T0251696

ADVANCED LABORATORY STRESS-STRAIN TESTING OF GEOMATERIALS

OUTCOME OF TC29 (TECHNICAL COMMITTEE NO.29) OF ISSMGE (INTERNATIONAL SOCIETY FOR SOIL MECHANICS AND GEOTECHNICAL ENGINEERING) FROM 1994 TO 2001

Advanced Laboratory Stress-Strain Testing of Geomaterials

Edited by

F. Tatsuoka
University of Tokio, Chairman of TC29, ISSMGE

S. Shibuya
Hokkaido University, Secretary of TC29, ISSMGE

R. Kuwano
*University of Tokio, Member of JGS (Japanese Geotechnical Society)
Domestic Committee for TC29*

Taylor & Francis
Taylor & Francis Group

LONDON AND NEW YORK

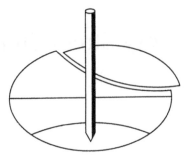

Sponsored by TC 29 on "Stress-Strain testing of Geomaterials in the Laboratory" of ISSMGE

The texts of the various papers in this volume were set individually by typists under the supervision of either each of the authors concerned or the editor.

Published by: Taylor & Francis
2 Park Square, Milton Park, Abingdon, Oxon, OX14 4RN
270 Madison Ave, New York NY 10016

Transferred to Digital Printing 2006

ISBN 90 2651 843 9

Publisher's Note
The publisher has gone to great lengths to ensure the quality of this reprint but points out that some imperfections in the original may be apparent

Printed and bound by CPI Antony Rowe, Eastbourne

Advanced Laboratory Stress-Strain Testing of Geomaterials, Tatsuoka, Shibuya & Kuwano (eds),
© *2001 Taylor & Francis, ISBN 90 2651 843 9*

Table of contents

Preface   VII

Recent advances in stress-strain testing of geomaterials in the laboratory;   1
background and history of TC-29 for 1994 - 2001
F.Tatsuoka, S.Shibuya & R.Kuwano

S-O-A Reports (National reports and international round-robin tests)

Report on the current situation of laboratory stress-strain testing of geomaterials in Italy   15
and its use in practice
A.Cavallaro, V.Fioravante, G.Lanzo, D.C.F.Lo Presti, S.Rampello,
A.d'Onofrio, F.Santucci de Magistris & F.Silvestri

A review of laboratory equipments for stress-strain testing in Greece   45
V.N.Georgiannou

Recent state of laboratory stress-strain tests on geomaterials in Japan   47
J.Kuwano & M.Katagiri

A review of Japanese standards for laboratory shear tests   53
J.Kuwano, M.Katagiri, K.Kita, M.Nakano & R.Kuwano

International round-robin test organized by TC-29   65
S.Yamashita, Y.Kohata, T.Kawaguchi & S.Shibuya

Report on applications of laboratory stress-strain test results of geomaterials   111
to geotechnical practice in Japan
J.Koseki, F.Tatsuoka, M.Yoshimine, M.Hatanaka, K.Uchida, N.Yasufuku & I.Furuta

Individual papers

Viscous deformation in triaxial compression of a dense well-graded gravel and its model simulation   187
L.Q.AnhDan, J.Koseki & F.Tatsuoka

Modelling of stress and strain relationship of dense gravel under large cyclic loading   195
K.Balakrishnaiyer & J.Koseki

Investigation of tied-back wall behavior and selection of geomechanical parameters   203
for numerical analysis
M.M.Berilgen, I.K.Ozaidin, & O.Inan

Small strain stiffness under different isotropic and anisotopric stress conditions   209
of two granular granite materials
A.Correia, L.Q.AnhDan, J.Koseki, & F.Tatusoka

Viscous and non viscous behaviour of sand obtained from hollow cylinder tests   217
H.Di Benedetto, H.Geoffroy & C.Sauzeat

Micaceous sands: stress-strain behaviour and influence of initial fabric 227
V.N.Georgiannou

Effect of ageing on stiffness of loose Fraser River sand 235
J.A.Howie, T.Shozen & Y.P.Vaid

Deformation characteristics of recompressed volcanic cohesive soil 245
M.Katagiri & K.Saitoh

Effects of ageing on stress-strain behaviour of cement-mixed sand 251
L.Kongsukprasert, R.Kuwano & F.Tatsuoka

Dependency of horizontal and vertical sugrade reaction coefficients on loading width 259
J.Koseki, Y.Kurachi, & T.Ogata

Low-strain stiffness and material damping ratio coupling in soils 265
C.G.Lai, O.Pallara, D.C.F.Lo Presti & E.Turco

A solid cylinder torsional shear device 275
J.Manzanas & V.Cuellar

Application of an elasto-viscous model to one-dimensional consolidation of clay under cyclic loading 279
T.Moriwaki, Y.E.Supranata, M.Sitoh & M.Hashinoki

Effects of stress path on the flow rule of sand in triaxial compression 287
H.Nawir, R.Kuwano & F.Tatsuoka

Simulation of viscous effects on the stress-strain behaviour of a dense silty sand 295
F.Santucci de Magistris, F.Tatsuoka & M.Ishihara

Interrelationship between the metastability index and the undrained shear strength of six clays 303
S.Shibuya, T.Mitachi, M.Temma & T.Kawaguchi

Case study on the practical use of elastic modulus in deformation analysis of soft clay ground 309
S.B.Tamrakar, S.Shibuya & T.Mitachi

Effects of fabric anisotropy of a sand specimen on small strain stiffness measured by 317
the bender element method
S.Yamashita & T.Suzuki

Strength and deformation characteristics of artificially cemented clay with dispersion 323
due to the mixing quality
N.Yasufuku, H.Ochiai, K.Omine & N.Iwamoto

Author index 329

Advanced Laboratory Stress-Strain Testing of Geomaterials, Tatsuoka, Shibuya & Kuwano (eds),
© *2001 Taylor & Francis, ISBN 90 2651 843 9*

Preface

In this book, recent developments in the measurement and interpretation of advanced laboratory stress-strain testing (i.e., static monotonic and cyclic loading tests, dynamic loading tests, wave propagation tests) of geomaterials (i.e., clay, sand, gravel, softrock and so on), substantial progress over the last decade or so, are described, together with a collection of case histories in applying the test results.

Most of the articles are produced after seven-year activities from 1994 to 2000 of the technical committee No. 29 (TC-29) on "Stress-Strain Testing of Geomaterials in the Laboratory" of International Society for Soil Mechanics and Geotechnical Engineering (ISSMGE), for which Tatsuoka, F. (University of Tokyo) and Shibuya, S. (Hokkaido University) have served as the chairman and the secretary, respectively.

The objectives (i.e., the terms of reference) of TC-29 are:
1) to promote co-operation and exchange of information about the recent developments in laboratory equipment and data acquisition systems;
2) to develop the recommendation for procedures referred to triaxial and torsional shear tests dealing with;
a) laboratory reconsolidation technique for cohesive soils, and
b) methods for measuring accurately, and in a reliable manner, stresses and strains of laboratory specimens for a strain range from 10^{-5} to 10^{-2}; and
3) to work out a uniform framework for comparison of stiffnesses measured by means of different laboratory techniques with emphasis on their relevance for the solution of problems of practical interest.

These objectives were set up by considering the following situation at the time of the establishment of TC-29 in 1994.
1) It had been one of important geotechnical engineering needs to accurately predict ground deformation and structural displacements, in particular those at working loads.
2) It had been often considered that, compared with field loading tests, laboratory stress-strain tests be less representative (so less useful) to meet this engineering need. Practising engineers were fully aware of the fact that the stiffness values obtained from conventional type laboratory stress-strain tests were often far low to explain actual field full-scale behaviour. It had been often considered that this discrepancy was attributed to serious effects of sample disturbance, among others. The link among the stiffness values from laboratory stress-strain tests, field loading tests and field full-scale behaviour was definitely out of thinking among practising engineers.
3) In addition, the reliable method for measuring deformation of soil specimen over a wide strain range had not been fully established.

Through the seven-year activities of TC-29, the perception of laboratory stress-strain testing has been greatly enhanced in respects that;
1) Accurate prediction of ground deformation and structural displacements has been, is and will be one of the important geotechnical engineering issues.
2) Laboratory stress-strain tests of geomaterials can provide useful and essential information to meet this engineering need, provided that the tests are performed adequately on high quality undisturbed samples, and that the results are properly interpreted by considering the every effect of controlling factors. In particular, accurate measurements of strain for a range from about 0.001 % to that at the peak,

for example, by using local axial strain measurements in the triaxial test, are crucially important, and the controlling factors should be first indentified, and then each effect be evaluated properly when applying the results to predict (or back-analyse) the relevant field behaviour.

3) The controlling factors include kinematic yielding, effects of recent stress-time history, anisotropy, structuration and destructuration, non-linearity by strain and pressure and the effects of cyclic loading. Careful distinctions ought to be made between elastic, plastic and viscous properties.

4) The stiffness values from different laboratory stress-strain tests, different field loading tests, field wave velocity measurements and field full-scale behaviour can be all linked to each other when taking into account the effects of these controlling factors. The stiffness values at very small strains measured dynamically and statically under otherwise the same conditions could be essentially the same among these different measurements, at least for fine-particle soils.

5) Detailed standard testing methods for laboratory stress-strain tests relevant to the engineering need described above are urgently needed to be developed with supportive back-up data.

The content of the book is as follows:

- detailed testing procedures of the advanced laboratory stress-strain testing of geomaterials (clay, sand, grave and soft rock), including the triaxial and torsional shear tests, which are relevant to the geotechnical design in general and the prediction of ground deformation and structural displacement;
- the results obtained from a series of international round robin tests (static and dynamic tests) on clay, sand and softrock and their interpretation, the link between the results from static tests (monotonic and cyclic loading) and dynamic tests (the resonant-column tests and the bender element tests) is discussed on experimental data.
- country reports on the stress-strain testing in the laboratory in several countries;
- country reports on case histories in which advanced laboratory stress-strain tests played an essential role in predicting ground deformation and structural displacement; and
- individual technical papers relating to the objectives of TC-29.

We believe that this book is a very useful reference for all researchers and practising engineers relating to the stress-strain property of geomaterial.

The editors would like to express their sincere thank to continuous cooperation in editing this book from the members of TC-29 of ISSMGE and the members of the domestic supporting committee of the Japanese Geotechnical Society.

April 2001

TATSUOKA, F.
University of Tokyo
Chairman of TC 29, ISSMGE

Advanced Laboratory Stress-Strain Testing of Geomaterials, Tatsuoka, Shibuya & Kuwano (eds),
© *2001 Taylor & Francis, ISBN 90 2651 843 9*

Recent Advances in Stress-Strain Testing of Geomaterials in the Laboratory – Background and History of TC-29 for 1994 - 2001–

F. TATSUOKA
University of Tokyo, Chairman of TC 29, ISSMGE

S. SHIBUYA
Hokkaido University, Secretary of TC-29, ISSMGE

R. KUWANO
University of Tokyo, Member of JGS Domestic Committee for TC29

ABSTRACT: The history, background and the main achievements for a period between 1994 and 2001 of TC-29 entitled "Stress-Strain Testing of Geomaterials in the Laboratory" of ISSMGE are reviewed by the authors.During the period, the devices, procedures and data acquisition systems for laboratory stress-strain testing of geomaterials have all shown significant developments, with which the in-situ stress-strain behaviour of geomaterials in use for predicting ground deformation and structural displacement can now be evaluated much more reliably and realistically than before. It is pointed out that the stress-strain behaviour over a 'practical' range of strain, which is usually less than about 0.5 % with stiff geomaterials, is often highly non-linear against strain, and very sensitive to pressure changes. It is emphasised that the stress-strain property of geomaterials for a strain range from less than 0.001 % to that at the peak can be evaluated by performing relevant laboratory stress-strain tests using a single specimen.

1 INTRODUCTION

The stress-strain testing of geomaterials in the laboratory has a long history. The primary objective of the conventional testing was used to be the evaluation of shear strength. This is because, except for the consolidation problem with soft clays, the conventional geotechnical engineering design of structures, natural slopes and geostructures (i.e., embankments, earth dams, rock-fill dams, retaining walls, piles and so on) was primarily based on the classical stability analysis assuming the isotropic rigid perfectly-plastic property of geomaterial (i.e., the limit equilibrium analysis, the upper-bound analysis, the stress characteristics method and so on). When ground deformations, structural displacements and the stress states in a soil mass and/or at the interface between a structure and a soil mass were to be evaluated, an additional over-simplified assumption that the geomaterial be linear-elastic with the fixed values of Young's modulus and Poisson's ratio was usually employed. It should be noted that even nowadays, these simplified assumptions with respect to the stress-strain property of geomaterials are employed in a simplified design approach.

As it becomes necessary to evaluate properly ground deformations, structural displacements and/or stress states, in particular those at working loads, the stress-strain testing in the laboratory is

demanded to yield the pre-failure stress-strain property of geomaterials more accurately than before. It has been found that the strains operating in the ground at working loads could be far smaller than we used to presume, i.e., typically less than about 0.5 % with stiff geomaterials, and that the stress-strain behaviour of geomaterials could be highly non-linear against strain even at these small strains. In addition, the stress-strain behaviour could be highly sensitive to changes in the pressure level, and it is significantly controlled by miscellaneous factors such as recent stress-time history, structure, loading rate and so on. In parallel to the above, an advancement for predicting ground deformations and structural displacements had been successively made by performing sophisticated non-linear numerical analysis. In performing realistic analysis of the failures of ground and structure by advanced numerical method such as the FEM, it was essential to establish a proper geomaterial model being capable of describing the pre-failure stress-strain property, the peak strength and the post-peak behaviour involved with strain localisation into a shear band(s). In link with the recent change of the demands for laboratory stress-strain testing of geomaterials, it was inevitable to see rapid developments in the device, procedure and data acquisition system.

In such a situation as described above, the establishment of a technical committee on "Stress-Strain

Testing of Geomaterials in the Laboratory" was proposed in 1994 by the Japanese Geotechnical Society (JGS) to the International Society for Soil Mechanics and Geotechnical Engineering (ISSMGE). This proposal was approved very positively by Prof. Jamiolkowski, M., who was elected as the President of ISSMGE at the 13th International Conference on SMFE, held in New Delhi, India. Technical committee No. 29 (TC-29) launched in January, 1994. The activities of TC29 for the first term (1994 – 1997) was rigorously evaluated, and the extension to the second term (1997 – 2001) was approved by the successor Prof. Ishihara, K., who was elected as the President at the 14th ICSMGE, held in September 1997 in Hamburg, Germany.

Throughout these two terms over a period between 1994 and 2001, the chairman and the secretary have been, respectively, Tatsuoka, F. and Shibuya, S. The members of TC-29 for the second term are listed in Table 1. Immediately after the launching of TC-29, a domestic committee was established within the JGS, which was the host member society of TC-29, in order to support the activities of TC-29. The members of the JGS domestic committee are listed in Table 2.

This report summarises the activities, the important outcomes, the present situation and the personal perspective of TC-29.

Table 1. List of TC-29 members for the second term (1997-2000)

Chairman:	Prof. F. Tatsuoka (JPN)	
Secretary:	Dr S. Shibuya (JPN)	
Core members:	Prof. E Flavigny (France)	
	Prof. D. Lo Presti (Italy)	
	Prof. R.J. Jardine (UK)	
	Prof. K.H. Stokoe (USA)	
Members:	Dr D. Airey (Australia)	
	Prof. A. Bolle (Belgium)	
	Prof. C.S. Pinto (Brazil)	
	Prof. O.M. Villar (Brazil)	
	Prof. Y. Vaid (Canada)	
	Dr J. Kurta (Cz/Slovakia)	
	Dr R.L. Verdugo (Chile)	
	Dr Lars Bo Ibsen (DMK)	
	Prof. Di Benedetto (FR)	
	Dr V. Georgiannou (Greece)	
	Dr A.V. Shroff (India)	
	Dr. M. Telesnick (Israel)	
	Dr V. Fioravante (Italy)	
	Dr F. Silvestri (Italy)	
	Prof. Won-Pyo Hong (Korea Republic)	
	Prof. E.J. Den Haan (NL)	
	Dr S. Nordal (Norway)	
	Dr M. Lipinski (Poland)	
	Prof. A. Correia (Portugal)	
	Dr V. Cueller (Spain)	
	Prof. M. Berilgen (Turkey)	
	Prof. D.W. Hight (UK)	
	Prof. J.T. Germain (USA)	
	Prof. P. Lade (USA)	

Table 2. List of members in Japanese domestic committee

Chairman:	Prof. F. Tatsuoka	
Secretary:	Dr S. Shibuya	
Core members:	Dr J. Kuwano	Dr Y. Kohata
	Dr S. Yamashita	Dr J. Koseki
Members:	Dr K. Kita	Mr. I. Furuta
	Dr M. Nakano	Mr. M.Nakajima
	Dr T. Kodaka	Mr. K. Hayano
	Mr. F. Fujiwara	Dr M.Yoshimine
	Mr. H. Uehara	Dr M.Katagiri
	Dr K. Oda	Dr T.Tsuchida
	Prof. K. Uchida	Dr Y.Tanaka
	Dr T. Moriwaki	Dr R.Kuwano
	Dr N. Yasufuku	Mr. T.Kawaguchi
	Dr M. Hatanaka	

2 TERMS OF REFERENCE AND THEIR BACKGROUND

TC-29 has been given with the following three terms of reference (i.e., the objectives);

1) To promote co-operation and exchange of information about the recent developments in laboratory equipment and data acquisition systems;

2) To develop the recommendation for procedures referred to triaxial and torsional shear tests dealing with;

   a) laboratory reconsolidation technique for cohesive soils, and

   b) methods for measuring accurately, and in a reliable manner, stresses and strains of laboratory specimens for a strain range from 10^{-5} to 10^{-2};

3) To work out a uniform framework for comparison of stiffnesses measured by means of different laboratory techniques with emphasis on their relevance for the solution of problems of practical interest.

As described in the following, these terms of reference reflected a situation at the time of 1994:

Engineering needs: It had been one of important geotechnical engineering needs to accurately predict ground deformations and structural displacements, in particular those at working loads (Fig. 1) (n.b., it is still the case at the present time). According to Tatsuoka et al. (1999a), the value of such predictions can be highlighted in the following:

i) to ensure no excessive foundation displacements, which in turn saves the cost of constructions of buildings, large bridges and so on;

ii) to ensure high serviceability levels for the completed structures;

iii) to avoid any damages to associated structures caused by ground movements (examples include low and high-rise buildings supported by a single raft foundation, and the effects of excavations on nearby masonry buildings, piles or tunnels); and

iv) predict the stresses acting on structural elements (rafts, retaining walls, tunnels, offshore structures etc).

2

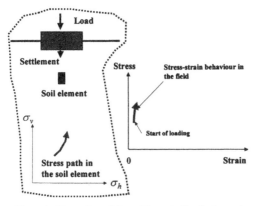

Figure 1. Illustration of the settlement of a footing subjected to working load, and stress path and stress-strain behaviour of a representative soil element in the ground.

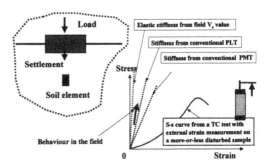

Figure 2. Illustration of inconsistent stress-strain behaviour among a laboratory stress-strain test, field loading tests and field full-scale behaviour.

Inconsistency between lab and field: The stiffness values evaluated by the conventional laboratory stress-strain tests as well as the field loading tests (such as plate loading tests and pressuremeter tests), which were conventionally considered as linear elastic properties, were far too low to explain many field full-scale behaviours, particularly with hard soils and soft rocks. Fig. 2 illustrates a typical example of inconsistency in stress-strain relationship between the field full-scale behaviour and the result from a conventional triaxial compression test. This kind of problem had been pointed out by Jardine et al. (1985, 1991), Burland (1989), Jamiolkowski et al. (1991), Atkinson and Sällfors (1991), Mair (1992) and Tatsuoka & Shibuya (1992), among others, before 1994, and was afterwards confirmed by Jardine (1995), Hight & Higgins 1995), Tatsuoka & Kohata (1995) and Tatsuoka et al. (1995b, 1999), among others. Such too low stiffness values obtained from conventional laboratory stress-strain tests were often considered due to serious effects of sample disturbance (n.b., this factor is still important for this discrepancy, but it may not be the exclusive cause).

Despite the above, a reliable and realistic procedure for such a prediction as above had not been established, or although it had been established in a limited number of leading laboratories, the practical implications were not widely appreciated.

On the other hand, as illustrated in Fig. 2, it had been commonly encountered to obtain substantially different stiffness values among conventional laboratory stress-strain tests, and also among conventional field loading tests assuming linear property of geomaterial. The stiffness values could also be largely different among different types of field loading tests. It had been often considered that, compared with field loading tests, laboratory stress-strain tests be less direct (so less useful) for such engineering needs. Corresponding to the above, the link among stiffness values from laboratory stress-strain tests, in-situ tests and field full-scale behaviour was rather missing, in particular among practicing engineers. It was also widely considered that the elastic stiffness from field shear wave velocity be utterly irrelevant when predicting ground deformations and structural displacements at static working loads.

Needs for more realistic soil-like constitutive models: It had also become clear that the in-situ pre-failure deformation behaviour could be significantly affected by the site-specifically formed geomaterial structure (or fabric), reflecting the recent stress-time history. In most cases, a simple geomaterial model comprising a density parameter coupled with consistency indices (such as plasticity index for clays and relative density for sands and gravels) and a simple stress history parameter such as OCR is not sufficiently representative to the real behaviour of geomaterial in the field. The fatal drawbacks of these stress-strain models developed after the results of laboratory test on reconstituted samples had been recognised seriously. Some leading researchers appreciated the need to predict the field behaviour by using a simpler model developed based on the results of relatively sophisticated stress-strain tests in which high-quality undisturbed samples are tested

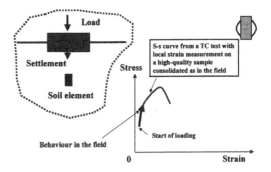

Figure 3. Illustration of consistent stress-strain behaviour between laboratory stress-strain tests and field full-scale behaviour.

3

by simulating the in-situ stress-strain history. Fig. 3 illustrates a satisfactory matching between the results from a laboratory stress-strain test and the field full-scale behaviour.

Needs for finding the link between static and dynamic tests, between laboratory and field tests and between tests and field full-scale behaviour: It was a classical practice to obtain the elastic deformation property of geomaterial by performing dynamic tests such as the resonant-column (RC) tests and the wave propagation tests (e.g. Hardin & Richart 1963; Hardin & Back 1968). Later, Hardin & Drnevich (1972), Iwasaki et al. (1978), Teachavorasinskun et al. (1991a & b), Tatsuoka & Shibuya (1991), Shibuya et al. (1992) and Tatsuoka et al. (1995a), among others, showed that the strain rate-dependency of the stiffness at small strains (less than about 0.001 %) of geomaterials in cyclic torsional shear was very low. The stress-strain behaviour was essentially reversible. Based on such experimental results as above, Woods (1991) and Tatsuoka & Shibuya (1992) pointed out that it is not necessary to distinguish between dynamically and statically measured elastic stiffness values, in particular for fine-graded soils, when they are measured under otherwise the same conditions.

In the meantime, it became possible in several leading geotechnical engineering laboratories to reliably measure strains for a range from less than about 0.001 % and associated small stresses in both triaxial and torsional shear tests. In triaxial tests, local measurement of deformation on the specimen lateral surface has become the standard method to reliably obtain the axial strain, particularly with stiff geomaterials that exhibit only very small deformation during consolidation (Tatsuoka et al. 1995a). It has therefore become rather popular to obtain the stiffness and damping values under cyclic loading conditions for a full range of concerned strain (usually from lower than 0.001 % to around 1 %) by cyclic static loading tests using a single specimen. It is only recent however that it has become possible to evaluate confidently the elastic deformation properties as well as the whole pre-peak stress-strain behaviour, including the peak strength by performing a monotonic loading test using a single specimen (e.g. Jardine 1995; Tatsuoka and Shibuya 1992; Tatsuoka 1994; Tatsuoka et al. 1995a & b; Jardine et al. 2001; and Lo Presti et al. 2001).

At the time of 1994, however, it was still popular in common engineering practice to treat dynamic and static stiffness values separately. For example, the deformation modulus from wave propagation tests was often called "the dynamic elastic modulus", while the deformation modulus from conventional laboratory stress-strain tests (such as unconfined tests and triaxial compression tests) or conventional field loading tests was called "the static elastic modulus" (see Fig. 2).

Needs for standard methods for relevant laboratory stress-strain tests: The practical strain levels operating in the ground at working loads could be relatively small; for example, less than about 0.5 % with stiff geomaterials (e.g., Tatsuoka & Kohata 1995). Besides, the stress-strain behaviour of geomaterials could be highly strain-non-linear even at these small strains, while the behaviour could be highly sensitive to the pressure change. In addition, the stress-strain behaviour at small strains could be affected largely by recent stress-time history (e.g., by different consolidation stress paths, different strain rate during consolidation, different ageing periods at the final consolidation stress state and so on). To evaluate such stress-strain properties of geomaterials as above, in addition to the use of high-quality undistorted samples sufficiently keeping the in-situ structure, the use of an advanced laboratory stress-strain testing system combined with relevant testing procedures becomes necessary. At the time of 1994, the standards for neither such sophisticated testing systems nor the testing procedures had been established, and therefore the testing of that kind was unpopular, in particular among practicing engineers.

Due mainly to these reasons described above, the activities of TC29 for the first and second terms were focused on the pre-failure stress-strain behaviour of geomaterials, covering a wide strain range from the elastic behaviour at strains less than about 0.001 % to the behaviour at larger pre-peak strains. The types of geomaterials concerned were soft and stiff clay, sand, gravel and soft rock.

3 ACTIVITIES INITIATED AND COMPLETED BY TC-29

In attempts to fulfill the terms of reference, the TC29 activities for the last two terms included the followings:
1) Recent developments in characterising the pre-failure deformation properties of geomaterials were reviewed and summarised.
2) The data required to predict ground deformations and structural displacements at working loads were collected.
3) Several practical methods of local strain measurements in the triaxial tests, together with dynamic methods such as the Bender Element (BE) and resonant column tests, were reviewed.
4) The relationships between static and dynamic experiments, between laboratory and field techniques, and between testing and field full-scale behaviour were examined, and it was attempted to give a framework for the above. In particular, it was shown that for fine-grained geomaterials, the elastic deformation characteristics are defined at strains less than about 0.001 %, and the values obtained by static and

dynamic experiments performed under otherwise the same test conditions are essentially the same, in particular for teh fine-grained geomaterials and rocks showing no sign of dominant discontinuities.

5) Important features of stress-strain properties of geomaterials were highlighted, including: kinematic yielding, effects of recent stress-time history, anisotropy, structuration and destructuration, non-linearity by strain as well as pressure and the effects of cyclic loading. Careful distinctions were made between elastic, plastic and viscous properties.

This book summarises in particular the following outcomes from the activities of TC-29 while completing three terms of reference:

T-O-R 1 *to promote co-operation and exchange of information about the recent developments in laboratory equipment and data acquisition systems*: As seen in Table 3, TC-29 sponsored two international conferences and one symposium, each with publication of proceedings and one workshop with pre-print, and will sponsor the third international conference to be held in 2003 in Lyon, France. In addition, a survey was made into the status quo of the relevant laboratory facilities in different countries, and the national reports on this issue were included in this book.

T-O-R 2 *to develop the recommendation for procedures referred to triaxial and torsional shear tests dealing with; a) laboratory reconsolidation technique for cohesive soils; and b) methods for measuring accurately, and in a reliable manner, stresses and strains of laboratory specimens for a strain range from 10^{-5} to 10^{-2}*: Several methods for laboratory stress-strain testing relevant to this objective (i.e., to provide necessary information for predicting ground deformation and structural displacements at practical strain levels), which had been developed in some leading laboratories were widely acknowledged in a number of member societies of ISSMGE. To confirm the applicability of these testing methods to general use, a series of international round robin (IRR) tests was initiated in 1997 under the initiative of the JGS domestic committee. The materials used in this IRR tests were; a) reconstituted clay (Fujinomori clay); b) reconstituted sand (Toyoura sand); and c) undisturbed sedimentary soft mudstone (Kazusa group of the early Pleistocene Era). The testing methods employed included; monotonic loading triaxial compression test, cyclic triaxial test, monotonic torsional test, cyclic torsional test, torsional resonant-column test and bender element test. A

Table 3.   International conferences, symposia, workshops and publications sponsored by TC-29

Year	Venue	Relevant meeting	Proceedings; Editors; Publisher
12-14, Sept., 1994	Sapporo, Hokkaido, Japan	International Conference on Pre-Failure Deformation of Geomaterials (IS Hokkaido '94)	Vol. 1 (1994) & Vol. 2 (1995); Shibuya, S., Mitachi, T. and Miura, S.; Balkema
4, Sept., 1997	London, UK	Géotechnique Symposium in Print (SIP) on Pre-failure Deformation Behaviour of Geomaterials	Pre-failure deformation behaviour of geomaterials (1998); Jardine, R.J., Davies, M.C.R., Hight, D.W., Smith, A.K.C. and Stallebrass, S.E.; Thomas Telford
26, Sept., 1999	Torino, Italy	International Workshop on Recent Advances Related to Terms of Reference of TC-29 of ISSMGE	Preprint
27-29, Sept., 1999	Torino, Italy	Second International Conference on Pre-Failure Deformation Characteristics of Geomaterials (IS Torino '99)	Vol. 1 (1999) and Vol. 2 (2001); Jamiolkowski, M., Lancelot, R. and Lo Presti, D.; Balkema
Aug., 2001	-	-	Advanced laboratory stress-strain testing of geomaterials (2001); Tatsuoka, F., Shibuya, S. and Kuwano, R. ; Balkema
22-24, Sept., 2003	Lyon, France	3rd International Symposium on the Deformation Characteristics of Geomaterials (DCG LYON '03)	Proceedings; Di Benedetto, H. et al. and Soils and Foundations (JGS)

number of volunteer members of TC-29 reviewed the standard test procedures that were suggested by the domestic committee in Japan. The test materials, together with the standard procedures for the testing, were distributed to a number of volunteer laboratories in the world. The members of TC29 examined the test results and evaluated the relevance of the proposed test procedures. Some of the main conclusions obtained from this activity are;

1) Rather consistent and reproducible results can be obtained in different laboratories when following the test procedures provided.

2) Local axial strain measurement is imperative to obtain accurately the stiffness values of relatively stiff geomaterials, such as sedimentary soft mudstone, in triaxial tests, while it may not be the case with soft clay, like reconstituted Fujinomori clay, that exhibits a relatively large compression during consolidation.

3) When measured at equally small strains under the same stress conditions, the elastic or quasi-elastic deformation characteristics can be obtained in common between the static and dynamic loading tests, particularly with fine-grained soils, such as Toyoura sand, Fujinomori clay and softrock comprising no major dominant discontinuities, as Kazusa group sedimentary soft mudstone.

As a result, the recommendation for the methods for stress-strain testing of geomaterials in the laboratory is proposed in this book.

T-O-R 3 *to work out a uniform framework for comparison of stiffnesses measured by means of different laboratory techniques with emphasis on their relevance for the solution of problems of practical interest*: To this end, information of related case histories was collected by the members of TC-29 and those of the domestic committee. Based on this information, a file of case histories was made. In each case history, the full-scale behaviour of ground, geo-structures and superstructures is predicted or back-analysed based on the stiffnesses from in-situ loading tests and/or laboratory stress-strain tests. The main conclusions that can be derived from this activity are as follows:

1) Relevant laboratory stress-strain test can provide information of stress-strain property at small strains of the cited geomaterial that is useful and essential in predicting reliably and realistically ground deformation and structural displacements at working loads.

2) The stress-strain behaviour of geomaterial is highly strain-non-linear at the operating small strains and highly pressure-sensitive. Considerations on both strain-non-linearity and pressure sensitivity are important, in particular, for drained unbound geomaterials.

3) The elastic deformation characteristics from field wave velocity measurement can be used in such predictions as described above after correcting for the

effects of relevant factors, including the strain-non-linearity and pressure-sensitivity.

4 CURRENT STATE-OF-THE-ART

The current state-of-the-art of the geotechnical engineering field which we are now discussing on could be illustrated as follows. Static loading could be either monotonic or cyclic at relatively low strain

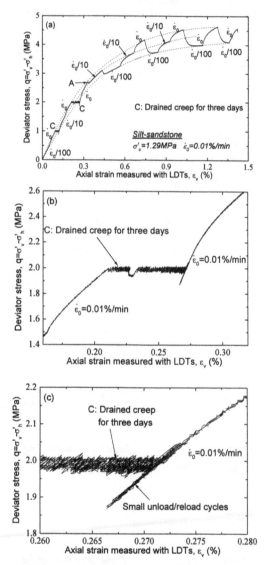

Figure 4. Stress-strain relationships from a drained TC test on a sedimentary silt-sandstone including stepwise changes in constant axial strain rate and creep stages (Hayano et al. 2001).

rates, while dynamic loading could also be either monotonic or cyclic, but at higher strain rates. It has been shown that a geomaterial element aged for some length of time, the initial stress-strain behaviour at small strains immediately after loading is restarted is essentially reversible. Fig. 4 shows the result from a CD triaxial compression test on an undisturbed sample of sedimentary soft siltstone. The sample was retrieved below the seabed at the mouth of the Tokyo Bay in relation to the design of a new suspension bridge. The specimen was isotropically re-consolidated to the field effective vertical stress, assuming that the field K_0 value be equal to unity. Axial strains were measured using a pair of LDTs (local deformation transducers, Goto et al., 1991). In this test, two creep tests, each left for three days, were performed while the strain rate was changed stepwise several times during otherwise the monotonic loading at a constant axial strain rate. At the end of each creep stage, several small unload/reload cycles were applied (Fig. 4c). It may

be seen that the stress-strain behaviour behaviour became essentially reversible immediately after loading was restarted following the ageing period. It is also the case with many other types of geomaterials (Tatsuoka et al. 2000). So, the factor that is most different between the static and dynamic tests is possibly attributed to the difference in the strain rate.

Fig. 5 shows a summary of data showing the rate effects of Young's modulus E_v defined at small strains of the order of 0.001 %. Other than some data of hard rock cores, concrete and mortar that were obtained from resonant-column tests and ultrasonic wave tests, the values of E_v were obtained from monotonic and cyclic triaxial tests. The references of these data are given in Tatsuoka et al. (1999a & b). It may be seen in Fig. 5 that the values of E_v are rather insensitive to changes in the strain rate, in particular at higher strain rates that are similar to those in unual dynamic tests. It is also true that with relatively soft geomaterials, as the strain rate becomes very low, the stress-strain relationship at strains less

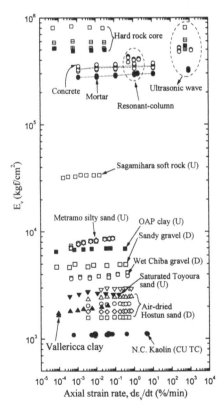

Figure 5. Relationships between the (quasi)-elastic Young's modulus E_v and the strain rate; except for some data from resonant-column tests and ultrasonic wave tests, the values of E_v were obtained from monotonic triaxial compression tests (only for kaolin) and cyclic triaxial tests; defined for an axial strain of the order of 0.001 % (Tatsuoka et al. 1999a & b).

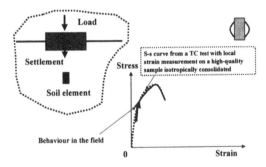

Figure 6. Illustration of inconsistent stress-strain behaviour between a TC test giving the specimen a stress-time history different from the one in the field and field full-scale behaviour.

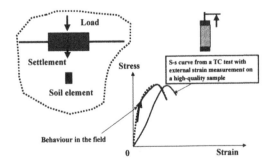

Figure 7. Illustration of inconsistent stress-strain behaviour between a TC test with external axial strain measurement on a high-quality sample (consolidated along a stress-time history different from the one in the field) and field full-scale behaviour.

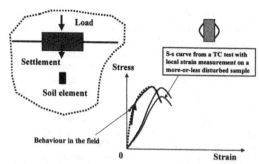

Figure 8. Illustration of inconsistent stress-strain behaviour between a TC test with local axial strain measurement on a more-or-disturbed sample (consolidated along a stress-time history different from the one in the field) and field full-scale behaviour.

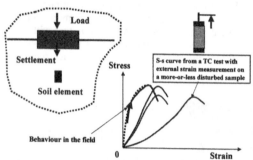

Figure 9. Illustration of inconsistent stress-strain behaviour between a TC test with external axial strain measurement on a more-or-less disturbed sample (consolidated along a stress-time history different from the one in the field) and field full-scale behaviour.

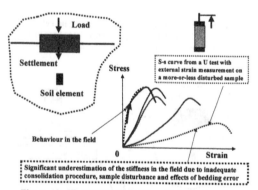

Significant underestimation of the stiffness in the field due to inadequate consolidation procedure, sample disturbance and effects of bedding error

Figure 10. Illustration of significantly inconsistent stress-strain behaviour between a unconfined compression test (on a more-or-less disturbed sample with a stress-time history different from the one in the field with external axial strain measurement) and field full-scale behaviour.

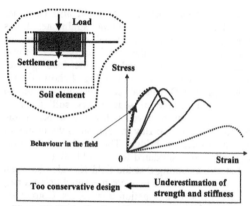

Too conservative design ← Underestimation of strength and stiffness

Figure 11. Illustration how the estimated stiffness of geo-material controls the design of foundation.

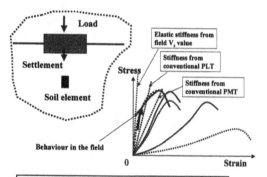

The missing link among results from laboratory and field tests and field full-scale behaviour

Figure 12. Inconsistent stress-strain behaviour between la-boratory and field experiments and between testing and field full-scale behaviour.

loading test performed at a relatively low strain rate could be smaller than the dynamically measured value. Therefore, the following can be concluded with geomaterials that are sufficiently homogeneous:
1) There is no reason to define separately dynami-cally and statically elastic modulus values measured under otherwise the same conditions.
2) When the strain rate and strain level are both simi-lar for a set of comparative tests, the stiffness values at strains less than about 0.001% should be similar between the dynamic and static tests performed un-der otherwise the same conditions.

Now, the main reasons for the discrepancy seen in Fig. 2 have been identified as:
a) with the exception of deformable materials such as soft clays or loose sands, the ground strains de-veloped by constructions are generally less than about 0.5 %, falling below the levels at which stiff-nesses could be defined reliably in conventional la-boratory or field tests; and

than 0.001 % becomes noticeably non-linear and be-comes noticeably strain rate-dependent. In such cases, the small strain stiffness E_v, defined for a strain of 0.001 %, that is measured by a static load-

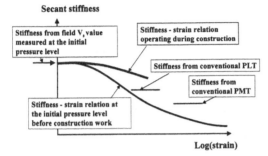

Secant stiffness

Stiffness from field V_s value measured at the initial pressure level

Stiffness - strain relation operating during construction

Stiffness from conventional PLT

Stiffness from conventional PMT

Stiffness - strain relation at the initial pressure level before construction work

Log(strain)

Figure 13. Interpretation of apparently inconsistent stress-strain behaviour between laboratory and field experiments and between testing and field full-scale behaviour only by considering strain-non-linearity of stiffness.

Secant stiffness

Stiffness from field V_s value measured at the initial pressure level

Stiffness - strain relation at a pressure level higher than that at the start of construction

Stiffness - strain relation operating during construction

Stiffness - strain relation at the initial pressure level before construction work

Log(strain)

Figure 14. Illustration of the interlink of secant modulus and strain relationship between the laboratory and field experiments and field full-scale behaviour considering both strain-non-linearity and pressure-sensitivity of geomaterial stiffness.

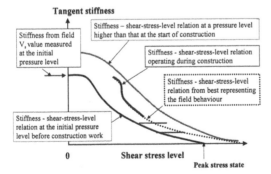

Tangent stiffness

Stiffness from field V_s value measured at the initial pressure level

Stiffness – shear-stress-level relation at a pressure level higher than that at the start of construction

Stiffness - shear-stress-level relation operating during construction

Stiffness - shear-stress-level relation from best representing the field behaviour

Stiffness - shear-stress-level relation at the initial pressure level before construction work

0   Shear stress level

Peak stress state

Figure 15. Illustration of the interlink of tangent modulus and strain relationship between laboratory and field experiments and field full-scale behaviour considering strain-non-linearity and pressure-sensitivity of geomaterial stiffness as well as effects of recent stress-time history.

b) the geomaterial stiffness is often highly strain-non-linear and pressure-sensitive over the 'practical' range of strain and pressure.

Fig. 3 illustrates the result of a laboratory stress-strain test being sufficiently consistent to the corresponding field behaviour. When the effects of recent stress-time history are ignored, for example when the specimen is isotropically consolidated, without being aged under the conditions of field initial stress state, the stress-strain behaviour in the laboratory would be different from the field behaviour even when strains are measured accurately (Fig. 6). When the axial strain is measured inadequately, even when using a high-quality sample, the stiffness in the field could be grossly underestimated (Fig. 7). Similarly, even when the axial strain is measured properly, the inconsistency as above would be significant when the sample is grossly disturbed (Fig. 8; e.g., Tatsuoka et al. 1995c). When the axial strain is measured externally, the discrepancy would further increase (Fig. 9). As for stiff clays and sedimentary softrocks, unconfined compression tests are often performed. When the effects of confining pressure are significant (as it is usually the case), when the stiffness is evaluated by such tests, the field stiffness would be significantly underestimated (Fig. 10; e.g., Tatsuoka and Kohata 1995). Fig. 11 illustrates schematically how underestimation of the stiffness of geomaterial results in a conservative design of foundation.

Fig. 12 illustrates inconsistent stress-strain behaviour between laboratory stress-strain tests and field loading tests (including shear wave velocity measurement) and between laboratory and field experiments and field full-scale behaviour. In Fig. 13, the inconsistency of stress-strain behaviour between laboratory and field experiments and between testing and field full-scale behaviour is interpreted by considering the strain-non-linearity of stiffness only. In this case, it is presumed that the pressure level increases during the construction. When the stiffness of geomaterial is pressure-sensitive, however, the above-mentioned way of interpretation would not be successful, but the pressure-sensitivity of stiffness (or more widely stress state-sensitivity of stiffness) as well as the effects of recent stress-time history should be taken into account (Fig. 14). That is, the pressure level in the subsoil during plate loading test and pressuremeter tests could become much higher than the initial level, while the pressure level during construction is also higher than the initial value, and the loading in the field case would start from an anisotropic stress state.

In such a case as above, it is not simple to relate the relationship between the secant stiffness and the strain in the field full-scale with those obtained from a laboratory stress-strain test performed at a constant pressure level. In such a case, the prediction of the relationship between the tangent stiffness and the shear stress level taking into account the effects of instantaneous pressure level and recent stress-time history would be much more relevant, as the tangent modulus is more objective than the secant modulus when the pressure level changes during loading (Fig. 15). This point is discussed in detail in Tatsuoka et al. (1999a).

5 CONCLUSIONS

Based on the activities for a period from 1994 to 2001 of TC 29, we can conclude the following:

1) Accurate prediction of ground deformation and structural displacements is one of the most important geotechnical engineering issues. We have now a much better ability than before to predict ground movements at working loads by;
a) achieving a shift from the use of conventional laboratory and field tests towards the use of the results of relatively sophisticated laboratory, field loading tests and field geophysics; and
b) moving from a relatively simple picture of geomaterial behaviour towards a more realistic one considering many relevant factors.

2) Laboratory stress-strain tests of geomaterials can provide useful and essential information to cope with this engineering need, provided that the tests are performed adequately on high quality undisturbed samples while taking into account the effects of controlling factors. In particular, accurate measurements of strain for a range from lower than about 0.001 % to that at the peak, for example, by local axial strain measurement in triaxial test, and due considerations on the strain-non-linearity and pressure-sensitivity of the stress-strain behaviour of geomaterials are important factors, among others.

3) The stiffness values from different laboratory stress-strain tests, different field loading tests, field wave velocity measurements and the field full-scale behaviour can be linked to each other only when taking into account the effects of controlling factors, including non-linearity of stress-strain behaviour by strain and pressure, effects of recent stress-time history and so on. In particular, under otherwise the same conditions at very small strains, dynamically and statically measured stiffnesses could be essentially the same at least with fine-particle soils and rocks without dominant discontinuities. The development of a more unified view which inter-links static and dynamic behaviour and laboratory and field testing has been achieved.

4) Standard testing methods for laboratory stress-strain tests that are relevant to the engineering need described above were investigated and the details of the above were suggested.

6 ACTIVITIES IN HAND AND SOME PERSPECTIVES

Some of the previous and incoming core members of TC29 (i.e., Tatsuoka,F., Shibuya,S., Jardine,R.J., Lo Presti,D. and Di Benedetto,H.) had a meeting in London in January 2001 to discuss the perspective of TC-29, and also to decide the activities for the next term starting from 2001 in case that the extension of TC-29 is approved by the incoming president who will be elected on the occasion of the 15th ICSMGE in Istanbul in August 2001. It was decided to propose the following:
1) The third international conference will be held in 2003 in Lyon, France, and the organising committee will be chaired by Prof. Di Benedetto, H., ENTPE. The main topics of this conference will be:
a) Soils and soft rocks;
b) Experimental investigations into deformation properties; from very small strains to those beyond failure;
c) Interpretation of laboratory, in-situ and field observations of deformation behaviour,
d) Characterising and modelling the stress-strain behaviour; and
e) Case histories.

Emphasis will be placed on exploring recent investigations into anisotropy and non-linearity, the effects of stress-strain-time history, ageing and time effects, yielding, failure and flow, cyclic and dynamic behaviour. However, equal weight will be given to reporting the application of advanced geotechnical testing to real engineering problems and to ways of synthesising information from a range of different sources when engaging in practical site characterisation studies.

The topics of the coming third international conference were determined by considering the fact that the terms of reference for the first and second terms cover a part of knowledge that is necessary for predicting the ground deformations and structural displacements at working loads. When the objective of the laboratory stress is expanded, for example, to provide information for stability analysis, the scope of TC-29 should be widened. So, the main interests and scope of the activity for the next term, starting from 2001 should be addressed to:

2) Main areas of interest
a) Experimental observations (laboratory and field)
b) Test procedures and recommendation
c) Characterisation of experimental findings
d) Synthesis of data from different sources for practical engineering with behaviour modelling
e) Practical significance of findings
f) Prototype behaviour, predictive validation, lessons learned from field observations
3) Topics
a) Small-strain stress-strain behaviour
b) Pre-failure stress-strain non-linearity

c) Stress and strain history-dependency of mechanical property

d) Sampling and sample disturbance

e) Hydraulic, temperature and mechanical coupling

f) Time effects (ageing, structure, rate effects etc)

g) Yielding, failure, flow, brittleness etc shear band and interface behaviour

h) Anisotropy

i) Cyclic and dynamic behaviour

j) Others (may include permeability)

4) Likely terms of reference for the term starting from 2001 of TC-29

a) To promote co-operation and exchange of information about recent developments in advanced geotechnical testing, including apparatus, techniques, data acquisition and interpretation (similar to the present T-O-R1, but no longer exclusively with laboratory based work).

b) To encourage the development of rational approaches to site characteristation, data interpretation and ground modelling that allow advanced testing to be applied usefully and effectively in practical geotechnical engineering (including suggestions and recommendations).

c) To advance the above aims through liaison with specialists working in the areas of laboratory and field testing, sampling, theoretical and numerical analysis in project engineering and full scale observation.

5) Likely new name of TC-29 for the term starting from 2001: Advanced mechanical geotechnical testing and its applications

REFERENCES

Atkinson,J.H. and Sällfors,G. (1991): Experimental determination of soil properties, Proc. 10th EC on SMFE, Firenze 3, 915-956.

Burland,J.B. (1989): Small is beautiful -the stiffness of soils at small strains, Canadian Geotechnical Journal 26, 499-516.

Di Benedetto,H,, Sauzeat,C. and Geoffroy,H. (2001b), "Modelling viscous effects and behaviour in the small strain domains", Proc. 2nd Int. Symp. on Prefailure Deformation Characteristics of Geomaterials, IS Torino (Jamiolkowski et al., eds.), Vol.2 (to be published).

Hardin,B.O. and Richart,F.E.Jr. (1963): Elastic wave velocities in granular soils, J. ASCE 89-SM1, 33-65.

Hardin,B.O. and Black,W.L. (1968): Vibration modulus of normally consolidated clay, J. SMF Div., ASCE 95-SM6, 1531-1537.

Hardin,B.O. and Drnevich,V.P. (1972): Shear modulus and damping in soils: design equations and curves, Jour. of SMF Div., ASCE, 98-SM7, 667-692.

Hardin,B.O. (1978): The nature of stress-strain behavior for soils, Proc. Geotech. Div. Specialty Conf. on Earthquake Eng. and Soil Dynamics, ASCE, Pasadena, I, 3-90.

Hardin,B.O. and Bladford,G.E. (1989): Elasticity of articulate materials, J. of Geotech. Engrg., ASCE, 115-6, 788-805.

Hayano, K. Matsumoto, M. Tatsuoka, F. and Koseki, J. (2001), "Evaluation of time-dependent deformation property of sedimentary soft rock and its constitutive modeling", Soils and Foundations, Vol.41, No.2, 21-38

Hight,D.W. and Higgins,K.G. (1995): An approach to the prediction of ground movements in engineering practice: Background and application, Proc. of Int. Symposium Pre-Failure Deformation of Geomaterials, IS Hokkaido '94 (Shibuya et al., eds.), Balkema, 2, 909-945.

Iwasaki,T., Tatsuoka,F. and Takagi,Y. (1978), "Shear moduli of sands under cyclic torsional loading", Soils and Foundations, 18-1, 39-56.

Jamiolkowski,M., Leroueil,S., and Lo Presti,D.C.F. (1991): Design parameters from theory to practice, Theme Lecture, Proc. Geo-Coast '91, Yokohama, 2, 877-917.

Jardine,R.J., Fourie,A.B., Maswoswse,J. and Burland,J.B. (1985): Field and laboratory measurements of soil stiffness, Proc. 11th ICSMFE, San Francisco, 2, 511-514.

Jardine,R.J., St.John,H.D., Hight,D.W. and Potts,D.M. (1991): Some practical applications of a non-linear ground model, Proc. 10th ECSMFE, Firenze, 1, 223-228.

Jardine,R.J. (1995): One perspective of the pre-failure deformation characteristics of some geomaterials, Proc. of Int. Sym. on Pre-Failure Deformation of Geomaterials, IS Hokkaido '94 (Shibuya et al., eds.), Balkema, 2, 855-885.

Jardine, R.J., Kuwano,R., Zdravkovic,L. and Thornton,C. 2001. Some fundamental aspects of the pre-failure behaviour of granular soils. Keynote Lecture, Preprint, Second Int. Conf. on Prefailure Deformation Characteristics of Geomaterials, 1999 (Jamiolkowski et al eds), Torino, Balkema, Vol.2 (to appear).

Lo Presti,D.C.F. Shibuya,S. and Rix, G.J. 2001. Innovation in soil testing. Keynote Lecture, Preprints of the 2nd Int. Conf. on Pre-Fairure Deformation Characteristics of Geomaterials, 1999 (Jamilkowski et al., eds.), September, Torino, Vol.2 (to appear).

Mair,R.J. (1993): Developments in geotechnical engineering research: application to tunnel and deep excavation; Unwin Memorial Lecture 1992, Proc. ICE 93, 27-41.

Shibuya,S., Tatsuoka,F., Teachavorasinskun,S., Kong,X.J., Abe,F., Kim,Y.S. and Park,C.-S. (1992), "Elastic Deformation Properties of Geomaterials", Soils and Foundations, Vol.32, No.3, pp.26-46.

Tatsuoka,F. and Shibuya,S. (1991), "Deformation characteristics of soils and rocks from field and laboratory tests", Keynote Lecture for Session No.1, Proc. of the 9th Asian Regional Conf. on SMFE, Bangkok, Vol.II, pp.101-170.

Tatsuoka,F. (1994), "Measurement of static deformation moduli in dynamic tests", Panel Discussion on Deformation of soils and displacements of structures, Panel Discussion, Proc. of the 10th European Conf. on S.M.F.E., Florence, Vol.4, pp.1219-1226.

Tatsuoka,F. and Kohata,Y. (1995), "Stiffness of hard soils and soft rocks in engineering applications", Keynote Lecture, Proc. of Int. Symposium Pre-Failure Deformation of Geomaterials (Shibuya et al., eds.), Balkema, Vol. 2, pp.947-1063.

Tatsuoka,F., Lo Presti,D.C.F. and Kohata,Y. (1995a), "Deformation characteristics of soils and soft rocks under monotonic and cyclic loads and their relationships", SOA Report, Proc. of the Third Int. Conf. on Recent Advances in Geotechnical Earthquake Engineering and Soil Dynamics, St Luois (Prakash eds.), Vol.2, pp.851-879.

Tatsuoka,F., Kohata,Y., Ochi,K. and Tsubouchi,T. (1995b), "Stiffness of soft rocks in Tokyo metropolitan area - from laboratory tests to full-scale behaviour", Keynote Lecture, Proc. Int. Workshop on Rock Foundation of Large-Scale Structures, Tokyo, Balkema, pp.3-17.

Tatsuoka,F., Kohata,Y., Tsubouchi,T., Murata,K., Ochi,K. and Wang,L. (1995c), "Sample disturbance in rotary core tube sampling of softrock", Conf. on Advances in Site Investigation Practice, Institution of Civil Engineers, London, pp.281-292.

Tatsuoka,F., Jardine,R.J., Lo Presti,D., Di Benedetto,H. and Kodaka,T. (1999a), "Characterising the Pre-Failure Deformation Properties of Geomaterials", Theme Lecture for the Plenary Session No.1, Proc. of XIV IC on SMFE, Hamburg, September 1997, Volume 4, pp.2129-2164.

Tatsuoka,F., Modoni,G., Jiang,G.L., Anh Dan,L.Q., Flora,A., Matsushita,M., and Koseki,J. (1999b): Stress-Strain Behaviour at Small Strains of Unbound Granular Materials and its Laboratory Tests, Keynote Lecture, Proc. of Workshop on Modelling and Advanced testing for Unbound Granular Materials, January 21 and 22, 1999, Lisboa (Correia eds.), Balkema, pp.17-61.

Tatsuoka,F., Santucci de Magistris,F., Hayano,K., Momoya,Y. and Koseki,J. (2000): "Some new aspects of time effects on the stress-strain behaviour of stiff geomaterials", Keynote Lecture, The Geotechnics of Hard Soils – Soft Rocks, Proc. of Second Int. Conf. on Hard Soils and Soft Rocks, Napoli, 1998 (Evamgelista and Picarelli eds.), Balkema, Vol.2, pp1285-1371.

Tatsuoka,F., Uchimura,T., Hayano,K., Di Benedetto,H., Koseki,J. and Siddiquee,M.S.A. (2001a); Time-dependent deformation characteristics of stiff geomaterials in engineering practice, the Theme Lecture, Proc. of the Second International Conference on Pre-failure Deformation Characteristics of Geomaterials, Torino, 1999, Balkema (Jamiolkowski et al., eds.), Vol. 2 (to appear).

Teachavorasinskun,S., Shibuya,S., Tatsuoka,F., Kato,H. and Horii,N. (1991a), "Stiffness and damping of sands in torsion shear", Proc. Second Int. Conf. on Recent Advances in Geotech. Earthquake Engnrg. and Soil Dynamics, March, St. Louis, Vol. I, pp.103-110.

Teachavorasinskun,S., Shibuya,S. and Tatsuoka,F. (1991b), "Stiffness of sands in monotonic and cyclic torsional simple shear", Proc. ASCE Geotechnical Engineering Congress, Boulder, Geotechnical Special Publication, Vol.27, pp.863-878.

Woods,R.D. (1991), "Field and laboratory determination of soil properties at low and high strains, Proc. Second. Int. Conf. On Recent Advances in Geotechnical Earthquake Engineering and Soil Dymanics, St. Lous, MO, Vol.II, pp.1727-1741.

S-O-A Reports
(National reports and international round-robin tests)

Advanced Laboratory Stress-Strain Testing of Geomaterials, Tatsuoka, Shibuya & Kuwano (eds),
© *2001 Taylor & Francis, ISBN 90 2651 843 9*

Report on the current situation of laboratory stress-strain testing of geomaterials in Italy and its use in practice

A. Cavallaro
University of Catania, Italy

V. Fioravante
University of Ferrara (formerly Ismes of Bergamo), Italy

G. Lanzo
University of Rome "La Sapienza"

D.C.F. Lo Presti, O. Pallara
Politecnico di Torino

S. Rampello
University of Messina (formerly University of Rome "La Sapienza")

A. d'Onofrio, F. S. de Magistris
University of Naples "Federico II"

F. Silvestri
University of Calabria (formerly University of Naples "Federico II")

1 CURRENT PRACTICE FOR GEOTECHNICAL INVESTIGATION IN ITALY

Laboratory testing for routine geotechnical design in Italy is mainly based on the results of oedometer, triaxial and direct shear tests. These tests provide compressibility, undrained shear strength and shear strength characteristics generally used in simplified design procedures. The undrained shear strength and/or shear strength parameters are inferred from UU or CU triaxial compression tests performed on specimens 38 mm in diameter and H/D≈2. The shear strength parameters are also inferred from direct shear tests performed under drained conditions.

A booklet prepared by the Italian Geotechnical Society (AGI, 1994) suggests the procedures recommended for laboratory tests such as: grain size analysis, odometer, direct shear and triaxial tests. Anyway, none of these suggested methods deal with the assessment of the stress-strain relationship of geomaterials. In a separate booklet (AGI, 1977) recommendations are provided in order to:
- program geotechnical investigations
- retrieve soil samples
- draw stratigraphic and geotechnical logging
- install piezometers

The above mentioned documents are written in Italian and mainly follow the recommendations of the International Society (ISSMGE). Associations of commercial laboratories (ALGI) and contractors (ANISIG) also guarantee that laboratory and in situ tests (including sampling and borehole logging) are carried out according to the AGI recommendations.

A comprehensive review of AGI recommendations on both in situ and laboratory tests is currently in progress. The reviewed version will closely refer to the principles addressed by Parts 2 and 3 of Eurocode 7 (Geotechnical design).

The Italian Ministry of Public Work (MLLPP) has recently established that Geotechnical Laboratories in Italy should guarantee some requirements and should follow given standards in performing laboratory and in situ tests on soils and rocks (MLLPP, 1999). According to the MLLPP the AGI recommendations will become mandatory. For those tests not considered by AGI it is suggested to follow the international standards such as ASTM, ISRM recommendations and BS 1377 - 5930.

Advanced stress-strain testing is in practice required only for special projects such as microzonation studies and/or design of very important infrastructures especially in seismic areas.

2 STANDARDS FOR STRESS-STRAIN TESTING

Accurate definition of the stress-strain relationships of geomaterials starting from very small strain levels are mainly performed in four different geotechnical laboratories, namely Enel.Hydro of Bergamo (formerly Ismes), the University of Naples "Federico II", Politecnico di Torino, and the University of Rome "La Sapienza". Some advanced tests can also be performed at the University of Florence (Ciulli 1998, Crespellani 1998) and the University of Catania (Cavallaro et al. 1998).

A common standard for determining laboratory stress-strain curves at small to medium strains does not exist. Yet, due to co-operation among the laboratories mentioned above, a common practice has been developed and accepted. Details of the commonly accepted testing procedures and apparatus requirements are summarised below. More details can be found in Fioravante et al. 1994, Fioravante 2000, d'Onofrio et al. 1999a,b, Silvestri 1991, Lo Presti et al. 1993, 1995, Aversa & Vinale 1995, Rampello 1991, Rampello et al. 1997, Callisto & Calabresi 1998, Lai et al. 2000.

1) The importance of local and accurate strain measurements is widely recognised. Local measurements of sample deformation are carried out using different kinds of transducers: submergible miniaturised LVDTs and LDTs (Naples), submergible miniaturised LVDTs and strain inclinometers (Ackerley et al. 1987) (Rome), or proximity transducers (Bergamo & Torino). Most of the sensors mentioned above are used for local measurements of axial displacements in triaxial tests; the proximity transducers are also used for local measurements of radial strain, very accurate evaluation of volumetric strain (Bergamo & Torino) and to measure the overall rotation of specimens subjected to torsional shear.

2) In order to improve the measurement resolution, most of the laboratories use specimens of non-standard sizes; usually, a ratio H/D = 2 is adopted with a diameter of 50 or 70 mm. Triaxial apparatus for coarse grained solis have been developed (Flora 1995, Lo Presti et al. 1999). These large triaxial cells enable one to use specimens 300 mm in diameter with H/D = 2.

3) Measurement of the shear wave velocity in triaxial or torsional shear tests is common practice in almost all the laboratories mentioned above. These measurements are usually performed using piezoceramic transducers called Bender Elements (BE); the transmitter and the receiver elements are installed into the top cap and the pedestal of the cell, respectively. Exceptionally, a couple of receivers have been used (Torino) as well lateral BE (Rome, Torino) of the type developed by Nash et al. (1999). The use of two receivers enables one to interpret the test on the basis of the so-called "true visual interval method" or cross-correlation method. The lateral BEs measure the velocities of vertically polarised shear waves propagating horizontally, V_{hv}, and of shear waves propagating in the horizontal planes, V_{hh}.

4) All the laboratories use a Resonant Column (RC) & Torsional Shear (TS) apparatus. It is quite common practice to perform both tests on the same specimen according to procedures that will be illustrated later.

5) All the laboratories use appropriate countermeasures to avoid swelling during specimen set-up. This problem is specially relevant in the case of stiff overconsolidated clays and dense compacted silty sands.

6) The adopted re-consolidation procedures try to avoid the occurrence of very large volume variations. For this purpose specimens are usually K_o consolidated to the in-situ geostatic stress. The reproduction in the laboratory of the in-situ stress-history is applied with some caution.

9) RC & TS tests are repeated on the same specimen. Some typical procedures are illustrated in the following sections:

Procedure A:
- specimen is firstly subjected to a multi staged isotropic or anisotropic (K_o) consolidation path during which it is possible to measure the initial shear modulus by low amplitude RC tests
- at the end of the consolidation path multi-step undrained cyclic torsional tests (CLTS) and a resonant column (RC) series are carried out at an increasing strain level until the maximum equipment capabilities are reached.

Procedure B:
- specimen is first subjected to an isotropic or anisotropic (K_o) compression stage. During this

stage the small strain modulus G_o is measured by low-amplitude RC tests.

- At the end-of this stage, a multi-step undrained shear load procedure is usually started with the following sequence:

1. a series of CLTS tests is carried out, at each shear strain step keeping the torque amplitude constant, and increasing the frequency from 0.01 Hz to 1 Hz (Naples) or from 0.01 Hz to the resonant frequency (Torino); torque, rotation and phase difference measurements are used to simultaneously obtain the shear modulus and damping ratio as a function of the frequency;
2. the CLTS sequence is followed by a resonant column (RC) test at the same strain amplitude;
3. steps 1 and 2 are repeated at increasing strain levels, at the onset of the non linear behaviour;
4. at the end of this sequence it is possible to restart the isotropic compression up to a defined value of mean effective stress.

As an example, the Tables shown below indicate the adopted procedures in some laboratories as far as triaxial and torsional shear (resonant column) tests are concerned.

REFERENCES

Ackerley S.K., Hellings J.E. & Jardine R.J. (1987) – Discussion on A new device for measuring local axial strains on triaxial specimens by C.R.I. Clayton and S.A. Khatrush, *Géotechnique* 37, No. 3: 413-417.

AGI 1994. *Raccomandazioni sulle prove geotecniche di laboratorio.*

AGI 1977. *Raccomandazioni sulla programmazione ed esecuzione delle indagini geotecniche.*

Aversa S.& Vinale F., 1995 Improvements to a stress-path triaxial cell *ASTM Geotecnhical tyesting journal* 1:116,120.

Callisto L. & Calabresi G. 1998. *Mechanical behaviour of a natural soft clay.* Géotechnique, 48, No. 4: 495-513.

Cavallaro A. Lo Presti D.C.F., Maugeri M. & Pallara O. 1998. *Strain Rate Effect on Stiffness and Damping of Clays.* Italian Geotechnical Review, Vol. XXXII, n. 4, pp. 30-49.

Ciulli B. 1998. *Caratterizzazione dinamica dei terreni e analisi della risposta sismica locale in alcuni siti di Fabriano finalizzate alla microzonazione sismica.* M.Sc. Thesis. University of Florence.

Crespellani T. (coord.) 1998. *Proprietà dinamiche dei terreni di Fabriano.* Poster, Convegno Nazionale GNDT, Roma.

Fioravante V., Jamiolkowski M. & Lo Presti D.C.F. 1994 *Stiffness of Carbonatic Quiou Sand.* Proceedings of the XIII ICSMFE, New Dehli, India, Vol 1, pp 163-167

Fioravante V., 2000 *Anisotropy of Small Strain Stiffness of Ticino and Kenya Sands from Seismic Wave Propagation Measured in Triaxial Tests.* Soils & Foundations, 40(4): 129-142.

Flora A. 1995. *Caratterizzazione Geotecnica e Modellazione Geotecnica dei Materiali a Grana Grossa* Ph.D. Thesis, Naples, University.

Lai C.G., Lo Presti D.C.F., Pallara O & Rix G.J. 1999. *Misura simultanea del modulo di taglio e dello smorzamento intrinseco dei terreni a piccole deformazioni.* Atti 9° Convegno Nazionale "L'Ingegneria Sismica in Italia", Torino 20-23 Settembre 1999. ANIDIS (formato CD)

Lo Presti D.C.F., Pallara O, Lancellotta R., Armandi M. & Maniscalco R. 1993 *Monotonic and Cyclic Loading Behaviour of Two Sands at Small Strains* Geotechnical Testing Journal, Vol 16, No 4, pp 409-424.

Lo Presti D.C.F., Pallara O. & Puci I. (1995) "A Modified Commercial Triaxial Testing System for Small Strain Measurements: Preliminary Results on Pisa Clay," Geotechnical Testing Journal, Vol. 18, No 1, pp. 15-31.

Lo Presti D.C.F., Pedroni S. & Froio F. 1999. *Rigidezza dei Terreni a Grana Grossa da Prove Triassiali Eseguite in una Cella di Grandi Dimensioni* XX Convegno Nazionale di Geotecnica, Parma 22-25 Settembre 1999, Pàtron Editore Bologna, pp. 149-153.

MLLPP 1999. *Circolare n. 349/STC.*

Nash D.F.T., Pennington D.S. & Lings M.L. (1999) The dependence of anisotropic G_o shear moduli on void ratio and stress state for reconstituted Gault clay. *Proc. 2nd Int. Symp. on Pre-Failure Deformation Characteristics of Geomaterials – IS Torino 99.* Torino, Balkema, vol. 1: 229-238.

d'Onofrio, A., F. Silvestri, and F. Vinale (1999a). Strain rate dependent behaviour of a natural stiff clay. *Soils and Foundations*, 39:2, 69-82.

d'Onofrio, A., F. Silvestri, and F. Vinale (1999b). A new torsional shear device. *ASTM Geotechnical testing journal*, 22:2, 101-111.

Rampello S. 1991. *Some remarks on the mechanical behaviour of stiff clays: the example of Todi clay.* Proc. Int. Workshop on "Experimental characterisation and modelling of soils and soft

Triaxial apparatus (University of Napoli "Federico II")

Type of test		Triaxial test	
Device	Model	Bishop & Wensley (1975) modified by Aversa & Vinale (1996)	
	Cell structure	External tie	
	Specimen/device connection	Low part: Rough porous stone Annular porous stone using BE Top part: Perspex top cap	
	Loading system	1) Hydraulic loading system stress controlled type by E/P converter 2) Hydraulic loading system strain controlled by Bishop ram and stepper motor	
Specimen	Diameter	38 mm	
	Height	76 mm	
	Accuracy in measuring specimen dimension	0.1 mm, 0.1 g	
	Specimen preparation	Sedimentation at 1.5 w_L Constipation in a Proctor mold and constant volume saturation	
	Specimen setting	Wet setting method	
	Reconsolidation of undisturbed samples	1) isotropic 2) automatic K_0 at the in situ stress following eventual overconsolidation path.	
	Period to stop consolidation	Ageing > 2 days at the chosen stress state	
Data control system		A/D D/A card + PC	
Control and measurement accuracy		Temperature	Room temperature ±1°C
		Resolution of the E/P converter	40 Pa
		Strain loading range	$4*10^{-5} \div 1.3$ mm/min (strain control) No limits (stress control)
		Resolution and sampling frequency of the data device	A/D= 16 bit D/A= 12bit; Sampling 0.25 ± 1 Hz (depending on digital filter); Sequential acquisition system

To be measured	Transducer type	Position	Resolution	Precision
Axial loading	Loading cell	inner	70 Pa	± 0.2 kPa
Pore pressure	Druck PDCR 810 PDCR 81	External Local	30 Pa 20 Pa	± 0.30 kPa ± 0.20 kPa
Cell pressure	Druck PDCR 810	External	30 Pa	± 0.30 kPa
Axial strain	1*LVDT ±12.5 mm 2*2 LVDT/ 2 LDT	External Local	0.33 μm 0.1 μm/0.25 μm	± 2 μm ± 0.2 μm/ ± 0.4 μm
Volume strain	IC volume gauge	External	0.0033 cm³	± 0.005 cm³

Resonant column and torsional shear apparatus (University of Napoli "Federico II")

test type		Resonant column	Torsional shear
designation			
specimen	outer diameter	36mm, 70mm, 100mm	
	inner diameter	hollow specimen possible but not used	
	wall thickness of hollow specimen		
	height	$h = 2 D$	
	accuracy in measuring specimen dimension	0.1 mm, 0.1gr	
	method of specimen preparation	trimming method	
	saturation of specimen	constant volume saturation outside the cell in a proper mould using hydraulic gradient equal to 20 kPa.	
	B value	B > 0.95 averaging values from increasing and decreasing cell pressure.	
	contact between specimen and device	rough platen screwed on top cap and porous stone screwed on the bottom base.	
loading device	type	electromagnetic motor	
control accuracy	cell pressure	control via E/P converter 40 Pa	
	back pressure	no control	
	axial stress	no control	
	torsional loading device	electric current control resolution $1.6*10^{-5}$ Nm (2 Pa on a $\varnothing$ 36 mm specimen) f.s. 5 Nm no back electromagnetic force	
	lateral strain		
	loading rate	no limits	
	temperature	no control	
measurement device type	cell pressure	high performance pressure sensor incorporating a high integrity silicon diaphragm	
	back pressure	miniaturised transducer with silicon diaphragm	
	pore water pressure	miniaturised transducer with silicon diaphragm	
	axial load		
	torque	strain gage torsional cell	
	axial deformation	D.C. LVDT	
	lateral deformation		
	volume change	volume gauge with A.C. LVDT	
	acceleration	piezoelectric accelerometer	
	rotational angle	proximitor probes microproximitor probes	
measurement accuracy	cell pressure	1.5 Pa	
	back pressure	0.7 Pa	
	pore water pressure	0.7 Pa	
	torque	$6x10^{-3}$ Nm	
	axial deformation	1 μm	
	lateral deformation	-	
	volume change	0.0014 cm^3	
	rotational angle	$2.5x10^{-8}$ rad (microprox) 2. $5x10^{-7}$ rad (prox)	
	acceleration	0.002 rad/s^2	
measurement frequency		> 100 data in one cycle	
period to stop consolidation		t_{100} determined on the (logt,w) curve	
test procedures		1 - isotropic compression with scheduled RC test at very small strain. 2 – Y1 surface probing at defined mean effective stress (Rc test followed by TS tests at the same stain level but at various frequencies carried out at increasing strain level up to the linear threshold strain (G = 0.95 Go) 3 – monotonic torsional shear test at the defined mean effective stress till reaching the maximum device capabilities	
others		during monotonic torsional shear test it is possible to apply any loading rate	

Triaxial apparatus (Politecnico di Torino)

Purpose		The strength and deformation characteristics of soil under undrained(CU) / drained(CD) triaxial compression after under K_o and K = constant consolidation.
Soil type		Mainly saturated cohesive soils.
Apparatus & tools	Load cell	The load cell is located inside the triaxial chamber just beneath the specimen pedestal. It is a commercially available, very rigid submergible load cell, which is insensitive against hydrostatic pressure. This transducer has a capacity of 5 kN and a resolution of 0.046 N (in the range of 0 + 600 N).
	Strain measurement	Three different measurements of the axial deformation ε_a are taken: • Local ε_a, in the central portion of the specimen, using a couple of high resolution submergible proximity transducers. Range, resolution and accuracy of these transducers are respectively 2.5 mm, 0.3 μm and 0.05 %FSO. • Internal ε_a, between the top cap and the base pedestal, using a couple of high resolution submergible LVDTs. Range, resolution and accuracy of these transducers are respectively 10 mm, 0.5 μm and 0.05 %FSO. • External ε_a, using a long stroke LVDT. Range, resolution and accuracy of this transducer are respectively 50 mm, 2.5 μm and 0.05 %FSO. • Radial deformation ε_r, are directly and locally measured by means of a couple of proximity transducers. Range, resolution and accuracy of these transducers are respectively 2.5 mm, 0.3 μm and 0.05 %FSO. A piece of thin aluminium foil is used as a target. The target is attached to the specimen surface by means of a few drops of water.
	Contact Between top cap and loading piston	The top cap is rigidly connected with the upper cell plate and the base cap is rigidly connected to the loading ram.
	Top and pedestal	Porous stones screwed on top cap and bottom base
	Stroke of loading piston	= 90 mm, 0.64 H
	Data sampling	A computer with a data acquisition board is used to convert all the analog signal of the transducers in digital output. The main characteristics of this converter are: 16 Single Ended analog inputs with 16 Bits of resolution on ±10 V range and a sample rate of 100000 Samples/s. It is possible to use two different data reading rates in order to reduce the noise of analog signal. The maximum sampling rates are: • ~0.5 Samples/s if every data reading is the average of 3000 samples (max noise reduction) • ~2.5 Samples/s if every data reading is equal to one sample (no noise reduction)
	Rubber membrane	~0.3 mm thick, ~0.95D
	Filter paper	0.2 + 0.3 mm thick, used for impermeable specimens
	O-ring	∅ internal ~0.7D, thick 3.53 mm
	Other requirements	The volume change is accurately measured by means of a special designed system, using a proximity transducer and a float bearing an aluminium target. The short burette is inside a little pressure chamber and is filled of water in contact with the specimen. The upper part of burette is filled with coloured silicon oil with a specific gravity equal to about 0.5. The internal burette does not deform with increasing values of the pore pressure, thus avoiding the occurrence of apparent volume change. The capacity of this gauge is 45 cm³ with a resolution of about 2 mm³.
Specimen	Diameter, D (mm)	D = 70 mm
	Height, H (mm)	H = 140 mm, H = 2D
	Accuracy in measuring specimen dimensions	see table below
	Method of specimen preparation	Trimming method
	Saturation of specimen	Dry setting method: • A flushing with de-aired water from bottom to top with an hydraulic gradient of 5 kPa is established for at least 24 h. During this stage, the vertical and horizontal stresses are independently increased to respectively prevent any axial and radial strains. The computer controls the following conditions: ΔH = 0 ±5 μm and ΔD = 0 ±2 μm. • A back pressurizing stage then starts to dissolve in the water any air bubbles that might still be trapped in the lines and in the space between the specimen and the membrane.
	Back pressure	Usually step of 50 kPa up to 200 kPa
	B-value	> 0.95
	Other requirements	A vertical stress σ_a = 5 kPa is applied under unconfined conditions and maintained during the sample preparation.
Test operation & control	Cell pressure, σ_r	$\Delta\sigma_r$ = ±0.5 kPa
	Back pressure, u_b	Δu_b = ±0.5 kPa

Accuracy in measuring specimen dimensions:

Dimension	Capacity	Resolution	Accuracy
H		0.01mm	±0.03 mm
D		0.01mm	±0.03 mm
Weight	1200g	0.1g	

accuracy	Axial load, P Axial stress, σ_a Axial deformation, ΔH	$\Delta\sigma_a = \pm0.5$ kPa
	Lateral strain, ε_r	$\Delta\varepsilon_r = \pm0.0015$ %
	Consolidation rate	Usually equal $\dot{\varepsilon}_a = 0.01$ %/min
	Consolidation period	Specimen consolidation is performed at constant strain rate under both K_o and $K =$ constant conditions. In both cases, the operator establishes the final axial stress σ_{af}, the strain rate, and the pausing time after σ_{af} is reached. Thus the loading rod is automatically raised at the established strain rate until the target vertical stress is reached. Thereafter σ_{af} is maintained constant while the horizontal stress is appropriately modified to keep a nil lateral strain. The tests pause at the consolidation stresses until the creep rate falls to 0.05 %/day
	Axial compression	Axial compression process is continued until $\varepsilon_a > 20$ %
	Loading rate, $\dot{\mathcal{E}}_a$	CU: ~0.1 %/min (silt), ~0.05 %/min (clay) CD: $< \varepsilon_{a(failure)} / 15t_c$ (t_c: consolidation period)
	Temperature	$< \pm2°$ C
Measurement accuracy	Cell pressure, σ_r Back pressure, u_b Pore water pressure, u	$\sigma_r = \pm0.5$ kPa $u_b, u = \pm0.5$ kPa
	Axial load, P	$P = \pm50$ N (max amplify: $P = \pm0.5$ N)
	Axial deformation, ΔH	Local $\varepsilon_a = \pm0.001$ %. Internal $\varepsilon_a, = \pm0.004$ %. External $\varepsilon_a, = \pm0.018$ %.
	Lateral deformation	$\varepsilon_r = \pm0.002$ %.
	Volume change, ΔV	$\Delta V = \pm2$ mm³
	Measurement frequency	*See data sampling*
Others		

Triaxial apparatus for cyclic tests (Politecnico di Torino)

Purpose		The deformation characteristics of geomaterials under an isotropic/anisotropic stress state when subjected to cyclic undrained/drained triaxial loading.
Soil type		Mainly saturated cohesive soils.
Apparatus & tools	Load cell	The load cell is located inside the triaxial chamber just beneath the specimen pedestal. It is a commercially available, very rigid submergible load cell, which is insensitive against hydrostatic pressure. This transducer has a capacity of 5 kN and a resolution of 0.046 N (in the range of 0 + 600 N). The max hysteresis of the load cell is 0.05% FS. The change of calibration factor are given in the following table:

Amplify [mV/V]	FS [kN]	Change of calibration factor [%FS]
0.2	0.6	0.01%
0.5	1.2	0.01%
1.0	2.4	0.01%
2.0	5.0	0.1%

	Strain measurement	Three different measurements of the axial deformation ε_a are taken: • Local ε_a, in the central portion of the specimen, using a couple of high resolution submergible proximity transducers. Range, resolution and accuracy of these transducers are respectively 2.5 mm, 0.3 μm and 0.05 %FSO. • Internal ε_a, between the top cap and the base pedestal, using a couple of high resolution submergible LVDTs. Range, resolution and accuracy of these transducers are respectively 10 mm, 0.5 μm and 0.05 %FSO. • External ε_a, using a long stroke LVDT. Range, resolution and accuracy of this transducer are respectively 50 mm, 2.5 μm and 0.05 %FSO. • Radial deformation ε_r, are directly and locally measured by means of a couple of proximity transducers. Range, resolution and accuracy of these transducers are respectively 2.5 mm, 0.3 μm and 0.05 %FSO. A piece of thin aluminium foil is used as a target. The target is attached to the specimen surface by means of a few drops of water. The change of calibration factor are given in the following table:

Tranducers	FS [mm]	Change of calibration factor [%FS]
Local ε_a,	2.5	0.1%
Internal ε_a,	10	0.04%
External ε_a,	50	0.01%
Radial deformation ε_r	2.5	0.1%

	Contact Between top cap and loading piston	The top cap is rigidly connected with the upper cell plate and the base cap is rigidly connected to the loading ram.
	Top and pedestal	Porous stones screwed on top cap and bottom base
	Stroke of loading piston	= 90 mm, 0.64 H
	Data sampling	A computer with a data acquisition board is used to convert all the analog signal of the transducers in digital output. The main characteristics of this converter are: 16 Single Ended analog inputs with 16 Bits of resolution on ±10 V range and a sample rate of 100000 Samples/s. It is possible to use two different data reading rates in order to reduce the noise of analog signal. The maximum sampling rates are: • ~0.5 Samples/s if every data reading is the average of 3000 samples (max noise reduction) • ~2.5 Samples/s if every data reading is equal to one sample (no noise reduction)
	Rubber membrane	~0.3 mm thick, ~0.95D
	Filter paper	0.2 + 0.3 mm thick, used for impermeable specimens
	O-ring	Ø internal ~0.7D, thick 3.53 mm
	Other requirements	The volume change is accurately measured by means of a special designed system, using a proximity transducer and a float bearing an aluminium target. The short burette is inside a little pressure chamber and is filled of water in contact with the specimen. The upper part of burette is filled with coloured silicon oil with a specific gravity equal to about 0.5. The internal burette does not deform with increasing values of the pore pressure, thus avoiding the occurrence of apparent volume change. The capacity of this gauge is 45 cm³ with a resolution of about 2 mm³.
Specimen	Diameter, D (mm)	D = 70 mm
	Height, H (mm)	H = 140 mm, H = 2D

Accuracy in measuring specimen dimensions	Dimension	Capacity	Resolution	Accuracy
	H		0.01mm	±0.03 mm
	D		0.01mm	±0.03 mm
	Weight	1200g	0.1g	

	Method of specimen preparation	Trimming method

	Saturation of specimen	Dry setting method: • A flushing with de-aired water from bottom to top with an hydraulic gradient of 5 kPa is established for at least 24 h. During this stage, the vertical and horizontal stresses are independently increased to respectively prevent any axial and radial strains. The computer controls the following conditions: $\Delta H = 0 \pm 5$ μm and $\Delta D = 0 \pm 2$ μm. • A back pressurizing stage then starts to dissolve in the water any air bubbles that might still be trapped in the lines and in the space between the specimen and the membrane.
	Back pressure	Usually step of 50 kPa up to 200 kPa
	B-value	> 0.95
	Other requirements	
Test operation & control accuracy	Cell pressure, σ_r Back pressure, u_b Axial stress σ_a	$\Delta\sigma_r = \pm 0.5$ kPa $\Delta u_b = \pm 0.5$ kPa $\Delta\sigma_a = \pm 0.5$ kPa
	Lateral strain, ε_r	$\Delta\varepsilon_r = \pm 0.0015$ %
	Consolidation rate	Usually equal $\dot{\varepsilon}_a = 0.01$ %/min
	Consolidation period	Specimen consolidation is performed at constant strain rate under both K_o and K = constant conditions. In both cases, the operator establishes the final axial stress σ_{af}, the strain rate, and the pausing time after σ_{af} is reached. Thus the loading rod is automatically raised at the established strain rate until the target vertical stress is reached. Thereafter σ_{af} is maintained constant while the horizontal stress is appropriately modified to keep a nil lateral strain. The tests pause at the consolidation stresses until the creep rate falls to 0.05 %/day
	Single amplitude cyclic axial strain $\varepsilon_{a(SA)}$ Cyclic axial load $P_C \& P_E$ Cyclic displacement $\Delta L_C \& \Delta L_E$	$\varepsilon_{a(SA)} = 0.005$ % ÷ 0.1 % (triangular wave)
	Cyclic loading rate, $\dot{\varepsilon}_a$	0.001 ÷ 0.5 %/min (slope of triangular wave)
	Temperature	$< \pm 2°$ C
	Cell pressure, σ_r Back pressure, u_b Pore water pressure, u	$\sigma_r = \pm 0.5$ kPa $u_b, u = \pm 0.5$ kPa
Measurement accuracy	Cell pressure, σ_r Back pressure, u_b Axial stress σ_a	$\Delta\sigma_r = \pm 0.5$ kPa $\Delta u_b = \pm 0.5$ kPa $\Delta\sigma_a = \pm 0.5$ kPa
	Axial load, P	$P = \pm 50$ N (max amplify: $P = \pm 0.5$ N)
	Axial deformation, ΔH	Local $\varepsilon_a = \pm 0.001$ %. Internal $\varepsilon_a = \pm 0.004$ %. External $\varepsilon_a = \pm 0.018$ %.
	Lateral deformation	$\varepsilon_r = \pm 0.002$ %.
	Volume change, ΔV	$\Delta V = \pm 2$ mm³
	Measurement frequency	> 40 data sets in each cycle
Others		E_{eq} and h are calculated for any cycle in each stage

Cyclic torsional shear test apparatus (Politecnico di Torino)

Purpose		The deformation characteristics of soils in an isotropic/anisotropic stress state when subjected to cyclic undrained/drained triaxial loading.
Soil type		saturated cohesive soils and sandy soils
Apparatus & tools	Load cell	A coil-magnet, resonant column actuator is used to apply torque to the specimen. It is an electric motor with eight coils fixed to support and four magnets rigidly connected to the specimen top cap; The weight is counterbalanced by a spring. The current passing through the coils (Volts) is measured during a test. A torque-voltage calibration factor is used to transduce the volts into torque and thereafter into shear stress. The maximum torque available is 1.1 N·m.
	Strain measurement	The axial strain is measured using a high-resolution LVDT to monitor the aluminium top-cap displacement. Range, resolution and accuracy of these transducers are respectively 4.0 mm, 0.2 μm and 0.02 %FSO. Radial strain is measured using two couples of proximity transducers located at the mid-height of the specimen and by monitoring both internal and external radius displacement. A small piece of aluminium foil, attached to the external membrane with silicon grease, is used as a target. Range, resolution and accuracy of these transducers are respectively 2.5 mm, 0.3 μm and 0.05 %FSO. Shear strain is measured by monitoring the top rotation with a couple of high-resolution proximity transducers. The steel target are fixed onto the top cap. Their correct alignment on the same diameter and their perpendicularity to the sensor is achieved by using an aluminium element appropriately machined. Range, resolution and accuracy of these transducers are respectively 1.5 mm, 0.06 μm and ±0.5 μm, corresponding to top rotation ($\Delta\theta$) of 0.05 rad, $2 \cdot 10^{-6}$ rad and $\pm 1.7 \cdot 10^{-5}$ rad
	Top and pedestal	Porous stones screwed on top cap and bottom base
	Rotational range	$\Delta\theta = 0.05$ rad
	Data sampling	*Cyclic loading rate: < 0.01 Hz* A data acquisition unit is used to convert all the analog signal of the transducers in digital output. The maximum sampling rate is: • 0.1 Samples/s for channel *Cyclic loading rate: 0.01 ÷ 10 Hz* Three HP 3421A digital multimeters are used to store the analog signal of the transducers for the shear strain measurements and for the current passing through the coils (Volts). The maximum sampling rate is: • 1000 Samples/s, data storage buffer: 512 readings
	Rubber membrane	~0.3 mm thick, ~0.95D
	Filter paper	0.2 ÷ 0.3 mm thick, used for impermeable specimens
	O-ring	Ø internal ~0.7D, thick 3.53 mm
	Other requirements	Commercial volume gauge An air cylinder fixed to the concrete anchor block provides a vertical consolidation stress. The cylinder transmits the vertical load to the top by means of a steal strand. The maximum obtainable vertical stress is about 300 kPa. A very thin strand does not affect either resonant column or torsional shear test results as is demonstrated by appropriate calibration.

		Specimen geometry		
Specimen	Diameter, D (mm)		Hollow		Solid
		$D_{external}$	$D_{internal}$	D
		50 mm	30 mm	50 mm
		70 mm	50 mm	70 mm

	Height, H (mm)	= 2D			
	Accuracy in measuring specimen dimensions	Dimension	Capacity	Resolution	Accuracy
		H		0.01mm	±0.03 mm
		D		0.01mm	±0.03 mm
		Weight	1200g	0.1g	
	Method of specimen preparation	• Trimming method • Air pluviation			
	Saturation of specimen	Applying high back pressure			
	Back pressure	Usually step of 50 kPa up to 200 kPa			
	B-value	> 0.95			
	Other requirements				
Test operation & control accuracy	Cell pressure, σ_r Back pressure, u_b Axial stress σ_a	$\Delta\sigma_r = \pm 0.5$ kPa $\Delta u_b = \pm 0.5$ kPa $\Delta P = \pm 0.002$ kN $\Rightarrow$ $\Delta\sigma_a = \pm 1.0 \div 1.5$ kPa (depending by hollow specimen geometry)			
	Lateral strain, ε_r				
	Consolidation rate				
	Consolidation period	The test pause at the consolidation stress until the creep rate falls to 0.05 %/day. (at least 24 hours for saturated cohesive soils)			

	Single amplitude cyclic shear strain $\gamma_{(SA)}$ Cyclic torque T_R&T_L	$\gamma_{(SA)}$ = < 0.001 % + 0.1 % (triangular wave)
	Cyclic loading rate, $d\gamma$	0.001 ÷ 10 Hz
	Temperature	< ±2° C
Measurement accuracy	Cell pressure, σ_r Back pressure, u_b Pore water pressure u	$\sigma_r = \pm0.5$ kPa $u_b, u = \pm0.5$ kPa
	Axial load, P	$P = \pm0.002$ kN
	Torque, T	$T = \pm2.5 \cdot 10^{-5}$ N·m
	Axial deformation, ΔH	$\Delta H = \pm0.8$ μm
	Lateral deformation	$\Delta R = \pm1.0$ μm
	Rotational angle	$\Delta\theta = \pm1.7 \cdot 10^{-5}$ rad
	Volume change, ΔV	
	Measurement frequency	> 40 data sets in each cycle
Others		G_{eq} and h are calculated for any cycle in each stage

Table 1 - *Range of values of some relevant physical properties and classification characteristics of Augusta clay specimens*

Depth of retrieved samples (m)	Unit weight γ (kN/m³)	Specific gravity G_s	Natural water content w_n (%)	Liquid limit w_L (%)	Plastic limit W_P (%)	Plasticity index PI	Liquidity index I_L	Void ratio e	Degree of saturation S_r (%)	Clay fraction CF (%)	Calcium Carbonate ($CaCO_3$) content (%)
13.5-46	18.2-19.4	2.70-2.78	29-38	57-74.5	22-31	31-46	0.08-0.34	0.68-1.03	94-100	56-70	20

rocks". University of Naples Federico II, Naples: 131-190.

Rampello S., Viggiani G.M.B. & Amorosi A. 1997. *Small-strain stiffness of reconstituted clay compressed along constant triaxial effective stress ratio paths*. Géotecnique, 47, No. 3:475-789.

Silvestri F. 1991, *Analisi del comportamento dei terreni naturali in prove cicliche e dinamiche di taglio torsionale*. Ph. D. thesis Università degli Studi di Napoli "Federico II".

3 ITALIAN ROUND ROBIN TEST AND CASE HISTORIES

3.1 Round Robin Test on Augusta clay

3.1.1 Outline
In 1994 a natural Italian clay, named Augusta clay, was selected as a test material for a collaborative laboratory investigation involving different universities and research teams. The primary purpose of this investigation was to evaluate the range in scatter of the deformation characteristics of natural clay, from very small to large strains, when tests were carried out on the same material in different laboratories and by different types of apparatus.

The laboratory tests were carried out from 1995 to 1999 by four research institutions, Politecnico di Torino, Enel.Hydro of Bergamo (former ISMES), and the and University of Catania all in Italy and the University of California at Los Angeles (UCLA) in the USA. The laboratory tests conducted in Italy were the resonant column (RC), monotonic loading torsional shear (MLTS), cyclic loading torsional shear (CLTS) and monotonic loading triaxial (MLTX) tests. The tests performed in the USA were cyclic simple shear tests, carried out in a double specimen direct simple shear (DSDSS) device (Doroudian & Vucetic, 1995).

Augusta clay is a marine pleistocene clay, coming from a quite homogeneous deposit located in the vicinity of the town of Augusta (southeastern Sicily, Italy). The characteristics of the soil profile at the site have been extensively presented elsewhere (Maugeri & Frenna, 1995; Cavallaro, 1997; Lo Presti et al., 1997, 1998). The clay is medium stiff, overconsolidated, with a medium to high plasticity index. Augusta clay samples were retrieved from depths of between approximately 15 and 46 m by means of an 86 mm diameter Shelby tube sampler. The general characteristics and index properties of the tested specimens are summarised in Table 1.

In Table 2 a summary of relevant information about the type of laboratory tests performed is provided, together with details of the index properties of the specimens and test conditions.

Well-established laboratory testing techniques, such as triaxial, resonant column and torsional shear, are extensively described in the literature. On the other hand, the double specimen direct simple shear device employed at the University of California at Los Angeles is a recently constructed non-commercial apparatus described in detail by Doroudian & Vucetic (1995).

The experimental activity involving all the research teams has been documented in several publications (see Table 2). In the following a synthetic description of the laboratory tests is presented with particular emphasis on: (i) comparison between stiffness and damping characteristics; (ii) the effect of strain-rate on the stiffness and damping ratio; (iii) the influence of specimen setting on the stress-strain curve and stiffness.

3.1.2 Comparison of G_0 values from different laboratory tests

Figure 1a shows a plot of the maximum shear modulus, G_0, against the mean effective confining stress, $\sigma'_m = (\sigma'_v + 2\sigma'_h)/3$, from the RC, MLTS, CLTS, TX and DSDSS tests. In the figure a clear trend of G_0 increasing with σ'_m can be recognised in all experimental studies, even if a certain scatter among the data can also be noticed. More specifically, it appears that the smallest G_0 values pertain to the DSDSS tests corresponding to the loading steps and to the MLTX tests corresponding to the wet setting method. The largest G_0 values pertain, with few exceptions, to the RC tests obtained by ISMES (1995). The data referring to the other RC, the MLTS, the CLTS, the MLTX corresponding to the wet setting method, and the DSDSS corresponding to the unloading and reloading tests plot in between the above mentioned set of data and show similar G_0 values at a comparable mean confining stress.

Table 2 - *Summary of Augusta clay data from RC, MLTS, CLTS, MLTX and DSDSS tests*

Reference	Type of test	Sample depth (m)	w_L	I_p	w_n (%)	e_0	e_c*	σ'_v (kPa)	σ'_h (kPa)	$\sigma'_m =$ $(\sigma'_v+2\sigma'_h)/3$ (kPa)
ISMES	RC	15	67	41	31.2	0.908	0.880	140	140	140
(1995)	RC	22.5	67	42	34.9	0.957	0.941	220	220	220
	RC	30	64	40	32.1	0.91	0.881	260	260	260
	RC	37.5	71	45	34.0	0.934	0.907	350	350	350
	RC	46	70	46	33.1	0.899	0.877	430	430	430
Cavallaro	RC	13.5	52	30	25	0.693	0.629	155	155	155
(1997)	CLTS	16.5	52	29	33	0.684	0.659	182	182	182
Lo Presti et al	RC	16.5	52	29	33	0.684	0.630	182	182	182
(1997,1998)	RC	21.5	67	39	30	0.829	0.790	253	253	253
	RC	28.5	68	40	32	0.889	0.709	259	259	259
	RC	29	52	24	28.9	0.847	0.768	250	250	250
	MLTS	35.5	62	38	25.3	0.972	0.834	377	377	377
	RC	35.5	62	38	25.3	0.972	0.811	377	377	377
	CLTS	35.5	62	38	25.3	0.968	0.768	398	398	398
Fioravante	CLTS	24.5	-	-	31.5	0.832	0.803	250	250	250
(1999)	RC	24.5	-	-	31.5	0.832	0.803	250	250	250
Lo Presti et al. (1999)	MLTX- Dry setting	21.5	67	40	35	0.954	0.914	200	129	153
		20.5	-	-	-	0.896	0.884	185	157	166
	MLTX-Wet setting	20.5	-	-	-	0.998	1.017	221	135	164
		20.5	69	41	36	0.998	0.974	187	159	168
Lanzo et al. (1999)	DSDSS	24.5	74.5	44	34.1	0.977	0.929	240	262[#]	254
							0.906	418	397[#]	404
							0.849	857	671[#]	733
							0.812	1254	919[#]	1031
							0.771	1672	1225[#]	1374
							0.775	1254	992[#]	1079
							0.784	857	751[#]	786
							0.809	418	444[#]	435
							0.834	240	292[#]	275
							0.796	857	751[#]	786
							0.757	1672	1225[#]	1374

* e_c = void ratio at the end of primary consolidation

\# σ'_h values were estimated with $\sigma'_h = K_0 \sigma'_v$, where K_0 was calculated from the Mayne & Kulhavy (1982) empirical correlation

In Fig. 1b, in order to eliminate the effect of the difference in density among the tests, the G_o values are normalised by the void ratio function F(e) proposed by Shibuya & Tanaka (1996), i.e. $F(e) = 1/(e^{1.5})$. The scatter is greatly reduced when the influence of the void ratio is taken into account, as the majority of the data fall into a fairly narrow band, with the exception of the data from ISMES (1995).

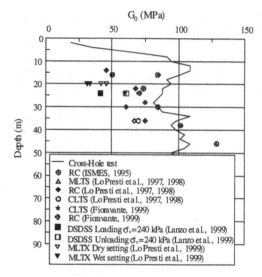

Figure 2   *Comparison between laboratory and field G_o data*

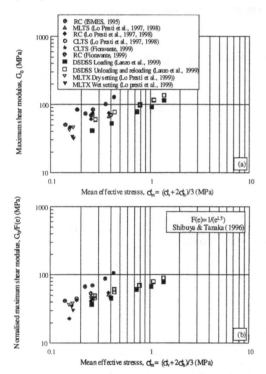

Figure 1   *Comparison between: (a) G_o-σ'_m and (b) $G_o/F(e)$-σ'_m, data obtained from different types of laboratory tests*

3.1.3 Comparison of G_o values from field and laboratory tests

In Fig. 2, the G_o data obtained in the laboratory are plotted with depth and compared to the G_o data obtained by measuring shear wave velocities in the field by the cross-hole method (Maugeri & Frenna, 1995). It can be seen that the G_o laboratory data are lower than the G_o field data. More specifically, differences between laboratory and field data are quite small in a few cases, while in other cases the ratio $G_o(lab)/G_o(field)$ is less than 0.5. A greater degree of sampling disturbance could be responsible for the lower values for the $G_o(lab)/G_o(field)$ ratio observed in some cases.

3.1.4 Comparison of G/G_o-γ and D-γ data from different laboratory tests

The normalised shear modulus, G/G_o, and the damping ratio, D, data obtained from different laboratory tests are plotted versus the shear strain amplitude, γ, in Figs. 3 and 4, respectively. The tests selected for comparison were all conducted on specimens retrieved from depths between about 25 and 35 m, consolidated to mean effective stress values varying between approximately 250 and 400 kPa .

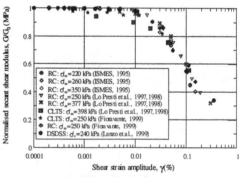

Figure 3   *Comparison of normalized shear modulus, G/G_o, vs. γ data obtained from RC, CLTS and DSDSS tests*

In Fig. 3, it appears that the G/G_o-γ data obtained in the RC, CLTST and DSDSS tests plot in a very narrow band, thus indicating a very good agreement among the different types of tests. The CLTS tests only exhibit G/G_o-γ values falling slightly below those determined in the other tests, at least up to γ=0.03%. The higher non-linearity observed in the CLTS tests has been attributed to rate effect (Lo Presti et al. 1997, 1998). In fact, the CLTS tests have been performed at a constant strain rate, while the other tests have been performed at a more or less constant frequency (i.e. the strain rate increases with strain level). In any case, the overall agreement leads to the conclusion that the G/G_o-γ data of high-plasticity Augusta clay are not very sensitive to factors that otherwise affect the G_o values.

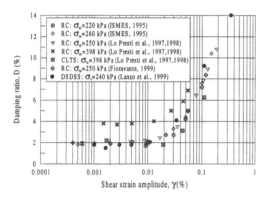

Figure 4   *Comparison of damping ratio, D, vs. γ data obtained from RC, CLTS and DSDSS*

In Fig. 4 the damping ratio, D, data obtained from different laboratory tests are plotted versus the shear strain amplitude, γ. With reference to the cyclic tests (CLTS and DSDSS) , the D values are those determined as an average of the experimental values from the 10th to 20th cycle. As for the normalised shear modulus, in Fig. 4 a very satisfactory agreement between the various D-γ data can be observed. More specifically, it appears that the D-γ data from the CLTS and DSDSS tests clearly form a narrow band, with relatively small scatter in the whole strain range investigated. The only exception is represented by the laboratory RC test with D values plotting considerably above the rest of the data, especially at small strain levels.

3.1.5 Influence of specimen setting and strain rate on the stress-strain curve and stiffness

Specimen setting and rate effect were evaluated by means of MLTX tests. More specifically, conventional wet-setting and dry setting (Lo Presti et al. 1998) methods were used. The stress-strain curves and stiffnesses of duplicated specimens subject to undrained compression in the triaxial apparatus were compared in order to evaluate the effect of the setting method. Naturally one specimen was set up wet and another dry.

The following conclusions were reached:
- The influence of dry setting on the small strain stiffness of Augusta clay is about 30 % as can be seen in Figure 5. A noticeable influence due to setting method was also observed at large strains (Fig. 5).
- The dry setting method showed equivalent results to other methods that inhibit soil swelling, such as the immediate application of a sufficiently large confining stress before the drainage opening (this kind of test was performed at Ismes).

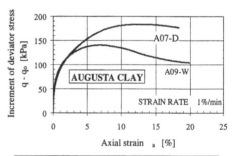

Test (*)	σ'_{vc} [kPa]	σ'_{hc} [kPa]	E_v [MPa]	G_{vh} [MPa]	C_{vh} [-]
A07-D	185	157	110	43	276
A09-W	187	158	75	32	236

(*)D = Dry Setting; W = Wet Setting

Figure 5   *Stress-strain curves obtained from dry and wet setting MLTX tests (Augusta clay)*

For a given strain level, the influence of the strain rate was evaluated by the ratio between the increase measured for the shear modulus in one log-cycle of strain rate and the stiffness obtained at a reference strain rate:

$$\alpha(\gamma) = \frac{\Delta G(\gamma)}{\Delta \log \dot\gamma \cdot G(\gamma, \dot\gamma_{REF})} \tag{1}$$

The above defined empirical coefficient of strain rate expresses the stiffness dependence on strain rate. The values obtained in the case of Augusta clay are plotted in Figure 6. The strain level dependence of the empirical coefficient is clearly seen.

3.1.6 Conclusions

The Italian round robin test on natural Augusta clay was performed between 1996 and 1999 in four geotechnical laboratories, three in Italy and one in the USA. The results of several static and dynamic tests, such as resonant column, monotonic and cyclic torsional shear tests, and monotonic triaxial and cyclic simple shear tests, were systematically analysed and presented. The bulk of the work focused on the comparisons between the shear modulus and damping ratio values of Augusta clay under a wide range of strains, from very small to large, obtained in different laboratories with various apparatus.

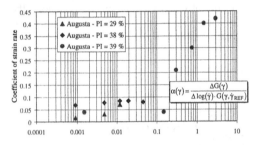

Figure 6   *Coefficient of strain rate α vs. γ from MLTX, CLTS and RC (Augusta clay)*

Despite the inherent differences in the boundary conditions, experimental procedures and interpretation criteria, a very satisfactory agreement can be generally recognised. The main results may be summarised as follow:

Only moderate to negligible differences exist among the values of G_o obtained for Augusta clay in static and dynamic tests when normalisation is made with an appropriate function of the void ratio.

The G/G_o-γ and the D-γ data obtained in resonant column, cyclic torsional shear and cyclic simple shear tests showed generally very consistent results with almost negligible scatter in the whole strain range investigated.

The dry setting method or similar countermeasures to avoid swelling turned out to be necessary.

The large strain stiffness and soil strength are mainly affected by the strain rate.

ACKNOWLEDGMENTS

The authors would like to thank Prof. Mladen Vucetic for allowing the DSDSS tests on Augusta clay to be conducted at the Soil Dynamics Laboratory of the University of California at Los Angeles (UCLA). The DSDSS testing was supported through a grant from the Pacific Earthquake Engineering Research (PEER) Center in the USA. The PEER Center is principally supported by the USA National Science Foundation (NSF).

REFERENCES

Cavallaro, A. 1997. *Influenza della velocità di deformazione sul modulo di taglio e sullo smorzamento delle argille*. PhD thesis. Università degli Studi di Catania (in Italian)

Doroudian, M. & Vucetic, M. 1995. *A direct Simple Shear Device for Measuring Small-Strain Behavior*. Geotechnical Testing Journal, GTJODJ, 18, No. 1, pp. 69-85.

Fioravante, V. 1999. *Personal communication*.

ISMES (1995). *Augusta (località Saline). Prove geotecniche di laboratorio*. Prog. DTA/7368; Doc. RTA-DTA-1442 Rev. 00.

Lanzo G., Vucetic M. & Doroudian M. 1999. *Small-strain cyclic behavior of Augusta clay in simple shear*. Proc. of the 2nd Int. Symposium on Pre-failure Deformations Characteristics of Geomaterials. IS-Torino99, Torino, Italy, 28-30 settembre 1999, Vol. 1, 213-220.

Lo Presti, D.C.F., Jamiolkowski, M., Pallara, O. & Cavallaro, A. 1997. *Shear modulus and damping of soils*, Géotechnique, 47, No. 3, pp. 603-617.

Lo Presti, D.C.F., Pallara, O., Maugeri, M. & Cavallaro, A. 1998. *Shear modulus and damping of stiff marine clay from in-situ and laboratory tests*, Proc. First Int. Conf. on Site Characterization, Atlanta, Georgia, 2, pp. 1293-1300.

Maugeri, M. & Frenna, S.M. 1995. *Soil-response analyses for the 1990 South-East Sicily earthquake*, Proc. Third Int. Conf. on Recent Adv. in Geotech. Earth. Eng. Soil Dynam., 2, pp. 653-658.

Shibuya, S. & Tanaka, H. 1996. *Estimate of elastic shear modulus in Holocene soil dposits*, Soils and Foundations, Vol. 36, 4, pp. 45-55.

3.2 Case histories

Table 3 lists some referenced research studies on the application of the proper assessment of the stress-strain relationship of geomaterials to full scale analyses of both static and seismic problems, most of which were addressed at the interpretation of an observed behaviour, rather than at blind predictions. The following applied research mainstreams can be individuated:

- 1D and 2D seismic response analyses addressed at microzonation studies;
- performance (mainly settlements) of shallow and deep foundations under working loading conditions;
- prediction of the deformation behaviour of earth dams.

Only two case histories are described in some detail in the following sections.

REFERENCES

Burghignoli A., D'Elia M., Miliziano S. & F.M. Soccodato 1999. *Settlements analysis of a silo founded on a cemented clayey soil.* Italian Geotechnical Journal, 33(3):23-36 (in Italian).

D'Elia M., Miliziano S., Soccodato F.M. & C. Tamagnini 1999. *Observed and predicted behaviour of a silo founded on a cemented soft clayey soil.* Proc. II Int. Symp. on Prefailure Deformation of Geomaterials, Torino, 1:741-748. Balkema, Rotterdam.

Lollino P., Cotecchia F. & L. Monterisi 1999. *Comportamento di una diga in calcarenite e della sua fondazione in argille pleistoceniche in base a dati di monitoraggio.* XX Convegno Nazionale di Geotecnica, Parma, 1:497-503. Patron, Bologna.

Mancuso C., Mandolini S., Silvestri F. & C. Viggiani 1997. *Non-linear analysis of the settlement of axially loaded piles.* Proc. XIV ICSMGE, Hamburg, 2:839-842. Balkema, Rotterdam.

Mancuso C., Mandolini A., Silvestri F. & C. Viggiani 1999. *Prediction and performance of axially loaded piles under working loads.* Proc. II Int. Symp. on Prefailure Deformation of Geomaterials, Torino, 1:801-808. Balkema, Rotterdam.

Mandolini, A. & C. Viggiani 1997. *Settlement of piled foundations.* Géotechnique, (47)4:791-816.

Olivares, L. & F. Silvestri 1995. *Observations on the pre- and post-cyclic compression and stiffness properties of a reconstituted high plasticity clay.* Proc. Int. Symp. on "Compression and consolidation of clayey soils", Hiroshima (Japan), 1:155-160. Balkema, Rotterdam.

Olivares, L. & F. Silvestri 2001. *A laboratory and numerical investigation on a post-seismic induced settlement in Southern Italy.* Proc. IV Int. Conf. on 'Recent Advances in Geotechnical Earthquake Engineering and Soil Dynamics', S. Diego (in press).

Rampello S. 1994a. *Observed behaviour of large diameter bored piles in medium to stiff clays.* Proc. Int. Workshop on Pile Foundations – Experimental investigations, analysis and design, 261-302. Hevelius Ed., Benevento.

Rampello S. 1994b. *Soil stiffness relevant to settlement prediction of the piled foundations a Pietrafitta.* Proc. Int. Workshop on Pile Foundations – Experimental investigations, analysis and design, 401-416. Hevelius Ed., Benevento.

Vinale F. 1988a. *Caratterizzazione del sottosuolo di un'area-campione di Napoli ai fini di una microzonazione sismica.* Rivista Italiana di Geotecnica, 22(2):77-100 (in italian). ESI, Napoli.

Vinale F. 1988b. *Microzonazione sismica di un'area-campione di Napoli.* Rivista Italiana di Geotecnica, 22(3):141-162 (in italian). ESI, Napoli.

CASE No. A (Seismic Microzonation of Castelnuovo Garfagnana in Central Italy)

Location
Castelnuovo Garfagnana (Tuscany)

Type of activity
The Tuscany regional government has started a comprehensive project for the evaluation of site effects in about 60 municipalities, located in the territories of Garfagnana and Lunigiana that are under its authority (Ferrini 2000). For this purpose a multidisciplinary task force has been working in order to assess the elastic response spectra taking into account both topographic and stratigraphic amplifications. The main objective is to provide a tool for land use planning, design of new structures and retrofitting of the existing ones. This kind of activity has recently been more or less completed for Castelnuovo Garfagnana, a town of about 6500 inhabitants and an area of about 28 Km^2.

The historical centre of Castelnuovo is located at the top of a hill and mainly consists of old masonry buildings with a maximum of four floors that have resisted the earthquakes thet have occurred in recent centuries. Offices and commercial activities are mainly concentrated in the historical centre. The alluvial valleys, which developed along the Serchio river and the Turrite Secca stream, mainly host new

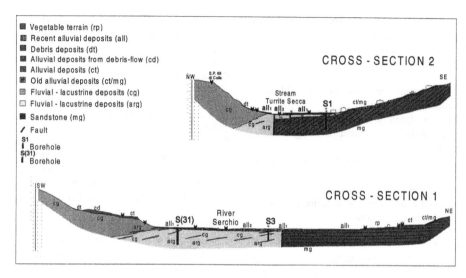

Figure A1   *Geological Cross-Sections (courtesy of Regione Toscana)*

residential areas and industries (paper mills to the value of US$ 200m.). Information about the seismicity in the valleys is not available. For the whole area, the national codes assume a maximum design acceleration of 0.07g for structures with a period not greater than 0.8 s. The national codes allow the definition of specific design spectra and amplification coefficients for a given area, based on the evaluation of local site effects.

Ground conditions

Figure A1 shows two geological cross sections that have been obtained from a geological survey and borehole logging. It is possible to recognise the following gelogical formations (Puccinelli 2000, Pochini 2000):
- Alluvial deposits (ALL) from the Holocene that mainly consist of well-graded gravels with sand and silt of variable thickness.
- Fluvial-lacustrine formations from the Plio-Pleistocene epoch (i.e. 1.5 to 3 million year old) that mainly consist of i) clay and grey sand, sandy clay and clayey sand of a thickness of up to 90 m (ARG) and ii) gravel and conglomerate in a clayey sandy matrix and pebbles of sandstone (CG, C/MG) of a thickness greater than 50 m.
- Macigno sandstone (MG) from the Oligocene epoch (i.e. 20 to 30 million year old) of a thickness of up to 2000 m. This formation is strongly

weathered at the top and becomes intact after a transition zone. The thickness of the weathered zone ranges from 10 to 20 m while that of the transition zone is much more variable.

The new development areas of Castelnuovo are mainly founded on the alluvial deposits, while the historical centre is located on the sandstone outcrop.

Laboratory tests

Laboratory tests have been performed on specimens retrieved by means of a Shelby tube sampler from the ARG formation and by means of a double core sampler from the MG (Lo Presti 2000).

In particular 6 resonant column (RC) and cyclic loading torsional shear tests (CLTS) were performed on ARG specimens using the same apparatus. CLTS tests were performed on the same specimens previously subjected to a RC test, after a rest period of 24 hrs with open drainage. CLTSs were performed under stress control conditions by applying a torque with a triangular time history at a frequency of 0.1 Hz. The maximum shear strain applied during CLTS was less than 0.05 %, because of the limited torque capability. One specimen was first subjected to a Cyclic Loading Triaxial test (CLTX), performed at a constant strain rate. Six different strain levels of progressive amplitude were imposed on the specimen. For each strain level 30 cycles were applied. The maximum applied axial strain (single amplitude) was about 0.07 %. The same specimen was subjected, after a rest period of 24 hrs with open drainage, to Monotonic Loading Triaxial test (MLTX). The shear wave velocity was measured by means of bender elements (BE) during both CLTX and MLTX in order to assess the "elastic" damage due to the progressive straining imposed on the specimen.

The normalised shear modulus G/G_o and damping ratio (D) obtained from RC and CLTS are shown in Figure A2. The same shear modulus decay is obtained from both types of test, but the damping ratio values provided by CLTS are much smaller than those measured in the RC tests. The results from CLTX and those from the RC and CLTS tests are compared in Figure A3. It is noticeable that theCLTX tests show a greater non-linearity in comparison to the other two types of test, while the damping ratio values from CLTX and those from CLTS are comparable. It should be remembered that

CLTX was performed at a constant strain rate equal to 0.01 %/min. Yet, the different deformation mechanism (different stress-path) could be responsible for the observed differences.

The effect of the number of loading cycles is shown in Figure A4 in terms of the degradation parameter ($t = -\log \delta / \log N$) and in Figure A5 by comparing the results of CLTX to those from MLTX. It is possible to see, at large strains, the progressive decrease of soil stiffness with the increase in N so that the stiffness due to CLTX becomes even smaller that that due to MLTX. It is possible to compute the following parameter ($n = 2\varepsilon_c / \varepsilon$), that is the ratio of the strain levels which provide the same stiffness due to CLTX and MLTX respectively. Table A1 summarises the values of the n parameter. It should be noticed that when n=2 the original 2nd Masing rule holds, while for n greater or smaller than 2 hardening or softening phenomena take place.

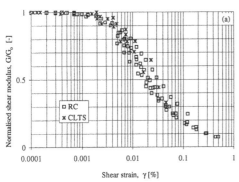
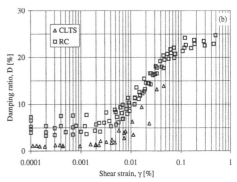

Figure A2 *Normalised shear modulus and damping ratio from RC and CLTS tests (Castelnuovo di Garfagnana)*

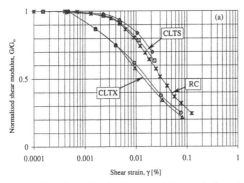
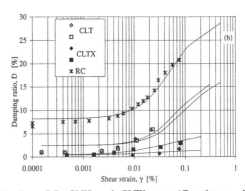

Figure A3 *Normalised shear modulus and damping ratio from RC, CLTS and CLTX tests (Castelnuovo di Garfagnana)*

Table 1 *Variation in the Masing scale-amplification factor with axial strain and number of loading cycles*

No. of cycles [-]	ε_c [%]	$\Delta\sigma_c$ [kPa]	n [-]
1 to 29	0.000405	3.67	4.46
1 to 29	0.001400	11.05	2.02
1 to 29	0.003495	23.74	2.27
1	0.008149	45.59	2.72
2	0.008222	45.19	2.62
3	0.008240	45.07	2.58
4	0.008386	45.56	2.57
5	0.008398	45.10	2.49
6	0.008361	44.75	2.45
10	0.008480	44.60	2.39
1	0.027273	93.66	2.62
2	0.027560	92.99	2.52
3	0.027745	93.77	2.56
4	0.028075	93.98	2.53
5	0.028068	93.24	2.49
6	0.028073	0.37	2.49
7	0.028180	93.30	2.48
8	0.028149	93.20	2.48
9	0.028213	92.24	2.41
13	0.028405	92.16	2.39
14	0.028375	91.58	2.35
15	0.028488	91.57	2.33
16	0.028498	90.73	2.29
1	0.06671	150.76	2.55
2	0.06773	146.92	2.31
3	0.06825	147.44	2.29
4	0.06879	146.66	2.17
5	0.06882	145.09	2.06
6	0.06891	144.10	1.98
7	0.06895	143.43	1.94
8	0.06921	142.71	1.86
9	0.06944	143.53	1.88
10	0.06941	142.08	1.80
11	0.06974	141.27	1.72
12	0.06956	140.76	1.70
13	0.06976	140.96	1.70
14	0.06980	141.32	1.71
15	0.06979	140.27	1.66
16	0.06970	138.79	1.57

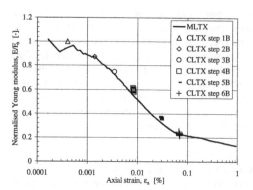

Figure A5 *Normalised Young modulus from MLTX and CLTX*

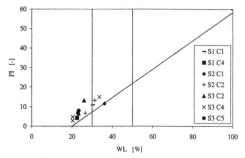

Figure A6 *Soil classification (Castelnuovo di Garfagnana)*

Classification test results on ARG specimens are summarised in Table A2 and in Figures A6 to A11. The soil samples are mainly classified as CL-ML and rarely as SM-SC. In practice, they are well graded, very dense soils of low plasticity.

Four K_o-Odometer tests were performed in order to assess the soil state. However, due to sample disturbance the assessment of a reasonable preconsolidation stress was possible only in one case. The G_o value, inferred from the reconsolidation stage to the in-situ vertical stress in the triaxial cell, was about 0.8, i.e. slightly larger than that observed in the odometer test.

Table 2 *Main physical and index properties of the ARG (Castelnuovo di Garfagnana)*

Borehole	S1	S1	S1	S2	S2	S3	S3	S3	S3
Sample	C1	C2	C4	C1	C2	C1	C2	C4	C5
Depth [-]	8.60 9.20	10.30 10.70	26.00 27.00	2.10 2.50	8.60 9.20	11.20 11.80	14.50 15.10	25.50 26.10	31.10 31.70
Type of tests	MLTX Class	Class	Class	MLTX CLTX Class	RC CLTS Class	ED RC CLTS Class	ED RC CLTS Class	ED RC MLTX CLTX CLTS Class	ED RC CLTS Class
Specific gravity, G_s [-]	2.672	2.673	2.684	2.689	2.740 2.726	2.766	2.693 2.741	2.736 2.709 2.712	2.711 2.723
Natural volume weight, γ_n [kN/m^3]	-	22.73	20.77	18.80	20.55	20.96	21.19	21.14	21.38
Void ratio, e_o [-]	-	0.225	0.425	0.779	0.518	0.485	0.435	0.457	0.427
Natural water content, W_N [%]	11.46	5.69	12.49	32.96	14.26 10.34	14.13	14.42 11.47	16.35 12.85 13.52	16.41 10.79
Liquid limit, W_L [%]	30.34	23.06	22.56	36.15	31.36 26.63	22.22	26.10 23.33	33.57 20.40 20.38	23.43 23.08
Plastic limit, W_P [%]	19.40	-	18.31	24.61	18.16 19.96	-	12.91 16.83	18.64 15.67 17.56	15.48 16.33
Plasticity index, IP [%]	10.94	-	4.25	11.54	13.20 6.67	-	13.18 6.50	14.93 4.73 2.82	7.95 6.75
Consistency index, IC [%]	1.73	-	2.37	0.28	1.30 2.44	-	0.89 1.82	1.15 1.60 2.43	0.88 1.82
Total carbonate $CaCO_3$ [%]	1.00	1.00	-	1.90	11.60 22.20	23.20	20.30 24.10	1.45 21.20 23.35	24.15 23.75

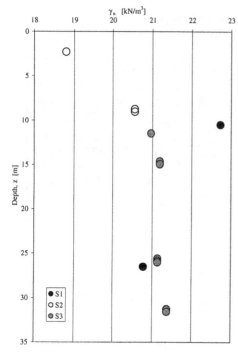

Figure A7   *Natural Volume weight (Castelnuovo di Garfagnana)*

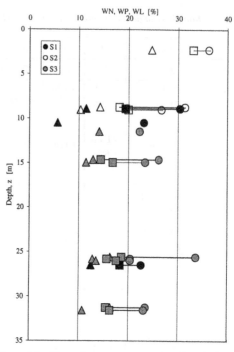

Figure A9   *Natural water content, liquid and plastic limits (Castelnuovo di Garfagnana)*

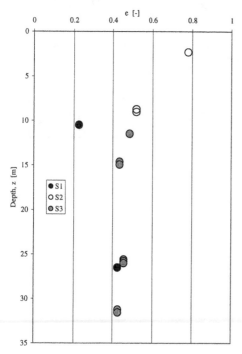

Figure A8   *Void ratio (Castelnuovo di Garfagnana)*

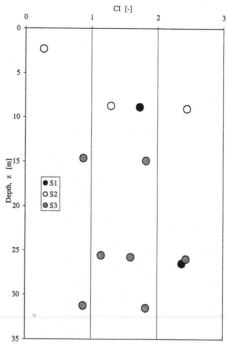

Figure A10   *Consistency   index   (Castelnuovo   di Garfagnana)*

36

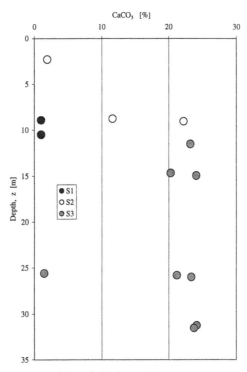

Figure A11 *Total carbonate content (Castelnuovo di Garfagnana)*

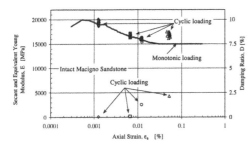
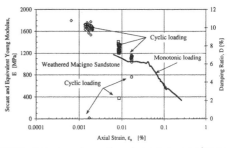

Figure A12 *Stiffness and damping ratio from MLTX and CLTX (Intact and weathered macigno sandstone, Castelnuovo di Garfagnana)*

One specimen of weathered MG and another of intact MG were firstly subjected to unconfined cyclic compression loading and then to unconfined monotonic compression loading tests. The normalised stiffness decay and damping ratio are shown in Figure A12.

Local axial and radial strain measurements were performed during triaxial tests performed on both the ARG and MG.

In-situ tests

Different kinds of geophysical tests were performed in the area under study in order to determine the shear wave velocity and hence the small strain shear modulus (Signanini 2000, Foti and Socco 2000). Two down hole (DH) tests were performed in the historical centre and another in the alluvial plain close to the paper-mill. DH tests have been performed using a pair of receivers and interpreting the wave traces by means of the visual true interval method. An SPT probe was done in the borehole, subsequently used for the down hole tests.

Seismic refraction tests using horizontally polarised shear waves (SH) and special SASW tests (Foti and Socco 2000) were performed at the same locations as the DH. An array of 24 geophones was used for the special SASW tests that were interpreted in the frequency-wave number domain (fk) (Foti and Socco 2000).

Figure A13 shows a comparison among the shear wave velocities inferred by means of different geophysical tests at the S3 borehole location. It is possible to see that the ARG has a very high initial shear modulus. On the whole there is good agreement among the different tests results, even though it is possible to observe a better agreement between the DH and SASW test results in the first few meters especially. It is also worthwhile remarking that the shear wave velocity profile from the DH and SASW is more accurate than that inferred from the SH test.

Comparison (laboratory-field)
The small strain shear modulus of ARG inferred from the RC and CLTX tests is plotted vs. depth in Figure A14. The G_o values inferred from the CLTS and MLTX tests are almost the same. The Figure also shows the small strain shear modulus inferred from DH tests at the S3 location. It is possible to observe that laboratory tests largely underestimate the initial stiffness probably because of sampling disturbance which can be very important when using Shelby samplers in very dense low plasticity soils. The poor sample quality should not influence the shear modulus decay curves if the sampling disturbance influences the soil stiffness at small and at large strains as well.

As far as the MG formation is concerned, the following values of shear wave velocity were inferred from the in-situ and laboratory tests:

Type of soil	$V_s[m/s]$ Situ	$V_s[m/s]$ Laboratory
Weathered sandstone	600	500 - 550
Transition zone	900	
Intact sandstone	1200-1500	1600-1700

The shear wave velocities inferred from in-situ and laboratory tests show very similar values. In the case of intact sandstone, those measured in the laboratory are even greater than those determined in situ. One may assume that the effect of sampling disturbance is not very important in the case of rock specimens. On the other hand, possible scale effects could explain why the shear wave velocity of intact sandstone determined from laboratory tests is larger than that obtained from in-situ tests.

The shear wave velocity was obtained from triaxial tests as $V_s = \sqrt{E_o/2(1+\nu)\rho}$ where: E_o = small strain Young modulus, ν = Poisson ratio assumed equal to 0.2 and ρ = mass density.

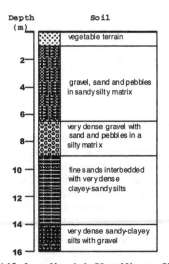
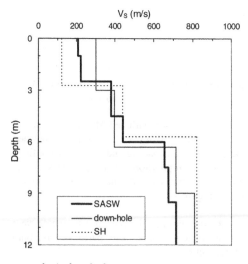

Figure A13 *Log of borehole S3 and Vs – profiles from different geophysical methods*

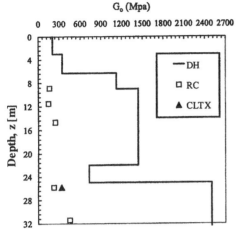

G_o (Mpa)

Figure A14 *Comparison between Go from in-situ and laboratory tests*

In the case of the RC or torsional shear tests the following relation has been used $V_s = \sqrt{G_o / \rho}$, where G_o = small strain shear modulus

Analysis results

On the basis of historical seismicity data, attenuation law (Sabetta & Pugliese 1996) and seismogenetic zone information (Scandone 1999), Pergalani et al. (2000) have obtained average response spectra. Input accelerograms on rock have been inferred from average response spectra on rock according to Sabetta & Pugliese (1996).

The analysis has not yet been completed. In any case, the stratigraphic amplification was evaluated by using a 1D linear equivalent code like SHAKE (Schnabel et al 1972), a 1D non linear code such as ONDA (Lo Presti et al. 2000) or a 2D code (Pergalani et al. 2000) similar to QUAD4M (Hudson et al. 1994), i.e. a 2D code with a compliant rock base. Topographic amplification has been evaluated comparing the results of the 1D analysis to that inferred by the 2D analysis.

ACKNOWLEDGEMENTS

The above described microzonation study was supported by the Tuscany Regional Government which also authorised the printing

REFERENCES

Ferrini M. 2000. *Il programma regionale VEL: le indagini ai fini della riduzione del rischio sismico negli strumenti urbanistici.* La riduzione del rischio sismico nella pianificazione del territorio: le indagini geologico tecniche e geofisiche per la valutazione degli effetti locali, CISM, Lucca 3-6 Maggio 2000.

Foti S. & Socco V. 2000. *Analisi delle onde superficiali per la caratterizzazione dinamica dei terreni.* La riduzione del rischio sismico nella pianificazione del territorio: le indagini geologico tecniche e geofisiche per la valutazione degli effetti locali, CISM, Lucca 3-6 Maggio 2000.

Hudson M., Idriss I.M. & Beikae M. 1994. *A computer program to evaluate the seismic response of soil structures using finite element procedures and incorporating a compliant base.* User's manual CGM, Davis, California.

Lo Presti D. 2000. *Indagini geotecniche per la valutazione degli effetti di sito.* La riduzione del rischio sismico nella pianificazione del territorio: le indagini geologico tecniche e geofisiche per la valutazione degli effetti locali, CISM, Lucca 3-6 Maggio 2000.

Lo Presti D., Lai C. & Camelliti Alessio 2000. *ONDA (One-dimensional Non-linear Dynamic Analysis): A Computer Code for the Non-linear Seismic Response Analysis of Soil Deposits.* Submitted to Italian Geotechnical Review.

Pergalani F. Luzi L. & Petrini V. 2000. *Definizione delle istruzioni tecniche e valutazione quantitativa degli effetti locali di alcune località (Castelnuovo Garfagnana).* Convenzione tra regione Toscana e IRRS.

Pochini A. 2000. *Applicazioni delle metodologie d'indagine ed illustrazione dei risultati per la valutazione delle caratteristiche dei terreni su alcuni Comuni dell'area.* La riduzione del rischio sismico nella pianificazione del territorio: le indagini geologico tecniche e geofisiche per la valutazione degli effetti locali, CISM, Lucca 3-6 Maggio 2000.

Puccinelli A. 2000. *Il contributo geologico e geomorfologico per la valutazione degli effetti di sito: criteri di indirizzo ed applicazioni.* La riduzione del rischio sismico nella pianificazione del territorio: le indagini geologico tecniche e geofisiche per la valutazione degli effetti locali, CISM, Lucca 3-6 Maggio 2000.

Sabetta F. & Pugliese A. 1996. *Estimation of response spectra and simulation of non stationary earthquake ground motion.* Bullettin of the Seismological Society of America, vol. 86, No. 2, pp. 337-352.

Scandone P. 1999. *Modello sismotettonico della penisola italiana.* Rapporto GNDT.

Schnabel P.B., Lysmer J. & Seed H.B. 1972. *A computer program for earthquake response analysis of horizontally layered sites.* User's manual, EERC, Berkley, California.

Signanini P. 2000 *Ricostruzione delle geometrie del sottosuolo mediante prove geofisiche e problematiche connesse con la determinazione del parametro Vs per la modellazione degli effetti di sito. Aspetti metodologici e risultati.* La riduzione del rischio sismico nella pianificazione del territorio: le indagini geologico tecniche e geofisiche per la valutazione degli effetti locali, CISM, Lucca 3-6 Maggio 2000.

CASE No. B Pietrafitta thermal power plant

Location
Pietrafitta (Central Italy)

Type of activity and ground conditions
The behaviour of two piled foundations has been briefly discussed by Rampello (1994a) who showed that a satisfactory agreement between observed and predicted settlement can be found via simplified approaches, assuming the soil to behave as a linear elastic medium with shear stiffness in the range of small to medium strains. The small-strain shear modulus has also proved to be appropriate when predicting the settlement of piled foundations (Mandolini & Viggiani 1997). Piled foundations were constructed at Pietrafitta, in central Italy, where the Italian State Electricity Agency (ENEL S.p.A) is building a new power plant. The site description and mechanical properties of the foundation soil are described in some detail by Rampello (1994b) and by Rampello et al. (1996). The soil profile is characterised by nearly horizontal layers and moving from the ground surface (≈ 225 m a.s.l.) downwards the following layers are encountered:

- clayey waste, medium to stiff, 15-25 m in thickness, with brown coal fragments and thin lenses of sandy silt;
- lignite, 2-4 m thick, which is the unexploited lowermost portion of the original layer;
- peaty clay, stiff, 2-4 m thick, with some brown coal fragments in the upper portion;
- blue silty clay, stiff, 13-18 m in thickness;
- clayey silt, stiff, down to a depth of 130-135 m, including layers of sandy silt. Under the chimney, a main layer of sandy silt is found at a depth of 55 to 69 m.

Laboratory and field tests
Accurate stress-strain relationships were measured in the laboratory on natural samples of Pietrafitta clay carrying out resonant column and triaxial compression tests with local measurements of axial strain (Georgiannou et al., 1991). In Fig. B1 values of the ratio E_u/C_u are plotted against the shear strain ε_s. For shear strains smaller than 0.01 % the non-dimensional undrained stiffness is seen to be nearly constant at $E_u/C_u = 2100$, while in the medium strain range (0.01 % $< \varepsilon_s <$ 0.1 %) a value of $E_u/C_u = 1200$ can be selected at $\varepsilon_s = 0.05$ %.

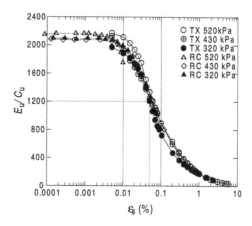

Figure B1   *Non-dimensional stiffness against shear strain*

In-situ measurements of the small-strain shear modulus G_o were obtained from a cross-hole test carried out along the axis of the chimney of the power plant. In Fig. B2 the values of G_o are plotted against the elevation above sea level together with the mean value weighted through the pile length. Dividing the values of G_o by the undrained shear strength evaluated by cone penetration tests, values of the ratio E_u/C_u in the range 2700 to 2900 were obtained for the in-situ measurements at small strains. Non-dimensional shear stiffness from in-situ tests about 30% higher than those measured in the laboratory may derive from both the scatter in the values of the undrained shear strength, and the interpretation of the cross-hole test.

Starting from the observation mentioned above, three different values of the ratio E_u/C_u were referred to when back-analysing the settlements of two piled foundations: namely $E_u/C_u = 2900$, 2100 and 1200.

Comparison between observed and predicted settlements

Settlement prediction was carried out for the productive area which included 4 buildings founded on both piled rafts and strips, and the chimney founded on a circular piled raft of 30.5 m in diameter. Figure B3 shows the piled foundations of the productive area; it consisted of 786 bored piles, 1.2 m in diameter, and covered an area of 108x104 m. The settlements were measured by precision levelling on 72 benchmarks installed in 13 sections of the area. The piled foundation of the chimney is shown in Fig. B4: it included 74 piles of diameter D = 1.2 m, and settlement was measured by precision levelling on 9 benchmarks installed along two perpendicular sections.

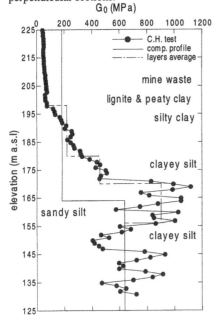

Figure B2 *Profile of G_o from the Cross-Hole test*

Settlement of the piled foundations may be evaluated using the simplified approaches of the equivalent raft or of the equivalent pier. According to Randolph & Clancy (1993), the degree to which a pile group behaves as a raft or a pier is governed by the parameter $R = \sqrt{(n \cdot s/l)}$, where n is the pile number, s is the pile spacing and l is the embedded pile length. For values of $R < 3$ the equivalent pier approach should be preferred, while for values of R > 3-4 the pattern of differential settlement resembles that of a fully flexible shallow foundation. The equivalent pier approach was then used for the chimney, for which $R = 2.2$, and the equivalent raft

approach was adopted for the productive area, for which $R = 5$.

The construction of the chimney took about 2 months and occurred in three stages: the circular raft was first constructed and the chimney core was then completed in 3 subsequent stages in which the height was increased to 1/3, 2/3, and then to its final value (140 m). The equivalent pier approach was used following the simplified procedure proposed by Poulos (1992) for piles in a layered soil. By this procedure, an approximate solution is obtained representing the piles as end bearing piles in an equivalent homogeneous stratum. The soil modulus assumed along the pier length and at the pier base are those shown in Fig. B2. In Fig. B5, the average settlements are plotted against the loads applied during construction of the chimney. The settlements were computed assuming undrained conditions and using three values of the ratio E_u/C_u. The best agreement with the observed behaviour was observed for a value of about E_u/C_u = 2100, which represents the non-dimensional shear stiffness of the soil measured at small strains in resonant column tests.

The settlement prediction of the productive area was carried out taking into account the different loads acting on the different buildings and assuming the loads applied down to a depth of 40 m, which corresponds to 2/3 of the pile length embedded in the in-situ soil. The loads applied at the time of the analysis were about 39 % of the final design load, and were applied over a time period of about 5 months. The best estimate of the observed settlement was obtained using a value of E_u/C_u = 1200, which is a measure of the stiffness in the range of medium strains (0.01% < ε_s < 0.1 %). A comparison between predicted and observed settlement is shown in Figure B6 for 4 analysed sections. The results are seen to be in reasonable agreement with the measurements, apart from the settlement computed close to the foundation edges where the foundation stiffness is likely to yield a measured settlement smaller than the computed one.

Using simplified approaches for estimating the immediate settlement of two piled foundations in medium to stiff clay, the small-strain shear modulus yielded a close agreement with the measured settlement in the case of the chimney, while the shear stiffness in the range of the medium strains was appropriate in the case of the productive area. In the latter case, consolidation settlement could have partially developed during construction, thus rendering the assumption of undrained conditions not entirely appropriate.

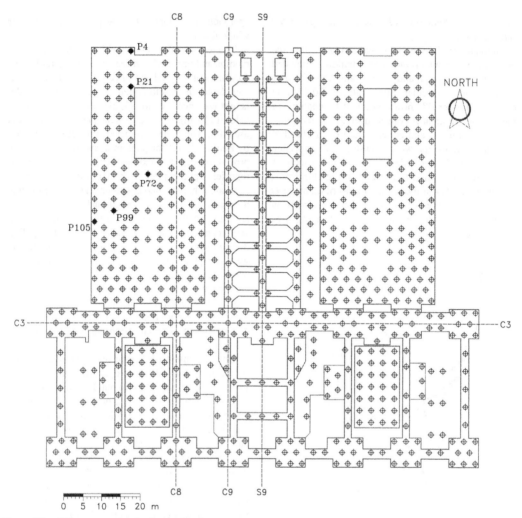

Figure B3   *Pile group under the Productive Area*

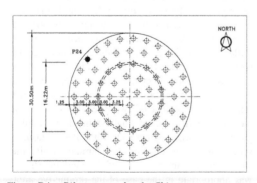

Figure B4   *Pile group under the Chimney*

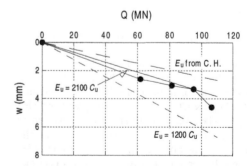

Figure B5   *Measured and predicted settlement of the Chimney*

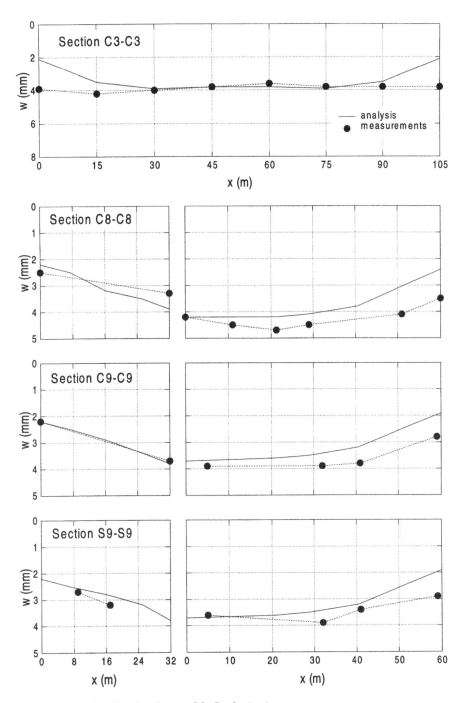

Figure B6 *Measured and predicted settlement of the Productive Area*

REFERENCES

Georgiannou V.N., Rampello S. & Silvestri F. (1991) - Static and dynamic measurements of undrained stiffness of natural overconsolidated clays. *Proc. X ECSMFE*, Florence, Balkema, 1:91-95.

Mandolini A. & Viggiani C. (1997) – Settlement of piled foundations. *Géotechnique*, 47, No. 4:791-816.

Poulos H.G. (1992) – Pile foundation settlement prediction – Hand and computer methods. *Research Rep. No. R661*. University of Sydney, Australia.

Rampello S. (1994a) – Soil stiffness relevant to settlement prediction of the piled foundations at Pietrafitta. *Int. Workshop on "Pile Foundations: Experimental Investigations, Analysis and Design"*, Università degli Studi di Napoli Federico II, Napoli: 407-415.

Rampello S. (1994b) – Observed behaviour of large diameter bored piles in medium to stiff clay. *Int. Workshop on "Pile Foundations: Experimental Investigations, Analysis and Design"*, Università degli Studi di Napoli Federico II, Napoli: 261-301.

Rampello S., Viggiani G. & Calabresi G. (1996) – Observed land subsidence due to the consolidation of a mine waste. *Proc. 2nd Int. Congress on Environmental Geotechnics*. Osaka, Balkema, 1:397-402.

Randolph M.F. & Clancy P. (1993) – Efficient design of piled rafts. *Proc. Conf. on Deep Foundation on Bored and Auger Piles*. Ghent, Balkema: 119-130.

Advanced Laboratory Stress-Strain Testing of Geomaterials, Tatsuoka, Shibuya & Kuwano (eds),
© *2001 Taylor & Francis, ISBN 90 2651 843 9*

A Review of Laboratory Equipments for Advanced Stress-Strain Testing in Greece

V.N. Georgiannou
National Technical University of Athens, Member of TC29, ISSMGE

ABSTRACT: The laboratory equipment currently in use in Greece refers to both static and dynamic testing. Test equipment and ancillary systems have been significantly developed in the recent years and particularly the advanced type of testing equipment. Both testing techniques and instrumentation characteristics associated with the later are presented herein.

1 EQUIPMENT

The experimental methods cover three different testing techniques: static, dynamic and pulse techniques. Greek, British and American standards are applied during shear testing of geomaterials in the laboratory. The testing equipment is described next:

1.1 *Triaxial tests:*

The equipment used consists of hydraulic triaxial stress path cells (Bishop & Wesley, 1975). These cells are automatically controlled to follow various stress paths and they can perform static and slow cyclic stress or strain controlled testing for extended periods. Apart from standard instrumentation such as strain gauged load cells and volume change transducers local instrumentation is mounted on the specimens. This comprises a mid-height pore pressure probe (Hight, 1982) and gauges for local small strain measurements (Burland and Symes, 1982) capable of measuring strain levels ranging from $5*10-4$ % to 10% for the investigation of soil stiffness. In addition, stiffness characteristics at very small strains can be obtained from measurements of the velocities of dynamic waves using piezoceramic transducers (Schulteiss, 1981). These are placed on the top and bottom platens.

Hydraulic pressure is regulated through high resolution digitally controlled manostat valves. Stresses can typically be resolved to within 0.2kPa.

1.2 *2. Cyclic Triaxial tests and Cyclic Torsional Shear tests on Hollow Cylindrical Specimens*

Cyclic tests, referred to also as dynamic tests to distinguish them from the slow cyclic tests which can be performed with the stress path cells, are usually carried out in Greece to determine strength and deformation characteristics of soils and liquefaction potential of saturated sands. The apparatuses used for the tests comply with the typical cyclic triaxial pressure cell as described in the ASTM (D3999-91) standards for cyclic triaxial tests. The internal measurement of strain is not usually available and the test is used for strain levels greater than about 0.01%.

Cyclic torsional shear tests on hollow cylindrical specimens can also be performed. The apparatus used is in compliance with the type described in the Japanese Standards JGS 0543-2000 (see also progress report on the activities of the Japanese domestic committee of the TC29, Kuwano et al, 2000). The inner and outer diameter of the specimen is $D_i=40mm$ and $D_0=70mm$ respectively, and the specimen height is 140mm. The torsional load capacity is 300kgf-cm with the achievable dynamic frequency ranging between 0.02 and 1Hz. Uniform and symmetric cyclic torque can be applied to induce shear strains, γ. The sensor measuring the rotational angle is placed inside the cell to measure the rotation of the cap directly. By repositioning the pneumatic actuator for vertical load at the upper center of the cell conventional vertical cyclic tests can be performed.

The apparatus can also be used for static tests to assess the response of the specimens to torsional shear under monotonic loading and to evaluate the effect of rotation of the principal stress (angle α between σ_1 and the vertical). The apparatus is fully automated and capable of applying strain or stess controlled loading conditions to isotropically and anisotropically consolidated specimens.

1.3 Stokoe Resonant Column Device

The resonant column equipment is the type in which the soil specimen is fixed at the bottom and free to oscillate at the top (Richart et al, 1970; Drnevich et al, 1978; Isenhower, 1979). A coil-magnet drive mechanism attached to the top of the specimen is used to vibrate the combined system at first-mode resonance in torsional motion. First-mode resonance is defined as the frequency at which maximum motion of the top of the specimen is obtained during a sweep of frequencies, usually starting at about 15 Hz. An accelerometer on the drive mechanism is used to monitor the torsional motion during each measurement, and a linear variable differential transformer (LVDT) is used to detect vertical height change of the specimen during the test.

Variations in the values of secant shear modulus, G, and viscous damping ratio, D, as functions of shearing strain amplitude, γ, are determined. The viscous damping ratio, D, can be directly related to the hysteretic damping ratio (λ).

Another resonant column device of the fixed-free type has been developed in Greece (Athanasopoulos, 2000). This device and its associated electronic equipment is used for the evaluation of dynamic moduli and damping of soil specimens under variable conditions of isotropic confinement and cyclic shear strain amplitude.

1.4 Direct Simple Shear Apparatus

The motorised direct simple shear apparatus is used to test undisturbed soils under conditions of simple shear and plane strain throughout the soil sample. The apparatus is designed for drained and undrained (constant volume) shear tests on undisturbed, normally consolidated and lightly overconsolidated clays, sands and silts. The sample cross-sectional area is $50cm^2$ and its height 20mm. The maximum vertical and shear load is 8kN and 4kN respectively, and the speed ranges from 0.01 to 10mm/min.

The apparatus is also capable of applying cyclic loading in the form of stress/strain load tests for the determination of dynamic soil parameters.

The cyclic loading system for the direct simple shear apparatus comprises an electronic pulse generator, an electronic-pneumatic converter and a pneumatic actuator. The pulse generator produces sinusoidal, triangular or square electronic pulses. The pulse form and frequency selected for the test is converted into a similar pneumatic pressure to pulse through a converter. The pressure pulse yields a similar cyclic load pulse. The maximum frequency is between 1 and 2Hz and the load cell capacity +/- 900N. The system is automatically controlled through especially designed software.

1.5 Seismic CPT (SCPT) System

SCPT soundings are carried out to determine the dynamic shear, G_0, and constrained modulus, M_0, for given density of the individual layers. The parameter of shear modulus at small strains, $G_0=G_{max}=\rho V_s^2$, has so far been the parameter mostly calculated for the evaluation of : liquefaction risk, site amplification parameters and performance of foundations for vibrating equipment.

Seismic CPT soundings are performed by pushing the probe to appropriate depths and recording the seismic travel times, from the surface to the SCPT cone adapter, after generating a seismic signal (polarized shear wave with a sledge hammer). The seismic channel adapter is equipped with a triaxial accelerometer (frequency ranges from 1 to 10kHz). The arrival times of the acoustic wave are measured with the SCPT Data Acquisition (SCPT-DAA) software. The application of this test, is based on the general Down-Hole methodology. Seismic CPT soundings can be carried out running concomitantly CPT(U) measurements.

ACKNOWLEDGEMENTS

Discussion and information provided by Mr Giannaros in connection to the equipment of the Greek Public Works Corporation is greatfully acknowledged.

REFERENCES

Athanasopoulos, G. (2000). Personal communication.

Bishop, A.W. and Wesley, L.D. (1975). A hydraulic triaxial apparatus for controlled stress path testing. Geotechnique 25. No. 4, 657-670.

Burland, J.B. and Symes, M. (1982). A simple axial displacement gauge for use in the triaxial apparatus. Geotechnique 32, No. 1, 62-65.

Drnevich, V.P., Hardin, B.O. and Shippy, D.J. (1978). Modulus and damping of soils by the resonant column method. Dynamic Geotechnical Testing, ASTM Special Technical Publication, STP 654, 91-125.

Hight, D.W. (1982). A simple piezometer probe for the routine measurement of pore pressure in triaxial tests on saturated soils. Geotechnique 32, No. 4, 396-401.

Isenhower, W.M. (1979). Torsional simple shear / Resonant Column properties of San Francisco Bay Mud. M.S. thesis, University of Texas at Austin.

Kuwano, J., Katagiri, M., Kita, K., Nakano, M. and Kuwano, R. (2000). A review of Japanese standards for laboratory shear tests. Progress report on the activities of Japanese domestic committee of TC-29; Stress-strain testing of geomaterials in the laboratory. The Japanese Geotechnical Society.

Richart, F.E., Hall, J.R., and Woods, R.D. (1970). Vibration of soils and foundations. Prendice Hall.

Schultheiss, P.J. (1981). Simultaneous measurement of P& S wave velocities during conventional laboratory soil testing procedures. Marine Geotechnology 4, No.4, 343-367.

Advanced Laboratory Stress-Strain Testing of Geomaterials, Tatsuoka, Shibuya & Kuwano (eds),
© 2001 Taylor & Francis, ISBN 90 2651 843 9

Recent state of laboratory stress-strain tests on geomaterials in Japan

J. Kuwano
Department of Civil Engineering, Tokyo Institute of Technology, Japan

M. Katagiri
Nikken Sekkei Nakase Geotechnical Institute (NNGI), Japan

ABSTRACT: This report summarizes the results of a Japanese nationwide questionnaire carried out in September 1995 to investigate the recent state of laboratory stress-strain tests on geomaterials in Japan

1 INTRODUCTION

Laboratory testing techniques on geomaterials have been making remarkable progress recently. Advanced techniques are utilized not only for academic research but also for practice. Besides the safety factor at failure, accurate estimations of deformation, displacement and stresses in the ground and in structures have become an important topic of study in geotechnical engineering. One of our major design concerns is whether the structure in question is adversely affected by ground deformation. Studies on applying laboratory test results to the prediction of in-situ full-scale deformation are being actively carried out, and their results are to be incorporated into the Eurocode.

Under the above-mentioned circumstances in 1994, the International Society for Soil Mechanics and Geotechnical Engineering established TC-29, Stress-Strain Testing of Geomaterials in the Laboratory, chaired by Prof. Tatsuoka. The Japanese Geotechnical Society as the chair country set up in 1995 a domestic committee to support TC-29. This report summarizes the results of a Japanese nationwide questionnaire carried out in September 1995 to investigate the recent state of laboratory stress-strain tests on geomaterials in Japan, which was part of the committee's preliminary study.

2 QUESTIONNAIRE

The questionnaires were limited to laboratory stress-strain tests in order to ascertain the state of laboratory shear tests in Japan in general. To avoid biased distribution of questionnaires, they were sent to more than 200 organizations in Japan, most of them are the organizations through which many of the Journals of the Japanese Geotechnical Society are

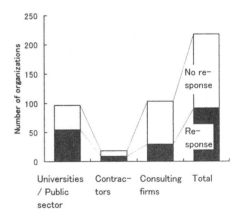

Figure 1. Responses to questionnaire.

distributed. Attention was paid to making the questionnaires simple so as not to impose a burden on the respondents.

92 questionnaires (42%) were collected out of 218 as summarized in Figure 1, where the organizations are divided into three groups, universities /public sector, contractors, and consulting companies including soil investigation companies. It must be mentioned here that in Japan, besides universities and public non-profitable research institutes, most of the major contractors and some of the consulting firms have their own geotechnical laboratories. Although a response rate of 42% may not be high, things are better than they appear, since some organizations, e.g. company in-house design sections, which probably do not possess their own laboratories, are included in the total.

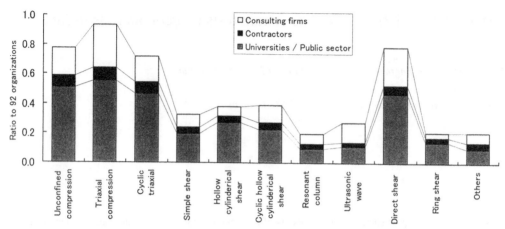

Figure 2. Proportion of organizations holding various apparatus

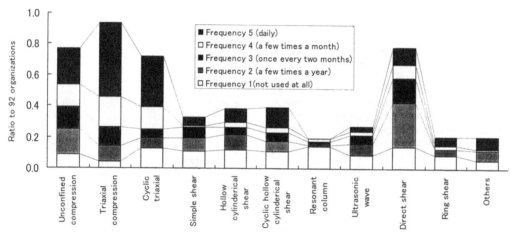

Figure 3. Frequency of use for various apparatus.

3 QUESTIONNAIRE RESULTS

3.1 *Laboratory shear test apparatus*

Figure 2 shows the proportion of organizations possessing various laboratory shear test apparatus for the 92 questionnaires collected. Triaxial test apparatus is owned by almost all (93%) of the organizations. Direct shear test apparatus, unconfined compression test apparatus and cyclic triaxial test apparatus also have a high penetration of 70 to 80%. These four types of apparatus appear to be the most used. The fifth-ranking hollow cylindrical shear test apparatus, is owned by only 39% of organizations. Although there is not much difference among organizations, consulting firms, as compared to universities, have fewer unconfined compression and cyclic triaxial test apparatus but more ultrasonic wave test apparatus.

Frequency of use of the respective apparatus is summarized in Figure 3. In this report, the frequency is determined as follows. Frequency 1: not used at all; 2: a few times a year; 3: once every two months; 4: a few times a month; 5: daily. As can be seen from this figure, triaxial test apparatus is used frequently by almost all organizations. On the other hand, direct shear test apparatus is used infrequently, though it is owned by about 80% of organizations. This is probably because the main concern of laboratory shear tests has shifted from simply determining the shear strength of the soil to obtaining deformation characteristics as well as strength under specified drainage conditions. It can also be seen in this figure that cyclic triaxial test, unconfined compression test and hollow cylindrical shear test apparatus are used rather more frequently. Ultrasonic wave test apparatus is perhaps still in its introductory stage. Frequency of use is relatively low for resonant col-

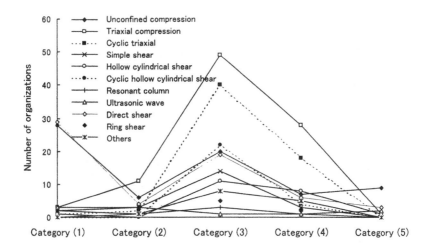

Figure 4. Controlling and measuring systems.

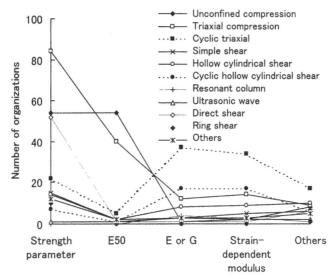

Figure 5. Purpose of laboratory shear tests.

umn test apparatus, suggesting that this apparatus has become less popular in Japan.

The reasons for choosing each particular piece of apparatus were also asked in the questionnaire. Triaxial test apparatus was selected because it is general and used for multi-purpose, cyclic triaxial test apparatus is used for soil liquefaction problems, unconfined compression test and direct shear test apparatus were selected because they were simple, hollow cylindrical shear test apparatus has high flexibility in stress control and is multi-purpose. A common response was that cyclic triaxial test and/or

cyclic hollow cylindrical shear test apparatus were used to determine deformation characteristics.

3.2 Controlling and measuring systems

Controlling and measuring systems for each apparatus are summarized in Figure 4. Systems are categorized from (1) to (5): (1) both controlling and measuring are mechanical and done by hand; (2) consolidation is carried out manually, shearing is controlled automatically with an electrical measurement system but recorded by hand; (3) consoli-

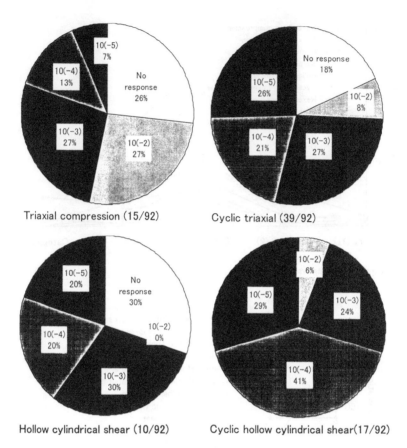

Triaxial compression (15/92)

Cyclic triaxial (39/92)

Hollow cylindrical shear (10/92)

Cyclic hollow cylindrical shear(17/92)

Figure 6. Minimum strain levels for strain-dependent moduli

dated manually and sheared and recorded automatically; (4) test is controlled and recorded automatically all the way from consolidation to shear; (5) other. Although the majority of unconfined compression tests and direct shear tests are still carried out under system (1), most of the other tests are performed under system (3). It can also be seen from Figure 4 that fully automated triaxial and cyclic triaxial shear test apparatuses under system (4) have come into wide use recently.

3.3 Purpose of laboratory shear tests

Figure 5 shows the purposes of laboratory shear tests. Strength parameters are obtained by most of the tests, but especially by direct shear tests. Unconfined compression tests are often used to obtain E_{50}. On the other hand, cyclic triaxial tests and cyclic hollow cylindrical shear tests are used most frequently to obtain the strain-dependent modulus E or G. Triaxial tests are not used very often to determine E or G.

Organizations performing tests to determine the strain-dependent modulus were asked about the strain levels. Figure 6 shows the results for the representative apparatus. In this figure, 10(-3), for example, means that the test is for a strain level of 0.1% or higher. The numerator beside the name of the apparatus indicates the number of organizations determining strain-dependent modulus with that particular piece of equipment. The denominator of 92 is the total number of organizations which replied to the questionnaire. Although 42% of organizations measure strain-dependent moduli by cyclic triaxial tests, less than 20% of organizations, not many at all, use other equipment. As answers concerning strain level were somewhat confusing, further study is needed on strain level through other research such as round-robin tests, for example.

3.4 Materials

Types of materials and their frequency of use are shown in Figure 7. Almost all types of material other than undisturbed gravel are used in triaxial tests with

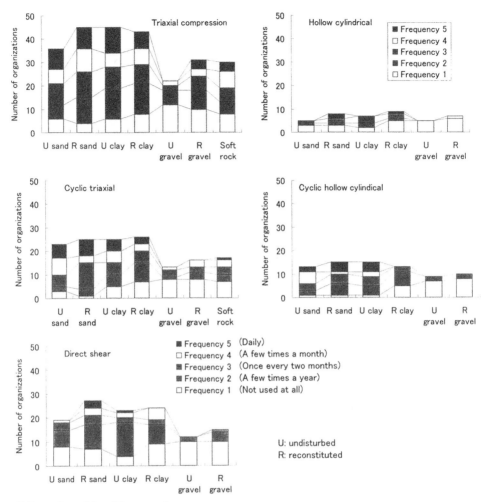

Figure 7. Types of materials and frequency of use.

considerable frequency. Undisturbed gravel is not used frequently presumably because it is difficult to sample and trim. Soft rock is used unexpectedly frequently. It is interesting to consider whether this is a recent trend or not. The cyclic triaxial test shows a similar tendency to the triaxial test. In the case of the (cyclic) hollow cylindrical shear test, the frequency of use for reconstituted sand is relatively high, while that of reconstituted clay and gravel is relatively low, probably because of difficulties in sample preparation. However, there are some organizations which use undisturbed clay frequently in hollow cylindrical tests. Direct shear test apparatus is used widely but not frequently for any type of material, a few times a month at most.

3.5 Test conditions

The frequency of use for various apparatus is summarized in Figures 8 (a) to (d) under different test conditions, i.e. UU, CU, $\overline{CU}$ and CD. As was expected, triaxial test apparatus is used most frequently under all test conditions. Direct shear tests are carried out frequently under CD conditions. A cyclic triaxial test under CD condition is probably the test used to study the long-term behavior of the ground under cyclic loads such as a wave load or a seasonal change in groundwater level, etc.

3.6 Utilization of test results

An important issue is how the obtained test results are utilized. Figure 9 shows how the test results obtained by the respective apparatus are used. The test results are used i) for evaluation of safety factors in

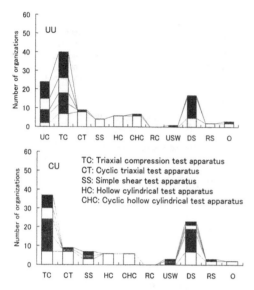

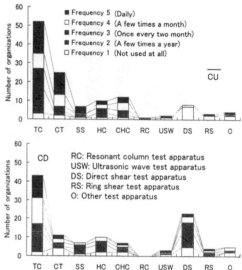

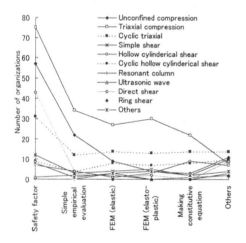

Figure 8. Frequencies of use for apparatus.

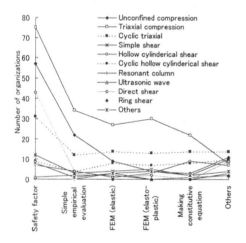

Figure 9. Utilization of test results.

ground failure problems such as a bearing capacity or slope failure; ii) for evaluation of ground deformation by a simple method such as an empirical equation; iii) for evaluation of ground deformation by FEM with elastic material properties; iv) for evaluation of ground deformation by FEM with elasto-plastic material properties; v) for making constitutive equations; and vi) other. It can be seen from the figure that triaxial test results are often used for all purposes, especially for FEM with elasto-plastic material properties. This is partly because the triaxial test has been widely used to develop elasto-plastic models and to determine their material properties, and partly because deformation analysis by FEM has become more popular in recent years. Cy-

clic triaxial tests are used less for stability problems, but more for the development of constitutive equations, together with hollow cylindrical and cyclic hollow cylindrical shear tests. They seem to be still used more for research purposes.

4 CONCLUSIONS

This report summarized the results of a Japanese nationwide questionnaire investigating the recent state of laboratory shear tests on geomaterials in Japan. As the questionnaires were made to cover various aspects of laboratory shear tests in general, the report may not be focused on the main aim of TC-29, i.e. stress-strain testing of geomaterials in the laboratory. However, it reflects the recent Japanese state to some extent, e.g. resonant column tests are being used less frequently in Japan. From the questionnaire, organizations conducting laboratory shear tests seem to be divided into two groups: those who try to obtain deformation characteristics from an extremely small strain level, and those who give priority to measuring strength for stability analysis.

Advanced Laboratory Stress-Strain Testing of Geomaterials, Tatsuoka, Shibuya & Kuwano (eds),
© 2001 Taylor & Francis, ISBN 90 2651 843 9

A review of Japanese standards for laboratory shear tests

J. Kuwano
Department of Civil Engineering, Tokyo Institute of Technology, Japan

M. Katagiri
Nikken Sekkei Nakase Geotechnical Institute (NNGI), Japan

K. Kita
Department of Marine Civil Engineering, Tokai University, Japan

M. Nakano
Department of Civil Engineering, Nagoya University, Japan

R. Kuwano
Department of Civil Engineering, University of Tokyo, Japan

ABSTRACT: Japanese standards for laboratory shear tests are introduced briefly with emphasis on pre-failure deformation characteristics. Summaries of the standards made by the WG1 of the Japanese domestic committee for TC29 are presented. Some of the features of these standards are compared with some others, such as the BS and ASTM standards.

1 INTRODUCTION

Testing techniques on geomaterials have been making remarkable progress lately. Advanced techniques are utilized not only for academic research but also for practice. This will inevitably require standardization of testing techniques. The Japanese Geotechnical Society (JGS) has published more than eighty issues of Standards since the 1960's, and has borne substantial responsibility for the Japanese Industrial Standards (JIS) related to geotechnology. These JGS Standards and JIS are being translated into English in the current situation of internationalization. The first English booklet "Standards of the Japanese Geotechnical Society for Soil Sampling" was published in May 1998 as a reference work. The second English booklet, "Standards of the Japanese Geotechnical Society for Laboratory Shear Tests" was published in May 1999. It contains 17 standards:

JGS 0520-2000: Preparation of soil specimens for triaxial tests.

JGS 0521-2000: Method for unconsolidated-undrained triaxial compression test on soils.

JGS 0522-2000: Method for consolidated-undrained triaxial compression test on soils.

JGS 0523-2000: Method for consolidated-undrained triaxial compression test on soils with pore water pressure measurements.

JGS 0524-2000: Method for consolidated-drained triaxial compression test on soils.

JGS 0525-2000: Method for Ko consolidated-undrained triaxial compression test on soils with pore water pressure measurement.

JGS 0526-2000: Method for Ko consolidated-undrained triaxial extension test on soils with pore water pressure measurement.

JGS 0527-1998: Method for triaxial compression test on unsaturated soils.

JGS 0530-2000: Preparation of specimens of coarse granular materials for triaxial tests.

JGS 0541-2000: Method for cyclic undrained triaxial test on soils.

JGS 0542-2000: Method for cyclic triaxial test to determine deformation properties of geomaterials.

JGS 0543-2000: Method for cyclic torsional shear test on hollow cylindrical specimens to determine deformation properties of geomaterials.

JGS 0550-1998: Preparation of hollow cylindrical soil specimens for torsional shear tests.

JGS 0551-1998: Method for torsional shear test on hollow cylindrical soil specimens.

JGS 0560-2000: Method for consolidated constant volume direct box shear test on soils.

JGS 0561-2000: Method for consolidated constant pressure direct box shear test on soils.

JIS A 1216-1998: Method for unconfined compression test of soils.

The WG1 of the Japanese domestic committee for TC29 has summarized some of the above standards. Their features will be briefly described following, together with other representative standards such as the BS and ASTM standards. The standards for laboratory shear tests whose main purpose is to determine shear strength will be discussed first, though the main concern of TC29 is the "Stress-Strain Testing of Geomaterials in the Laboratory" or the "Pre-failure Deformation Characteristics of Geomaterials". However, this report does not cover the

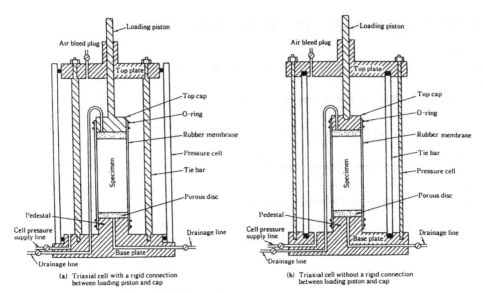

(a) Triaxial cell with a rigid connection between loading piston and cap

(b) Triaxial cell without a rigid connection between loading piston and cap

Figure 1. Typical examples of triaxial cell (JGS0523-2000)

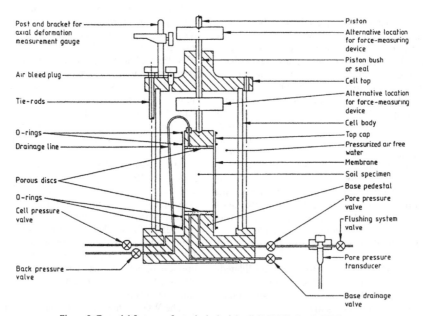

Figure 2. Essential features of a typical triaxial cell (BS1377: Part8: 1990)

standards for "Unconsolidated-Undrained Triaxial Test", "Triaxial Test on Unsaturated Soil", "Direct Box Shear Test" and "Unconfined Compres-sion Test". Next, the standards for tests used to determine the deformation properties of geomaterials will be discussed.

2 STANDARDS USED MAINLY TO DETERMINE SHEAR STRENGTH

Table 1 summarizes features of the JGS standards for $\overline{CU}$ tests (consolidated-undrained triaxial compression tests with pore pressure measurement) and CD tests (consolidated-drained triaxial compression tests). These tests are regarded as standard tests for

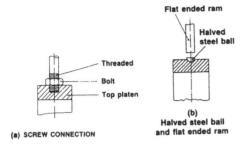

(a) SCREW CONNECTION

(b) Halved steel ball and flat ended ram

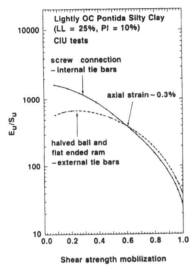

Figure 3. Comparison of stiffness measurements with different types of test details (Baldi et al.1988)

obtaining the shear strength of soils both in research and daily practice. Typical examples of triaxial cells are shown in Figure 1 (JGS 0523-2000). The corresponding British Standards (BS) are summarized in Table 2 and a typical example of a triaxial cell is shown in Figure 2 (BS 1377: Part8: 1990). Although the diameters are almost the same in both the standards, the height of the Japanese specimen is higher than that of the British specimen. One of the remarkable differences between the JGS standards and BS is the contact between the top cap and the loading piston. JGS standards for $\overline{CU}$ tests and CD tests presuppose a rigid connection between the top cap and the loading piston. Therefore, the JGS standards state "if the loading piston is not connected to the top cap rigidly...". On the other hand, British Standards seem to presuppose a free connection. However, it should be mentioned here that rigid connection and free connection sometimes give different test results as pointed out by many researchers. For example, Figure 3 is a comparison of stiffness measurements with different types of test details given by

Baldi et al. (1988). A higher initial stiffness is achieved with the cell with the internal tie bards and screw connection. Another remarkable difference is the method of saturating the specimen. The JGS standards suggest four different methods of saturation as shown in Table 1, whereas the BS recommends the wet method in which even a rubber membrane is pre-soaked overnight as shown in Table 2 and applied back pressure is usually higher than 300 kPa. 50 to 200 kPa is recommended as the back pressure in the JGS standards. Another feature of the BS is that the pressure chamber is filled with de-aired water, whereas the JGS standards do not mention filling the cell with water. Needless to say, filling the pressure chamber with de-aired water helps to keep the specimen at a high degree of saturation for a long period. All the details of the BS seem to suggest that they presuppose saturated stiff clay such as London clay and that therefore this is suitable for such types of soil. On the other hand, various types of soil from fine-grained soils to coarse granular materials need to be considered in the JGS standards.

Ko consolidated-undrained triaxial compression and extension tests are standardized by the JGS as summarized in Table 3. As seen from the table, the accuracy for control and measurement in Ko tests is severer than that in ordinary tests such as the CU and CD tests. To measure the lateral strain, ε_r, the JGS 0525-2000 introduces three methods in its notes:

1) to measure the volume change of the specimen by differential pressure transducer etc.,

2) to measure the change of water level of the inner cell by proximeter etc. when the triaxial apparatus has a double cell and a free water surface,

3) to measure the lateral displacement of the specimen.

The top cap is usually rigidly connected to the loading piston, especially in the case of the KoE test.

3 STANDARDS USED MAINLY TO DETERMINE DEFORMATION PROPERTIES

Cyclic triaxial tests and cyclic torsional shear tests on hollow cylindrical specimens are usually carried out in Japan to determine the deformation properties of soils. Standards for these are summarized in Tables 4 and 5, together with the standards for the corresponding ordinary tests. Typical apparatus and test results are shown in Figures 4 to 7. Deformation property tests cover a wide strain range from less than 0.001% to more than 0.1%. In these test methods, the internal measurements should be used for the deformation as well as the load. In the small strain range especially, where the error becomes significant, the axial strain should be measured locally on the lateral surface of the specimen.

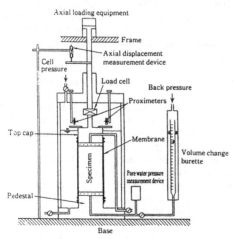

Figure 4. Typical triaxial apparatus for cyclic loading test (JGS 0542-2000)

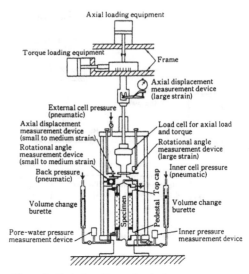

Figure 6. Typical cyclic torsional shear apparatus (JGS 0543-2000)

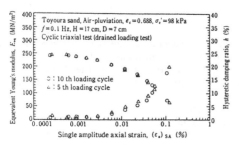

Figure 5. Example of cyclic triaxial test results (JGS 0542-2000)

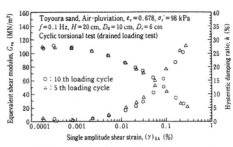

Figure 7. Example of cyclic torsional shear test results (JGS 0543-2000)

It can be seen that the requirements for control accuracy of an axial load is severer in the ordinary cyclic triaxial test than in the cyclic triaxial test to determine deformation properties, because the cyclic undrained strength of sand (liquefaction strength) is sensitive to various factors (Tatsuoka et al. 1986) and therefore the cyclic load should be quite uniform and symmetric.

The ASTM standards for the determination of modulus and damping properties are summarized in Table 6. Although the measurable strain range in the cyclic triaxial test is not described clearly in the standards (ASTM D3999-91), it is presumed that a cyclic triaxial test is usually used for a strain greater than about 0.01%. For a small strain level, a resonant-column test would be utilized. A typical cyclic triaxial apparatus is shown in Figure 8. As seen in the figure, internal measurement is not presupposed in the ASTM standard. It is stated that "Accurate deformation measurements require that the transducer be properly mounted to avoid excessive mechanical

system compression between the load frame, the triaxial cell, the load cell, and the loading piston".

As mentioned in ASTM D 3999-91, "Top and bottom platen alignment is critical to avoid increasing a non-uniform state of stress in the specimen". This alignment is prescribed in both the JGS standards and the ASTM standards. ASTM D 3999-91 shows clearly the limits on acceptable platen and load rod alignment, as shown in Figure 9. It also shows connections between the top cap and the loading piston, as shown in Figure 10. A "Hard lock system" seems to be the most appropriate to fulfill the requirements shown in this figure. However, a "vacuum system" might make it difficult to satisfy the requirements. The effect of system compliance on test results is schematically illustrated in ASTM D 3999-91 as shown in Figure 11. It is stated that "as compliance increases in the triaxial test system the deviation from the resonant column results increases". It seems to the authors that the difference in the mode of shearing between the cyclic triaxial

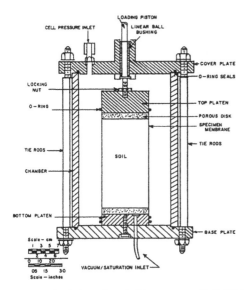

Figure 8. Typical cyclic triaxial pressure cell (ASTM D3999-91)

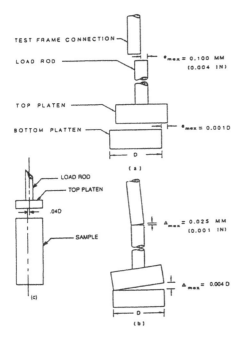

Figure 9. Limits on acceptable platen and load rod alignment: (q) eccentricity, (b) parallelism, (c) eccentricity between top platen and sample (ASTM D3999-91)

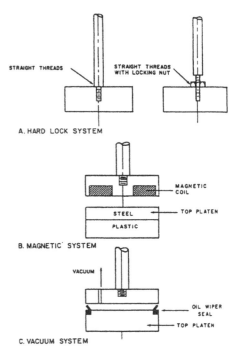

Figure 10. Top platen connection (ASTM D3999-91)

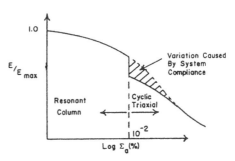

Figure 11. Variation in modulus curves caused by system compliance (ASTM D3999-91)

test and the resonant column test may also increase the deviation. On the contrary, deformation properties are obtained by the same test method in the Japanese standards from very small strain levels to large strain levels. Therefore, deviation as shown in Figure 11 is not encountered.

One of the characteristic features of the JGS standards for deformation property tests is the prescription of apparent damping caused by hysteresis of the load cell and/or the displacement measurement device. The time lag in the measurements between stress and strain also causes the apparent damping as illustrated in Figure 12 (Kohata et al. 1993). If the recording of the strain, ε, lags behind the recording of the stress, σ, by Δt as shown in Figure 12(a) and therefore ε at t-Δt is recorded with σ at t, the recorded stress-strain curve (hysteresis loop) becomes larger than the true loop. Apparent damping appears

Table 1. Summary of JGS standards for triaxial compression tests.

Type of test		$\overline{CU}$ test (effective stress)	CD (consolidated-drained) test
Designation		JGS 0523-2000	JGS 0524-2000
Specimen	Diameter, D (mm)	>35~100, >20D_{max}	
	Height, H (mm)	H=2~2.5D	
	Other requirements		
	Accuracy in measuring specimen dimension	0.1mm, 0.1g	
	Method of specimen preparation	Trimming method / Suction method (after reconstitution)	
	Saturation of specimen	a) flushed with de-aired water. b) high back pressure is applied. c) pore air is replaced with CO_2 gas then a) and b). d) vacuumed then a) and b).	
	B-value		
	Contact between top cap and loading piston	Connected rigidly / Not connected rigidly	
Control accuracy	Cell pressure, σ_r Back pressure, u_b	< ±4kPa (σ_r / u_b < 200kPa) or < ±2% of σ_r / u_b (> 200kPa).	
	Axial load, P Axial stress, σ_a Axial deformation, ΔH		
	Lateral strain, ε_r		
	Loading rate	0.1%/min (silt), 0.05%/min (clay)	$\dot{\varepsilon}_a = \varepsilon_{af} / 15t_c$ (t_c: consolidation period)
	Temperature	<±2°C	
Measurement accuracy	Cell pressure, σ_r Back pressure, u_b Pore water pressure, u	< ±2kPa (σ_r / u_b / u < 200kPa) or < ±1% of σ_r / u_b / u (> 200kPa).	
	Axial load, P	< ±0.01P_{max}	
	Axial deformation, ΔH	< ±0.001H	
	Lateral deformation		
	Volume change, ΔV	< ±0.001V	
Measurement frequency		every 0.2mm (up to P_{max}), every 0.5mm (after P_{max})	
Period to stop consolidation		3t-method or >100min. (D=35mm) / >150min. (D>50mm)	
Others		The volume change of the pore water pressure measurement system should meets the following conditions: {($\Delta V/V$)/Δu}<5×10^{-6}m^2/kN.	

even for linear elastic material. This increases with the increase in recording time lag and loading frequency if stress recording lags behind strain recording. Apparent damping is negative and decreases with recording time lag and loading frequency if strain recording lags behind stress recording.

4 CONCLUDING REMARKS

The Japanese standards for laboratory shear tests have been introduced briefly with emphasis on pre-failure deformation characteristics. Summaries of the standards made by the WG1 of the Japanese domestic committee for TC29 were presented. Some of the features of these standards have been compared with other standards, such as the BS and ASTM. Standards from different sources are similar in some senses. However, each set of standards has its own characteristic features due to their different backgrounds.

Table 2. Summary of British Standards for triaxial tests.

Type of test		$\overline{CU}$ test (effective stress)	CD test
Designation		BS 1377 : Part8 : 1990	BS 1377 : Part8 : 1990
Specimen	Diameter, D (mm)	usually $38 \sim 100$, $>15D_{max}$	
	Height, H (mm)	$H \approx D \sim 2D$	
	Other requirements		
	Accuracy in measuring specimen dimension	The bulk density to be calculated to an accuracy of ±1%	
	Method of specimen preparation	Trimming Compacted / Pluviated (reconstituted)	
	Saturation of specimen	Specimen should be prepared in the wet conditions (membrane is pre-soaked over-night with de-aired water), then a) the cell pressure is raised. b) high back pressure (usually >300kPa) is applied.	
	B-value	>0.95	
	Contact between top cap and loading piston	Usually a free connection, or a self-aligning seating (a central conical recess with a half-angle of 60° to accommodate a steel ball or the hemispherical end of the piston) is recommended between the top cap and the loading piston	
Control accuracy	Cell pressure, σ_r Back pressure, u_b Axial load, P Axial stress, σ_a Axial deformation, ΔH	The cell and back pressure applications are capable of remaining constant to within ±0.5% of the reading indicated.	
	Lateral strain, ε_r		
	Loading rate	Should be slow enough to ensure adequate equalization of excess pore pressures. The time to failure should not be less than 2hours.	Should be slow enough to ensure that pore pressure changes due to shearing are negligible (<4% of the effective confined pressure).
	Temperature	Should be maintained constant within ±2°C.	
Measurement accuracy	Cell pressure, σ_r Back pressure, u_b Pore water pressure, u		
	Axial load, P		
	Axial deformation, ΔH	Readable to 0.01mm	
	Lateral deformation		
	Volume change, ΔV	Readable to 0.2cc	
Measurement frequency			
Period to stop consolidation		Until at least 95% of excess pore water pressure is dissipated. √t-method (isotropic)	
Others		The cell is completely filled with de-aired tap water.	

Table 3. Summary of JGS standards for Ko compression / extension triaxial tests.

Type of test		Ko $\overline{CU}$ C test (effective stress)	Ko $\overline{CU}$ E test (effective stress)
Designation		JGS 0525-2000	JGS 0526-2000
Specimen	Diameter, D (mm)	>35~100, >20D_{max}	
	Height, H (mm)	H=2~2.5D	H=1.5~2D
	Other requirements		
	Accuracy in measuring specimen dimension	0.1mm, 0.1g	
	Method of specimen preparation	Trimming method / Suction method (after reconstitution)	
	Saturation of specimen	a) flushed with de-aired water. b) high back pressure is applied. c) pore air is replaced with CO_2 gas then a) and b). d) vacuumed then a) and b).	
	B-value		
	Contact between top cap and loading piston	Connected rigidly / Not connected rigidly	Usually connected rigidly
Control accuracy	Cell pressure, σ_r Back pressure, u_b	<±2kPa (σ_r / u_b < 200kPa) or <±1% of σ_r / u_b (> 200kPa).	
	Axial load, P Axial stress, σ_a Axial deformation, ΔH		
	Lateral strain, ε_r	<±0.05%	
	Loading rate	0.1%/min (silt), 0.05%/min (clay)	-0.1%/min (silt), -0.05%/min (clay)
	Temperature	<±2°C	
Measurement accuracy	Cell pressure, σ_r Back pressure, u_b Pore water pressure, u	<±2kPa (σ_r / u_b / u < 200kPa) or <±1% of σ_r / u_b / u (> 200kPa).	
	Axial load, P	<±0.01P_{max}	
	Axial deformation, ΔH	<±0.0002H	
	Lateral deformation	<±0.00025D	
	Volume change, ΔV	<±0.0005V	
Measurement frequency			
Period to stop consolidation		3t-method or >100min. (D=35mm) / >150min. (D>50mm)	
Others		The volume change of the pore water pressure measurement system should meet the following conditions: $\{(\Delta V/V)/\Delta u\}<5\times10^{-6}$m^2/kN.	

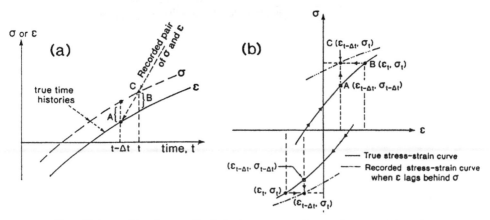

Figure 12. Apparent damping caused by the lag in measurements between the stress and the strain

Table 4. Summary of JGS standards for cyclic triaxial tests.

Type of test			Cyclic undrained triaxial test on soils	Cyclic triaxial test to determine deformation properties
Designation			JGS 0541-2000	JGS 0542-2000
Specimen	Diameter, D (mm)		>35 (clay), >50 (sand), >20D_{max}	>35 (clay), >50 (sand), >20D_{max}, >10D_{max} (coarse material)
	Height, H (mm)		H=1.5~2.5D	
	Other requirements			Both ends should be flat enough with a roughness of $\Delta H/H$<0.5%.
	Accuracy in measuring specimen dimension		0.1mm, 0.1g	
	Method of specimen preparation		Trimming method / Suction method (after reconstitution)	
	Saturation of specimen		a) flushed with de-aired water. b) high back pressure is applied. c) pore air is replaced with CO_2 gas then a) and b). d) vacuumed then a) and b).	
	B-value		>0.95	
	Contact between top cap and loading piston		Usually rigid connection	Rigid connection
Control accuracy	Cell pressure, σ_r Back pressure, u_b Axial load, P Axial stress, σ_a Axial deformation, ΔH Cyclic axial load, P_C and P_E Cyclic displacement, ΔL_C and ΔL_E		< ±2kPa (σ_r / u_b / σ_a < 200kPa) or < ±1% of σ_r / u_b / σ_a (> 200kPa). 0.98 < σ'_a/ σ'_r <1.02 at the end of consolidation. Cyclic axial load should meet the following conditions (ε_{aDA}<2%); 1) fluctuation of (P_C+P_E)<10% 2) 0.9 < P_C/P_E < 1.1	< ±2kPa (σ_r / u_b / σ_a < 200kPa) or < ±1% of σ_r / u_b / σ_a (> 200kPa). For ε_{aSA}=(~)0.001%~0.1%(~), [axial load control] 1) fluctuation of (P_C+P_E)<10% 2) 0.8 < P_C/P_E < 1.2 [axial strain control] 1) fluctuation of (ΔL_C+ΔL_E)<10% 2) 0.8 < ΔL_C/ΔL_E < 1.2
	Lateral strain, ε_r			
	Loading rate		0.1~1Hz (not be rectangular nor trapezoid).	0.05~1Hz (Sinusoidal or serrated wave).
	Temperature			
Measurement accuracy	Cell pressure, σ_r Back pressure, u_b Pore water pressure, u		< ±2kPa (σ_r / u_b / σ_a < 200kPa) or < ±1% of σ_r / u_b / σ_a (> 200kPa).	
	Axial load, P		< ±0.005P_{DA} (friction<0.02σ'_0A_0, if the load cell is outside the cell)	< ±0.01P_{DA} for ε_{aSA}>0.01% (should be measured inside the cell)
	Axial deformation, ΔH		< ±0.0002H (consolidation) or < ±0.0005H (cyclic loading)	< ±0.01ΔH (ε_{aSA}>0.01%)
	Lateral deformation			
	Volume change, ΔV		< ±0.0005V	
Measurement frequency			> 40 data in one cycle	
Period to stop consolidation			>30min (sand with FC<10%) or 3t-method	3t-method (isotropic) dε_a/dt<0.1%/min (anisotropic)
Others			The volume change of the pore water pressure measurement system should meet the following conditions: {($\Delta V/V$)/Δu}<5×10^{-6}m^2/kN.	Angle between top and bottom planes should be <0.5°. Eccentricity between top cap and sample < 0.01H. ε_a is to be measured internally at the cap for ε_{aSA}<0.1%. ε_a is to be measured locally on the lateral surface of the specimen, when either ε_a or h measured at the cap is expected to differ by more than 5% from the relevant value. Calculate E_{eq} and h at 5th and 10th of 11 cycles in each stage.

Table 5. Summary of JGS standards for torsional shear tests on hollow cylindrical soil specimens.

	Type of test	Torsional shear test on hollow cylindrical soil specimens	Cyclic torsional shear test on hollow cylindrical specimens to determine deformation properties
	Designation	JGS 0551-1998	JGS 0543-2000
Specimen	Outer diameter, D_o Inner diameter, D_i (mm)	D_o: >50 (clay), >70 (sand) D_i: >20 (clay), >30 (sand)	
	Wall thickness of hollow specimen, D_o-D_i	$(D_o\text{-}D_i)>10D_{max}$ $(D_o\text{-}D_i)>5D_{max}$ (soil with wide grain size distribution)	
	Height, H (mm)	$H=1\sim2D_o$	
	Accuracy in measuring specimen dimension	0.1mm (tolerance: ±1% for diameter, ±0.5% for height) 0.1g	
	Method of specimen preparation	Trimming method / Suction method (after reconstitution)	
	Saturation of specimen	a) flushed with de-aired water. b) high back pressure is applied. c) pore air is replaced with CO_2 gas then a) and b). d) vacuumed then a) and b).	
	B-value		>0.95
	Contact between top cap and loading piston		
Control accuracy	Cell pressure, σ_r Back pressure, u_b Axial stress, σ_a	$<\pm2$kPa (σ_r / u_b / $\sigma_a<200$kPa) or $<\pm1\%$ of σ_r / u_b / σ_a (> 200kPa).	$<\pm2$kPa (σ_r / u_b / $\sigma_a<200$kPa) or $<\pm1\%$ of σ_r / u_b / σ_a (> 200kPa).
	Torsional loading Torque in clockwise and in counterclockwise, T_R and T_L	The apparatus should be capable of continuously applying shear strain on the horizontal plane until it exceeds 22.5%.	Uniform and symmetric cyclic torque should be applied to induce the range of strain, γ_{SA}, from less than 0.001% to greater than 0.1%. Sinusoidal or serrated wave (should not be rectangular nor trapezoid). 1) fluctuation of $(T_R + T_L)<10\%$ 2) $0.8 < T_R / T_L < 1.2$
	Lateral strain, ε_r		
	Loading rate		0.05~1Hz
	Temperature		
Measurement accuracy	Cell pressure, σ_r Back pressure, u_b Pore water pressure, u	$<\pm2$kPa (σ_r / u_b / $\sigma_a<200$kPa) or $<\pm1\%$ of σ_r / u_b / σ_a (> 200kPa).	
	Axial load, P	$<\pm1\%$ of the axial load	
	Torque, T	$<\pm0.01T_{max}$ (load cell should be inside)	$<\pm0.01T_{DA}$ for $\gamma_{SA}>0.01\%$ (load cell should be inside the cell)
	Axial deformation, ΔH	$<\pm0.0002H$	
	Lateral deformation		
	Volume change, ΔV	$<\pm0.0005V$	
	Rotational angle, θ	$<\pm0.01\theta_{max}$ System compliance should be less than 1%, if the sensor is mounted outside the cell.	$<\pm0.01\theta_{DA}$ for $\gamma_{SA}>0.01\%$ Sensor should be inside to measure the rotation of the cap directly for strain level of $\gamma_{SA}<0.1\%$.
	Measurement frequency	Frequent enough to draw smooth curves	> 40 data in one cycle
	Period to stop consolidation	3t-method	
	Others		Calculate G_{eq} and h at 5th and 10th of 11 cycles in each stage.

Table 6. Summary of ASTM standards for the determination of modulus and damping properties.

Type of test		Cyclic triaxial test for modulus and damping	Resonant-column test for mudulus and damping
Designation		D 3999-91	D 4015-92
Specimen	Diameter, D (mm)	>36, >6D_{max}	>33, >10D_{max} (solid cylinder)
	Height, H (mm)	H=2~2.5D	H=2~7D (2~3D when σ_a>σ_r)
	Other requirements		
	Accuracy in measuring specimen dimension	D and H: 0.025mm (D<150mm) 0.25mm (D>150mm) Weight: 0.01g (D<63.5mm) 0.1g (D>63.5mm), Water content: 0.1%	0.1%
	Method of specimen preparation	Trimming method / Suction method (after reconstitution)	
	Saturation of specimen	a) flushed with de-aired water. b) high back pressure is applied. c) pore air is replaced with CO_2 gas then a) and b). d) vacuumed then a) and b).	
	B-value	>0.95	
	Contact between top cap and loading piston	Rigid connection	
Control accuracy	Cell pressure, σ_r Back pressure, u_b Axial load, P Axial stress, σ_a Axial deformation, ΔH	<±14kPa for σ_r and u_b. Piston friction<0.02P_{SA} for no correction Cyclic axial stress should meet: 1) fluctuation < 0.005σ_{DA} 2) 0.9 < P_E/P_C < 1.1 3) 0.9 < t_E/t_C < 1.1	
	Lateral strain, ε_r		
	Loading rate	0.1~2Hz	< 0.5% of excitation frequency
	Sine wave generation		Total distortion < 3%
	Temperature	<±4°C	
Measurement accuracy	Cell pressure, σ_r Back pressure, u_b Pore water pressure, u	<±0.5% of full scale	
	Axial load, P	<±0.25% of full scale	
	Axial deformation, ΔH	<±0.25% of full scale, 0 hysteresis	$2.5×10^{-6}$ m with 10% accuracy
	Rotational angle, $\Delta\theta$		10^{-6} rad with 10% accuracy
	Lateral deformation		
	Volume change, ΔV	<±0.0005V	
Measurement frequency		> 40 data in one cycle	
Period to stop consolidation		1 log cycle of time or 1 overnight after 100% primary consolidation by $\sqrt{t}$ method. 2 hours after 100% primary consolidation determined by log$t$ method.	
Others		Calculate E_{eq} and D for each individual hysteresis loop of 40 cycles. Eccentricity between top cap and sample < 0.004D. The volume change of the pore water pressure measurement system should meet the following conditions: {($\Delta V/V$)/Δu}<$3.2×10^{-6}$m^2/kN. Closure error of loop: $\Delta\varepsilon_a$ < 0.2%.	Reference on ambient stress: D2166: unconfined compressive strength of cohesive soil. D2850: unconsolidated, undrained compressive strength of cohesive soil. E: determined from longitudinal vibration G: determined from torsional vibration

REFERENCES

Baldi, G., D.W. Hight & G.E. Thomas 1988. A reevaluation of conventional triaxial test methods. Advanced Triaxial Testing of Soil and Rock, ASTM STP 977: 219-263.

Kohata, Y., S. Teachavorasinskun, T. Suzuki, F. Tatsuoka, T. & T. Sato 1993. Effecsts of time lag on damping in cyclic loading tests. Proc. 28th Japan National Conference on Geotechnical Engineering, Kobe, JGS 1: 887-890.

Tatsuoka, F., S. Toki, S. Miura, Y. M. Okamoto, S. Yamada, S. Yasuda & F. Tanizawa 1986. Some factors affecting cyclic undrained triaxial strength of sand. Soils and Foundations 26(3): 99-116.

Advanced Laboratory Stress-Strain Testing of Geomaterials, Tatsuoka, Shibuya & Kuwano (eds),
© *2001 Taylor & Francis, ISBN 90 2651 843 9*

International Round-Robin Test Organized by TC-29

S.Yamashita
Kitami Institute of Technology, Kitami, Japan

Y.Kohata
Muroran Institute of Technology, Muroran, Japan

T.Kawaguchi
Hakodate National College of Technology, Hakodate, Japan

S.Shibuya
Hokkaido University, Sapporo, Japan

(Japanese Domestic Committee for TC-29)

ABSTRACT: An international round-robin (IRR) test programme was launched in 1996 as a central facet of the TC-29 activities. The programme aimed at providing a valuable framework of reference. Based on the outcome from the programme, recommendation for procedures referred to triaxial and torsional shear tests may be developed. Nineteen laboratories, spreading over six nations, participated into the IRR tests, performed reconstituted Toyoura sand, reconstituted Fujinomori clay and sedimentary soft rock. This report briefly describes the IRR test programme, prepared by a working group in the Japanese domestic committee for TC-29. The reported results are analysed by the authors paying attentions to the stress-strain behaviour over a wide strain range from 10^{-5} to 10^{-2}. It has been manifested that the test results are consistent among different laboratories worldwide only when the specified procedures for testing are properly followed using a properly instrumented apparatus.

1 INTRODUCTION

1.1 Outline

In 1994, the International Society for Soil Mechanics and Geotechnical Engineering (ISSMGE) established TC-29 on Stress-Strain Testing of Geomaterials in the Laboratory, chaired by Prof. F. Tatsuoka. In the following year, the Japanese Geotechnical Society (JGS) set up a domestic committee to support the activities of TC-29 as a chair country. Through international cooperative studies, TC-29 aims at exchanging the latest knowledge and experiences of laboratory test techniques and deformation characteristics of geomaterials in a wide rage of strains, from very small strains to large ones at failure.

One of the terms of reference is to develop the recommendation for procedures referred to triaxial and torsional shear tests dealing with;
1) methods for measuring accurately, and in a reliable manner, stresses and strains of laboratory specimens over a strain range from 10^{-5} to 10^{-2}; and
2) laboratory reconsolidation techniques for cohesive soils.

In the first term from 1994 to 1997, research related to the above-mentioned terms of reference was carried out by volunteered members. In the light of the outcome from the above, a set of drafts for the standard test programme were prepared by a task group in the domestic committee in Japan, and these were then subjected to review by TC-29 members.

The following test materials were selected and prepared by the Japanese members;
1) one kind of sand (n.b., the test specimens were to be reconstituted by pluviating air-dried sand particles in each laboratory);
2) one kind of clay (n.b., the test specimens were to be reconstituted from clay powder in each laboratory); and
3) one kind of undisturbed sedimentary soft rock (n.b., several specimens retrieved at a single depth at a common site were distributed to the laboratories that participated in the programme).

The execution of the IRR test programme has been underway since 1997. The test materials and the standard test procedures were distributed to volunteer laboratories in the world. The types of laboratory tests that were employed included triaxial tests, torsional tests, resonant-column tests and bender elements tests.

Finally, the results of tests from eighteen laboratories, spreading over five countries, were reported to the task group (Table 1-1). In this report, the testing procedures, together with the summary of test results, are presented. Interpretation of the test results is also made by the authors.

1.2 Stiffness parameters

Figure 1-1 illustrates a typical stress-strain relationship of geomaterials in monotonic loading (ML) and cyclic loading (CL) tests. Young's modulus (E) can

Table 1-1. List of participating laboratories in the round-robin test.

Country	Organization	Lab. representative
Greece	Aristotle University of Thessaloniki	Dr. Theodora Tika
Spain	CEDEX	Dr. V. Cuéllar
Japan	Central Research Institute of Electric Power Industry	Dr. Y. Tanaka
Japan	Dia Consultants Company Limited	Mr. M. Nakajima
Japan	Geo-Research Institute	Mr. T. Hongo
Japan	Hokkaido University	Dr. S. Shibuya
Italy	ISMES	Dr. V. Fioravante
Japan	Kitami Institute of Technology	Dr. S. Yamashita
Korea	Korea Advanced Institute of Science and Technology	Dr. Dong-soo Kim
Japan	Kyushu University	Dr. N. Yasufuku
Japan	Nippon Koei Company Limited	Dr. H. Shimokura
Japan	OYO Corporation	Mr. I. Furuta
Japan	Takenaka Corporation	Dr. M. Hatanaka
Portugal	Technical University of Lisbon	Prof. A. G. Correia
Italy	Technical University of Turin	Prof. D. Lo Presti
Japan	Tokyo Institute of Technology	Dr. J. Kuwano
Italy	University of Naples 'Federico II'	Prof. F. Silvestri
Italy	University of Padua	Dr. P. Simonini
Japan	Yokohama National University	Dr. T. Pradhan

be obtained from a relationship between the deviator stress (q) and the axial strain (ε_a) by triaxial (or uniaxial) tests. Similarly, shear modulus (G) can be directly obtained from a relationship between the shear stress (τ) and the shear strain (γ) by torsional (or simple) shear tests.

The non-linear stress-strain response can be descrived in terms of the following stiffness parameters;

Secant shear and Young's moduli;

$$G_{sec} = \tau / \gamma \text{ and } E_{sec} = q / \varepsilon_a \qquad (1.1)$$

Tangent shear and Young's moduli;

$$G_{tan} = d\tau / d\gamma \text{ and } E_{tan} = dq / d\varepsilon_a \qquad (1.2)$$

Equivalent (peak-to-peak) shear and Young's moduli;

$$G_{eq} = \tau_{SA} / \gamma_{SA} \text{ and } E_{eq} = q_{SA} / \varepsilon_{aSA} \qquad (1.3)$$

Hysteretic damping ratio;

$$h = (1 / 2\pi) \cdot (\Delta W / W) \qquad (1.4)$$

where ΔW is the area of hysteretic stress-strain loop and W is $\tau_{SA} \times \gamma_{SA}$ or $q_{SA} \times \varepsilon_{aSA}$.

The subscript "SA" in Equation 1.3 means the "single amplitude", which is equal to a half of double amplitude of cyclic strain or stress. The stress parameters, together with the corresponding strains, are defined at the initially equilibrium state prior to start of shearing.

1.3 Standard Test Specification for TC-29 Round-Robin Tests

The Appendixes A to C are the standard test specifications distributed to the participating laboratories.

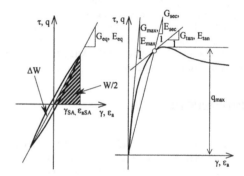

Figure 1-1. Definitions for Young's and shear moduli, together with hysteretic damping ratio.

As stated earlier, the draft was prepared by the task group of the JGS committee, and was endorsed by TC-29 members.

2 RESULTS OF ROUND-ROBIN TESTS ON TOYOURA SAND

2.1 Participating laboratories and test conditions

Table 2-1 summarizes the type of tests performed, together with the number of data reported in each laboratory. Table 2-2 summarizes the details of test conditions. As seen in this table, 112 sets of test data have been reported from 16 laboratories in six countries.

In the triaxial tests, the majority of laboratories employed in-cell device for measuring axial deformation of the specimen. Instrumentation for measuring the axial deformation of the specimen may be conveniently categorized into the following three groups; a) a method of measuring the movement of

Table 2-1. Summary of the reported test results on Toyoura sand.

Lab. No.	Monotonic Test				Cyclic Test						Bender Element Test	Total
	Triaxial		Torsional		Triaxial		Torsional		Resonant Column			
	Drained	Undrained	Drained	Undrained	Drained	Undrained	Drained	Undrained	Drained	Undrained		
1	3						2		2		3	10
2	2				3							5
3	3	2	2				2		1			10
4	2				2							4
5	2						2					4
6	2		2		2	2	2	2			3	15
7	2											2
8	2	2	2		2	2	2					12
9	2	2	2	2	2	2	2	2				16
10	2				2		2					6
11											2	2
12									9			9
13						2				2		4
14											2	2
18							3		3			6
19									3	2		5
Total	22	6	8	2	13	8	17	4	18	4	10	112

loading piston outside the cell by using LVDT, strain gauge type dial gauge (DG) or others (external); b) a method of directly measuring the axial movement of top cap by using a proximity transducer (PT), LVDT or others (internal); and c) a method of measuring the local axial deformation on the lateral surface of specimen using local displacement transducer (LDT) (Goto et al. 1991), PT or LVDT (local). Note that in all the cases, the axial load was measured by using a load cell mounted inside the triaxial cell.

2.2 Monotonic loading triaxial tests

2.2.1 Small-strain behaviour
Figure 2-1 plots the secant Young's modulus E_{sec} against the axial strain ε_a in the logarithmic scale obtained from drained monotonic loading triaxial compression (MTX) tests. In this figure, to eliminate the effects of the difference in the dry density among the tests, the Young's modulus was normalized by using a void ratio function. The adapted void ratio function is $F(e)=(2.17-e_c)^2/(1+e_c)$ (Hardin & Richart 1963; Iwasaki & Tatuoka 1977). Figure 2-2 shows the relationships between the deviator stress and the axial strain at small strains in tests where axial strains were measured inside the cell. In the case of internal measurement of axial strain, some scatter can be seen in the data for strain levels smaller than about 0.001 %. This could be attributed to a fluctuation in reading of the axial displacement devices. However, the scatter reduced as the strain level became larger than about 0.001 %. It may also be noted that the difference between the local strain and the internal strain is very small. Therefore, it may be surmised that the effects of bedding error at the boundary be-

tween the specimen and the rigid platens are insignificant in these tests on Toyoura sand.

Figure 2-3 shows the results based on external measurement of axial strain. Significant scatter is seen, which is possibly due to a poor resolution and response of the strain-gauge type DG in comparison with the load cell. Because of it, Young's modulus for strain levels smaller than about 0.01 % shows unrealistically large values.

Figure 2-4 shows the relationship between the void ratio and the normalized secant Young's modulus at four different strain levels based on internal and external measurements. Despite a relatively large scatter, it is clearly seen that bedding error is relatively small in tests on Toyoura sand at all the strain levels.

2.2.2 Large-strain behaviour
Figures 2-5 and 2-6 show the relationships between the deviator stress and the axial strain and those between the deviator stress and the volumetric strain from the tests where the axial deformation was measured outside the cell (which was the case in most of the tests). The scatter may be due partly to the difference in the dry density of the specimens (see Table 2-2).

Figures 2-7 through 2-9 show the angle of shear resistance, the axial strain and volumetric strain at peak strength, plotted against the relative density of the specimens. Note that approximately 90 % of the test data is within about ±10 percent from the mean value in each relationship.

Table 2-2. Test conditions of Round-robin test on Toyoura sand;

(a) Drained Monotonic Triaxial Compression Test (MTX-D Test)

Test No.	saturate or dry	σ_c (kPa)	u_b (kPa)	σ_c' (kPa)	axial strain rate (%/min)	t_m (mm)	D_{oc} (mm)	H_c (mm)	e_c	D_{rc} (%)	B	axial strain measurement		
												local	internal	external
1-1	saturated	298	201	98	0.01	0.35	71	139	0.787	53.5	1.000	O		
1-2	saturated	298	199	99	0.01	0.35	71	139	0.745	64.9	0.960	O		
1-3	saturated	398	200	98	0.01	0.35	71	139	0.685	81.4	0.970	O		
2-1	saturated	298	200	98	0.3	0.2	70	132	0.793	51.8	0.950			O
2-2	saturated	298	200	98	0.3	0.2	70	132	0.687	80.8	0.950			O
3-1	air-dry	101	0	101	0.0020 - 0.014	0.3	70	139	0.782	54.8	N/A	LVDT	PT	LVDT
3-2	air-dry	99	0	99	0.0054 - 0.014	0.3	70	139	0.777	56.0	N/A	LVDT	PT	LVDT
3-3	saturated	291	189	101	0.001 - 0.014	0.3	70	140	0.648	91.4	-	LVDT		LVDT
4-1	saturated	200	100	100	0.5	0.3	100	159	0.837	39.7	0.906	PT		DG
4-2	saturated	200	100	100	0.5	0.3	100	160	0.695	78.7	0.934	PT		DG
5-1	saturated	200	100	100	0.05	0.25	50	98	0.785	53.9	0.980	O		O
5-2	saturated	200	100	100	0.05	0.25	50	99	0.693	79.1	0.980	O		O
6-1	saturated	200	100	100	0.05	0.2	70	169	0.810	47.1	0.984		PT	DG
6-2	saturated	200	100	100	0.05	0.2	70	170	0.697	78.1	0.978		PT	DG
7-1	saturated	200	100	100	0.05	0.25	50	102	0.776	56.5	0.900			DG
7-2	saturated	200	100	100	0.05	0.25	50	102	0.718	72.3	0.900			DG
8-1	saturated	200	100	100	0.05	0.2	51	103	0.802	49.2	0.970	LDT		
8-2	saturated	200	100	100	0.05	0.2	52	105	0.705	75.9	0.960	LDT		
9-1	saturated	200	100	100	0.05	0.25	50	100	0.817	45.2	0.990			O
9-2	saturated	200	100	100	0.05	0.25	50	100	0.702	76.8	0.990			O
10-1	saturated	200	100	100	0.05	0.2	50	101	0.822	43.9	-			O
10-2	saturated	200	100	100	0.05	0.2	50	100	0.672	84.9	0.850			O

PT: proximity transducer, DG: dial gauge, LDT: local displacement transducer

(b) Undrained Monotonic Triaxial Compression Test (MTX-U Test)

Test No.	saturate or dry	σ_c (kPa)	u_b (kPa)	σ_c' (kPa)	axial strain rate (%/min)	t_m (mm)	D_{oc} (mm)	H_c (mm)	e_c	D_{rc} (%)	B	axial strain measurement		
												local	internal	external
3-1	saturated	440	389	51	0.075 - 0.138	0.3	70	141	0.783	54.5	-	LVDT	PT	LVDT
3-2	saturated	238	188	51	0.0602-0.1384	0.3	70	150	0.769	58.3	-	LVDT	PT	LVDT
8-1	saturated	200	100	100	0.05	0.2	51	104	0.809	47.5	0.980		PT	
8-2	saturated	200	100	100	0.05	0.2	51	105	0.708	75.1	0.960	LDT		
9-1	saturated	200	100	100	0.05	0.25	50	100	0.800	49.9	0.990			O
9-2	saturated	300	200	100	0.05	0.25	50	100	0.695	78.6	0.960			O

(c) Drained Monotonic Torsional Shear Test (MTS-D Test)

Test No.	saturate or dry	σ_c (kPa)	u_b (kPa)	σ_c' (kPa)	shear strain rate (%/min)	t_m (mm)	D_{oc} (mm)	D_{ic} (mm)	H_c (mm)	e_c	D_{rc} (%)	B	shear strain measurement		
													local	internal	external
3-1	air-dry	98	0	98	0.0005-0.0019	0.3	71	49	138	0.800	49.9	N/A		PT	
3-2	air-dry	101	0	101	0.00028-0.0016	0.3	71	49	1400	0.730	69.1	N/A		PT	
6-1	saturated	200	100	100	0.05	0.2	100	60	199	0.794	51.5	0.957		PT,PM	
6-2	saturated	200	100	100	0.05	0.2	100	60	200	0.680	82.7	0.967		PT,PM	
8-1	saturated	200	100	100	0.05	0.2	73	42	170	0.789	52.8	0.970	O		
8-2	saturated	200	100	100	0.05	0.2	73	42	170	0.695	78.6	0.960	O		
9-1	saturated	300	200	100	0.05	0.25	70	29	140	0.800	49.8	0.980		PT	PM
9-2	saturated	300	200	100	0.05	0.25	70	30	140	0.691	79.8	0.970		PT	PM

PM: potentiometer

(d) Undrained Monotonic Torsional Shear Test (MTS-U Test)

Test No.	saturate or dry	σ_c (kPa)	u_b (kPa)	σ_c' (kPa)	shear strain rate (%/min)	t_m (mm)	D_{oc} (mm)	D_{ic} (mm)	H_c (mm)	e_c	D_{rc} (%)	B	shear strain measurement		
													local	internal	external
9-1	saturated	300	200	100	0.05	0.25	69	30	140	0.800	49.8	0.970		PT	PM
9-2	saturated	300	200	100	0.05	0.25	70	30	140	0.700	77.4	0.950		PT	PM

(e) Drained Cyclic Triaxial Test (CTX-D Test)

Test No.	saturate or dry	σ_c (kPa)	u_b (kPa)	σ_c' (kPa)	f (Hz)	t_m (mm)	D_{oc} (mm)	H_c (mm)	e_c	D_{rc} (%)	B	axial strain measurement local	Internal	external
2-1	saturated	558	460	98	1	0.2	70	132	0.792	52.1	0.950			
2-2	saturated	468	370	98	1	0.2	70	132	0.689	80.3	0.950			
2-3	saturated	468	370	98	1	0.2	70	132	0.689	80.3	0.950			
4-1	saturated	200	100	100	0.1	0.3	100	159	0.821	44.2	1.000	PT		
4-2	saturated	200	100	100	0.1	0.3	100	160	0.693	79.3	0.961	PT		
6-1	saturated	196	98	98	0.1	0.2	70	168	0.808	47.6	0.974		PT	
6-2	saturated	196	98	98	0.1	0.2	70	168	0.688	80.5	0.973		PT	
8-1	saturated	200	100	100	0.1	0.2	52	103	0.803	49.0	0.985		O	
8-2	saturated	200	100	100	0.1	0.2	50	103	0.698	77.8	0.963		O	
9-1	saturated	300	200	100	0.1	0.25	50	100	0.795	51.2	0.950		LVDT	
9-2	saturated	200	100	100	0.1	0.25	50	100	0.697	78.1	0.990		LVDT	
10-1	saturated	200	100	100	0.1	0.2	50	100	0.811	46.9	0.960		O	
10-2	saturated	200	100	100	0.1	0.2	51	100	0.692	79.5	0.860		O	

(f) Undrained Cyclic Triaxial Test (CTX-U Test)

Test No.	saturate or dry	σ_c (kPa)	u_b (kPa)	σ_c' (kPa)	f (Hz)	t_m (mm)	D_{oc} (mm)	H_c (mm)	e_c	D_{rc} (%)	B	axial strain measurement local	Internal	external
6-1	saturated	196	98	98	0.1	0.2	70	168	0.801	49.7	0.979		PT	
6-2	saturated	196	98	98	0.1	0.2	70	168	0.694	78.8	0.972		PT	
8-1	saturated	200	100	100	0.1	0.2	50	99	0.795	51.1	0.985		O	
8-2	saturated	200	100	100	0.1	0.2	52	104	0.691	79.7	0.975		O	
9-1	saturated	300	200	100	0.1	0.25	50	100	0.811	46.9	0.950		LVDT	
9-2	saturated	300	200	100	0.1	0.25	50	100	0.686	81.1	0.950		LVDT	
13-1	saturated	441	343	98	1	0.3	51	102	0.818	44.9	-			
13-2	saturated	441	343	98	1	0.3	51	102	0.689	80.3	-			

(g) Drained Cyclic Torsional Shear Test (CTS-D Test)

Test No.	saturate or dry	σ_c (kPa)	u_b (kPa)	σ_c' (kPa)	f (Hz)	t_m (mm)	D_{oc} (mm)	D_{ic} (mm)	H_c (mm)	e_c	D_{rc} (%)	B	shear strain measurement local	internal	external
1-1	air-dry	98	0	98	0.2	0.3	50	N/A	106	0.827	42.5	N/A			
1-2	air-dry	98	0	98	0.2	0.3	50	N/A	107	0.698	77.8	N/A			
3-1	air-dry	98	0	98	0.1	0.3	71	49	138	0.800	49.9	N/A		PT	
3-2	air-dry	101	0	101	0.1	0.3	71	49	1400	0.730	69.1	N/A		PT	
5-1	saturated	200	100	100	0.1	0.3	70	30	70	0.788	53.1	0.980			
5-2	saturated	200	100	100	0.1	0.3	70	30	70	0.681	82.5	0.980			
6-1	saturated	196	98	98	0.1	0.2	100	60	200	0.793	51.7	0.946		PT,PM	
6-2	saturated	196	98	98	0.1	0.2	100	60	200	0.678	83.4	0.950		PT,PM	
8-1	saturated	200	100	100	0.1	0.2	75	42	170	0.812	46.7	0.960			
8-2	saturated	200	100	100	0.1	0.2	74	43	170	0.698	77.9	0.950			
9-1	saturated	300	200	100	0.1	0.25	70	29	139	0.793	51.7	0.970		PT	PM
9-2	saturated	300	200	100	0.1	0.25	70	30	140	0.684	81.7	0.960		PT	PM
10-1	saturated	200	100	100	0.1	0.3	101	60	105	0.807	48.0	0.560			
10-2	saturated	200	100	100	0.1	0.3	101	60	99	0.678	83.3	0.960			
18-1	air-dry	50-200	0	50-200	0.5	0.3	70	N/A	146	0.644	92.7	N/A		PT	
18-2	air-dry	50-200	0	50-200	0.5	0.3	70	N/A	146	0.680	82.8	N/A		PT	
18-3	air-dry	50-200	0	50-200	0.5	0.3	70	N/A	146	0.755	62.3	N/A		PT	

(h) Undrained Cyclic Torsional Shear Test (CTS-U Test)

Test No.	saturate or dry	σ_c (kPa)	u_b (kPa)	σ_c' (kPa)	f (Hz)	t_m (mm)	D_{oc} (mm)	D_{ic} (mm)	H_c (mm)	e_c	D_{rc} (%)	B	shear strain measurement		
													local	internal	external
6-1	saturated	196	98	98	0.1	0.2	100	60	200	0.799	50.2	0.966		PT,PM	
6-2	saturated	196	98	98	0.1	0.2	100	60	200	0.704	76.1	0.950		PT,PM	
9-1	saturated	300	200	100	0.1	0.25	69	30	140	0.794	51.6	0.960		PT	PM
9-2	saturated	300	200	100	0.1	0.25	70	30	140	0.691	79.8	0.950		PT	PM

(i) Drained Resonant Column Test (RC-D Test)

Test No.	saturate or dry	σ_c (kPa)	u_b (kPa)	σ_c' (kPa)	t_m (mm)	D_{oc} (mm)	D_{ic} (mm)	H_c (mm)	e_c	D_{rc} (%)
1-1	air-dry	98	0	98	0.3	50	N/A	106	0.802	49.3
1-2	air-dry	98	0	98	0.3	50	N/A	107	0.677	83.6
3-1	air-dry	101	0	101	-	71	49	140	0.730	69.1
12-1	air-dry	0	-50	50	-	70	N/A	100	0.744	65.2
12-2	air-dry	0	-80	80	-	70	N/A	100	0.744	65.2
12-3	air-dry	0	-50	50	-	70	N/A	100	0.740	66.3
12-4	air-dry	0	-80	80	-	70	N/A	100	0.740	66.3
12-5	air-dry	0	-80	80	-	70	N/A	100	0.722	71.2
12-6	air-dry	40	-10	50	-	70	N/A	100	0.717	72.6
12-7	air-dry	70	-10	80	-	70	N/A	100	0.717	72.6
12-8	air-dry	0	-50	50	-	70	N/A	100	0.715	73.2
12-9	air-dry	80	0	80	-	70	N/A	100	0.715	73.2
18-1	air-dry	50-200	0	50-200	0.3	70	N/A	146	0.644	92.7
18-2	air-dry	50-200	0	50-200	0.3	70	N/A	146	0.680	82.8
18-3	air-dry	50-200	0	50-200	0.3	70	N/A	146	0.755	62.3
19-1	air-dry	104	0	104	0.4	70	N/A	143	0.793	51.7
19-2	air-dry	104	0	104	0.4	70	N/A	143	0.738	66.8
19-3	air-dry	105	0	105	0.4	69	N/A	142	0.700	77.1

(j) Undrained Resonant Column Test (RC-U Test)

Test No.	saturate or dry	σ_c (kPa)	u_b (kPa)	σ_c' (kPa)	t_m (mm)	D_{oc} (mm)	D_{ic} (mm)	H_c (mm)	e_c	D_{rc} (%)
13-1	saturated	392	294	98	0.3	38	N/A	76	0.794	51.5
13-2	saturated	392	294	98	0.3	38	N/A	76	0.677	83.6
19-1	saturated	507	410	97	0.35	71	N/A	142	0.659	88.6
19-2	saturated	505	408	97	0.375	71	N/A	140	0.716	72.9

(k) Drained Bender Element Test (BE-D Test)

Test No.	saturate or dry	σ_c, σ_v (kPa)	u_b (kPa)	σ_c', σ_v' (kPa)	t_m (mm)	D_{oc} (mm)	H_c (mm)	e_c	D_{rc} (%)	B	NB
1-1	saturated	298	201	98	0.35	71	139	0.787	53.5	1.000	triaxial
1-2	saturated	298	199	99	0.35	71	139	0.745	64.9	0.960	triaxial
1-3	saturated	398	200	98	0.35	71	139	0.685	81.4	0.970	triaxial
6-1	saturated	150-500	100	50-400	0.20	70	150	0.829	41.9	0.960	triaxial
6-2	saturated	150-500	100	50-400	0.20	70	150	0.757	61.6	0.960	triaxial
6-3	saturated	150-500	100	50-400	0.20	70	150	0.692	79.6	0.960	triaxial
11-1	air-dry	20-400	0	20-400	-	60	50	0.798	50.5	-	oedometer
11-2	air-dry	20-400	0	20-400	-	60	50	0.688	80.5	-	oedometer
14-1	saturated	-	-	98	-	-	-	0.683	81.9	-	triaxial
14-2	saturated	-	-	98	-	-	-	0.649	91.2	-	triaxial

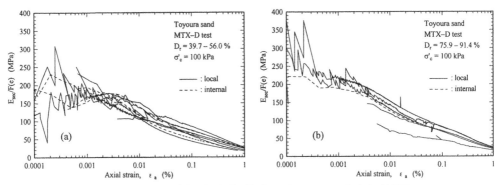

Figure 2-1. Normalized Young's modulus vs. axial strain obtained by internal measurement in drained MTX tests;
(a) Dr = 39.7 - 56.0 %, (b) Dr = 75.9 - 91.4 %.

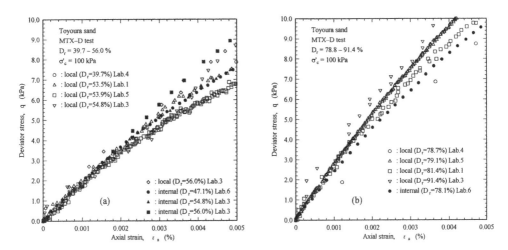

Figure 2-2. Deviator stress vs. axial strain at small strains obtained by internal measurement in drained MTX tests;
(a) Dr = 39.7 - 56.0 %, (b) Dr = 78.8 - 91.4 %.

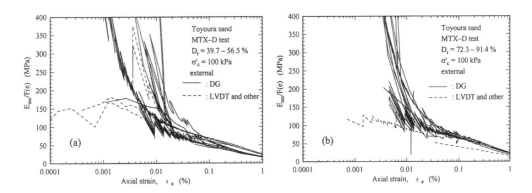

Figure 2-3. Normalized Young's modulus vs. axial strain obtained by external measurement in drained MTX tests;
(a) Dr = 39.7 - 56.5 %, (b) Dr = 72.3 - 91.4 %.

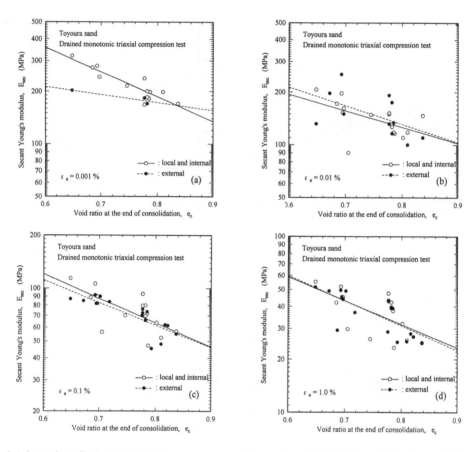

Figure 2-4. Comparison of internal and external measurements in axial strain; (a) ε_a = 0.001 %, (b) ε_a = 0.01 %, (c) ε_a = 0.1 %, (d) ε_a = 1.0 %.

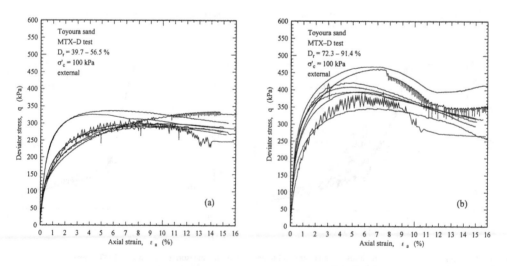

Figure 2-5. Deviator stress vs. axial strain obtained by external measurement in drained MTX tests; (a) Dr = 39.7 - 56.5 %, (b) Dr = 72.3 - 91.4 %

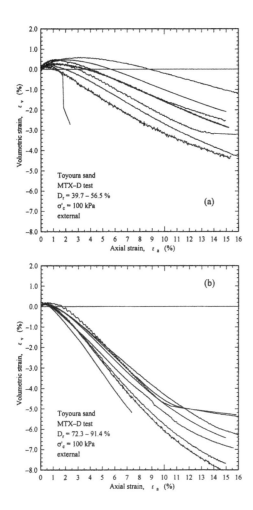

Figure 2-6. Volumetric strain vs. axial strain in drained MTX tests; (a) Dr = 39.7 - 56.5 %, (b) Dr = 72.3 - 91.4 %.

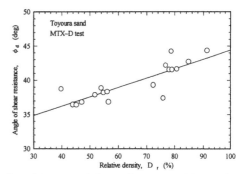

Figure 2-7. Angle of shear resistance vs. relative density in drained MTX tests.

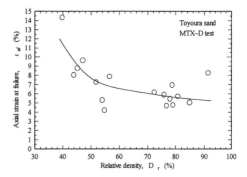

Figure 2-8. Axial strain at failure vs. relative density in drained MTX tests.

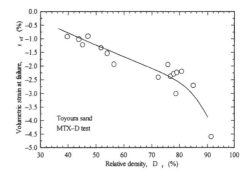

Figure 2-9. Volumetric strain at failure vs. relative density in drained MTX tests.

2.3 Monotonic loading torsional tests

Figure 2-10 plots the normalized secant shear modulus $G_{sec}/F(e)$ against the shear strain γ in the logarithmic scale from drained monotonic torsional shear (MTS) tests. Figure 2-11 shows the relationship between the shear stress and the shear strain at small strains from these MTS tests.

The amount of scatter seen in the MTS test data is remarkably small when compared to the scatter in the triaxial test data at strain levels smaller than about 0.001 %. Note that the test data from laboratories where small strains could not be measured deviates from the other data, implying that the consistency of the test data among different laboratories is high only when the standard test procedures are properly fulfilled using a properly instrumented apparatus.

Figure 2-12 shows the relationship between the shear stress and the shear strain over the entire strain range. Figure 2-13 shows the peak shear stress plotted against the relative density. It can be seen that the observed scatter in the stress-strain relationship may be due partly to the difference in the dry density among the specimens. Indeed, the peak strength shows a clear trend to increase with the relative density.

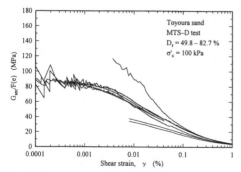

Figure 2-10. Normalized shear modulus vs. shear strain in drained MTS tests.

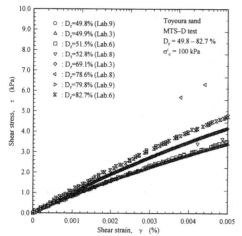

Figure 2-11. Shear stress vs. shear strain at small strains in drained MTS tests.

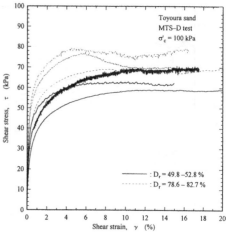

Figure 2-12. Shear stress vs. shear strain at large strains in drained MTS.

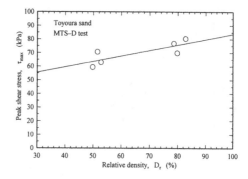

Figure 2-13. Peak shear stress vs. relative density in drained MTS tests.

In the light of the test results previously presented, it may be concluded that the test results in MTS test are more consistent (or stable) when compared to the comparable MTX test.

2.4 Cyclic triaxial tests

The results from the drained and undrained cyclic triaxial (CTX) tests are shown in Figs. 2-14 and 2-15. In these figures, the variation in the normalized equivalent Young's modulus $E_{eq}/F(e)$ and the hysteretic damping ratio h evaluated at the tenth cycle in each cyclic loading stage are plotted against the single amplitude axial strain $(\varepsilon_a)_{SA}$. It can be seen that the scatter of the data is very small except for the data from one laboratory. Generally, the reproducibility in this RRT programme is high as it has already been demonstrated in the similar round-robin tests performed in Japan (Toki et al. 1995).

2.5 Cyclic torsional tests

Figures 2-16 and 2-17 summarize the results from the drained and undrained cyclic torsional shear (CTS) tests. Similarly, the variation in the equivalent shear modulus G_{eq} and the hysteretic damping ratio h at the tenth cycle in each cyclic loading stage are plotted against the single amplitude shear strain $(\gamma)_{SA}$. In these figures, the shear modulus is also divided by $F(e)$. Note that the difference in the void ratio as well as the Young's modulus among the data set is not significant as compared to that in the triaxial test data. The scatter of CTS test data is also smaller than that in the CTX tests. It should be recalled that the trend is similar to that of the results from the monotonic tests. It should be noted that the scatter of damping ratio is in particular small in comparison with that with the results of CTX test.

2.6 Resonant-column and bender element tests

Figures 2-18 and 2-19 show the results of resonant column (RC) and bender element (BE) tests, re-

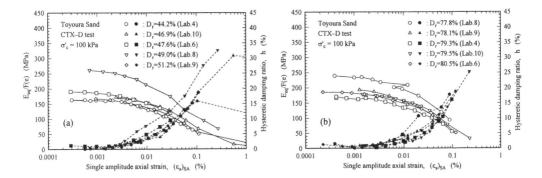

Figure 2-14. Normalized equivalent Young's modulus and hysteretic damping ratio vs. axial strain in drained CTX tests; (a) Dr = 44.2 - 51.2 %, (b) Dr = 77.8 - 80.5 %.

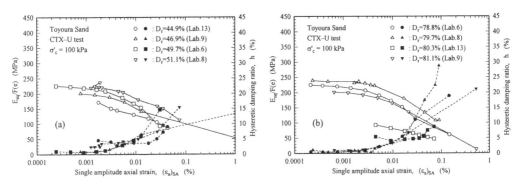

Figure 2-15. Normalized equivalent Young's modulus and hysteretic damping ratio vs. axial strain in undrained CTX tests; (a) Dr = 44.9 - 51.1 %, (b) Dr = 78.8 - 81.1 %.

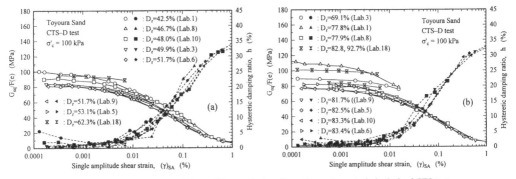

Figure 2-16. Normalized equivalent shear modulus and hysteretic damping ratio vs. shear strain in drained CTS tests; (a) Dr = 42.5 - 62.3 %, (b) Dr = 69.1 - 92.7 %

respectively. A detailed examination into the scatter of the test data is not easy, since the number of reported data is limited and the test conditions differ among the participating laboratories. Despite that, it seems that the scatter of the resonant-column test data is relatively small.

2.7 Comparison of monotonic and cyclic test data

Figures 2-20 and 2-21 show comparisons of the stiffness-strain relationships between the monotonic and cyclic tests for the cases of triaxial and torsional tests, respectively. In spite of some noticeable scatter observed at very small strains, it can be seen that the secant and equivalent Young's or shear moduli at small strains are virtually the same for the two types

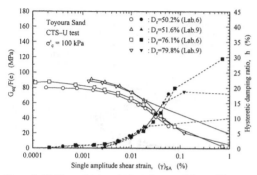

Figure 2-17. Normalized equivalent shear modulus and hysteretic damping ratio vs. shear strain in undrained CTS tests.

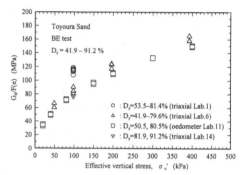

Figure 2-19. Normalized shear modulus vs. effective vertical stress in BE tests.

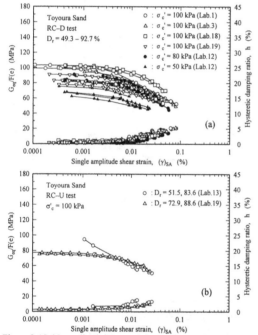

Figure 2-18. Normalized shear modulus and hysteretic vs. shear strain in RC tests; (a) drained test, (b) undrained test.

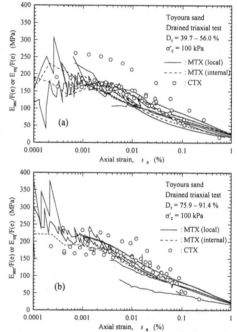

Figure 2-20. Comparison of drained monotonic and cyclic triaxial test data; (a) Dr = 39.7 - 56.0 %, (b) Dr = 75.9 - 91.4 %.

(monotonic and cyclic) of tests for both triaxial and torsional tests.

Figures 2-22 and 2-23 show the variation in the Young's and shear moduli with the void ratio at three different levels of strain in triaxial and torsional tests, respectively. In torsional shear test, the degradation of secant shear moduli in monotonic tests is much more pronounced than that observed in cyclic tests. In triaxial tests, it is the case until a intermediate strain of 0.01 %. At a large strain of 0.1 %, however, the Young's modulus becomes similar between the monotonic and cyclic loading tests. This behaviour could be attributed to the fact that the axial stress continuously increased in the monotonic

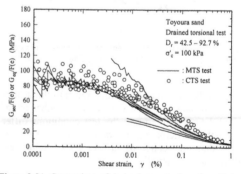

Figure 2-21. Comparison of drained monotonic and cyclic torsional test data.

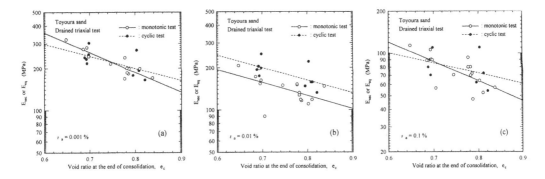

Figure 2-22. Comparison of drained monotonic and cyclic triaxial test data; (a) $\varepsilon_a = 0.001$ %, (b) $\varepsilon_a = 0.01$ %, (c) $\varepsilon_a = 0.1$ %.

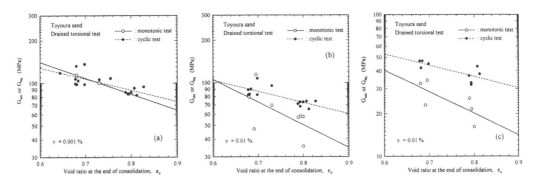

Figure 2-23. Comparison of drained monotonic and cyclic torsional test data; (a) $\gamma = 0.001$ %, (b) $\gamma = 0.01$ %, (c) $\gamma = 0.1$ %

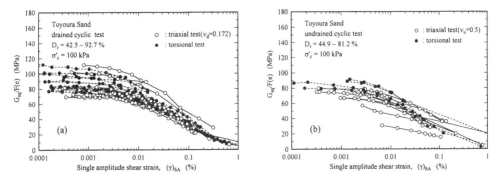

Figure 2-24. Comparison of G_{eq} obtained by cyclic triaxial and torsional test data; (a) drained test, (b) undrained test.

triaxial tests, while the axial stress increased and also decreased in the cyclic triaxial test.

2.8 Comparison between cyclic triaxial and torsional test data

Figures 2-24 and 2-25 show the relationships between G_{eq} and h and $(\gamma)_{SA}$ from the drained and undrained CTS and CTX tests, respectively. In these figures, G_{eq} and $(\gamma)_{SA}$ in CTX tests are defined by the following equations, respectively;

$$E = 2(1+v)G \tag{3-1}$$

$$\gamma = (1+v)\varepsilon_a \tag{3-2}$$

where the drained Poisson's ratio v_d was assumed to be 0.172, base on the results from drained CTX tests (Hoque & Tatsuoka 1998), whereas the undrained Poisson's ratio v_u for the CTX tests was assumed to be 0.5. In the undrained tests, G_{eq}(CTS) is slightly higher than G_{eq}(CTX). The main reason for this could be a decrease of Poisson's ratio below 0.5

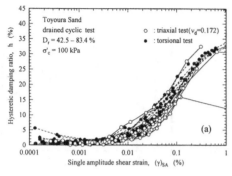

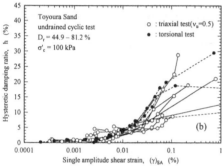

Figure 2-25. Comparison of h obtained by cyclic triaxial and torsional test data; (a) drained test, (b) undrained test.

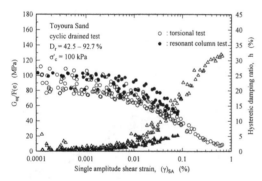

Figure 2-26. Comparison of drained torsional and resonant column test data.

under undrained cyclic triaxial conditions due to membrane penetration effects (Yamashita et al. 1996).

It is also found that the damping ratio at small strains is similar between the triaxial and torsional shear tests. However, the damping ratio at strains larger than about 0.01 % is slightly smaller in the triaxial tests than in the torsional tests. It seems that the difference of damping ratio between the triaxial and torsional tests at strains larger than about 0.01 % is caused by asymmetry of the stress-strain curves un-

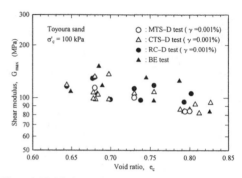

Figure 2-27. Maximum shear modulus obtained by torsional, resonant column and bender element tests.

der compression and extension conditions in triaxial tests.

2.9 Comparison of cyclic torsional and resonant-column and bender element test data

Figure 2-26 compares the values of G_{eq} and h, plotted against, $(\gamma)_{SA}$, from the drained CTS and RC tests. It can be seen that the value of G_{eq}(RC) is slightly higher than the value of G_{eq}(CTS). On the other hand, the damping ratio of RC test is nearly the same as that from the CTS test at strains less than about 0.01 %. However, the damping ratio of RC test is smaller than that of CTS test at strains larger than about 0.01%. It should be pointed out that these damping ratio values were evaluated by different methods, and that the frequency and number of loading cycle are different between the two types of tests. In the CTS test, the damping ratio was obtained from the hysteresis loop of stress-strain relationship. On the other hand, the h value of the reported RC test was obtained from the free vibration decay curve. In addition, the number of loading cycles on CTS test is 11 cycles at each stage, whereas the number of cycles in the RC test may be at least more than 1,000. Therefore, it is likely that a large part of the difference at strains larger than about 0.01 % between h(CTS) and h(RC) is mainly due to the difference in the number of loading cycles, as reported by Tatsuoka et al. (1978) and Kim & Stokoe (1995).

Figure 2-27 compares the maximum shear modulus G_{max} values at small strains plotted against the void ratio from the monotonic and cyclic torsional tests, bender element and resonant-column tests. The small strain shear moduli from theses different types of tests are located with in a narrow range irrespective of the type of loading. The shear modulus from the RC and BE tests is slightly larger than the G_{max} from torsional test. Collection of more data from RC and BE tests is certainly needed to confirm this speculation.

78

Table 3-1. Summary of the reported test results on Fujinomori clay.

Lab. No.	Undrained Monotonic Test				Undrained Cyclic Test				Bender Element Test	Total
	Triaxial		Torsional		Triaxial		Torsional			
	I.C.	A.C.	I.C.	A.C.	I.C.	A.C.	I.C.	A.C.		
3									1	1
5	1						1			2
8					1	1*				2
11									1	1
16	1	1*							1	3
17	1				1					2
Total	3	1	0	0	2	1	1	0	3	11

* the reconsolidation procedures differed from the specification

Table 3-2. Test conditions of Round-robin test on Fujinomori clay;

(a) Undrained Monotonic Triaxial Compression Test (MTX Test)

Test No.	u_b (kPa)	σ'_c (kPa)	σ'_{ac} (kPa)	σ'_{rc} (kPa)	axial strain rate	D_0 (mm)	H_0 (mm)	e_0	D_c (mm)	H_c (mm)	e_c	B	axial strain measurement	
													internal	external
5-1	100	130	130	130	0.05	50.2	99.6	1.538	48.9	96.21	1.323	0.99	O	O
16-1	202	133	133	134	0.04	49.8	100.0	1.632	48.0	97.4	1.388	0.99	O	
16-2	202	199	100	133	0.04	50.1	99.2	1.568	50.2	89.3	1.324	0.95	O	
17-1	196	130	130	130	0.05	49.8	99.9	1.508	49.0	97.8	1.368	0.93	O	O

(b) Undrained Cyclic Triaxial Compression Test (CTX Test)

Test No.	u_b (kPa)	σ'_c (kPa)	σ'_{ac} (kPa)	σ'_{rc} (kPa)	cyclic loading wave form	f	D_0 (mm)	H_0 (mm)	e_0	D_c (mm)	H_c (mm)	e_c	B	axial strain measurement	
														internal	external
8-1	200	133	200	100	sinusoidal	0.1	49.5	100.8	1.584	49.7	92.7	1.393	0.97	O	
8-2	200	130	130	130	sinusoidal	0.1	49.3	100.8	1.652	48.4	99.2	1.426	0.98	O	
17-2	196	130	130	130	sinusoidal	0.1	49.6	100.2	1.507	48.9	97.8	1.373	0.91	O	O

(c) Undrained Cyclic Torsional Shear Test (CTS Test)

| Test No. | u_b (kPa) | σ'_c (kPa) | σ'_{ac} (kPa) | σ'_{rc} (kPa) | cyclic loading wave form | f | D_{o0} (mm) | D_{i0} (mm) | H_0 (mm) | e_0 | D_{oc} (mm) | D_{ic} (mm) | H_c (mm) | e_c | B |
| 5-2 | 100 | 130 | 130 | 130 | sinusoidal | 0.1 | 69.9 | 30.1 | 70.0 | 1.559 | 66.7 | 28.7 | 67.9 | 1.261 | 0.98 |

(d) Bender Element Test (BE Test)

Test No.	u_b (kPa)	σ'_c (kPa)	σ'_{ac} (kPa)	σ'_{rc} (kPa)	D_0 (mm)	H_0 (mm)	e_0	D_c (mm)	H_c (mm)	e_c	B	NB
3-2	55	129	203	92	70.0	140.0	1.501	70.1	126.5	1.259	0.96	triaxial test
11-1	N/R	N/R	196	N/R	60.0	50.0	1.451	60.0	48.8	1.247	N/R	consolidometer test
16-3	200	132	199	99	49.5	99.4	1.533	49.5	91.7	1.336	0.95	triaxial test

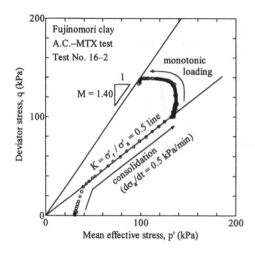

Figure 3-1. Anisotropic consolidation path of Fujinomori clay sample in a triaxial cell.

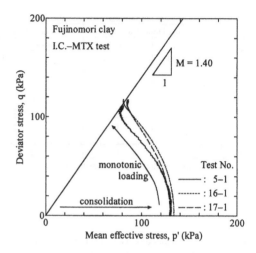

Figure 3-3. Undrained effective stress paths of Fujinomori clay samples in monotonic triaxial test.

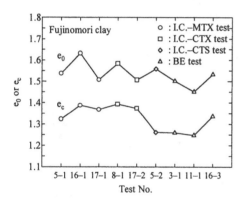

Figure 3-2. Scatter of void ratio of Fujinomori clay samples.

3 RESULTS OF ROUND-ROBIN TESTS ON FUJINOMORI CLAY

3.1 Participating laboratories and test conditions

Six laboratories participated into the IRR test programme on Fujinomori clay. Table 3-1 summarizes the types of tests performed, together with the number of tests reported. Table 3-2 summarizes the details of test conditions.

In the triaxial tests, all of the participating laboratories employed internal axial strain measurement using PT. The nominal diameter of 50 mm with the height/diameter ratio of two was the common feature of the cylindrical specimens.

Two different kinds of consolidation path; i.e., isotropic consolidation (I.C.) and anisotropic consolidation (A.C.) with $K=\sigma'_h/\sigma'_v$ of 0.5 were im-

posed. An undrained CTS test was performed in a laboratory, whereas MTX and CTX tests were performed in the rest of the laboratories. Note that the pore pressure parameter B was more than 0.95 in most of the tests.

3.2 Consolidation

The rate of axial stressing during consolidation was prescribed as 5 kPa per min in the original specification for test procedures (see Appendix B). The rate was obviously too fast for the clay specimen to dissipate excess pore water pressure during consolidation. In fact, some laboratories reported unexpected failure during anisotropic consolidation.

Figure 3-1 shows an anisotropic consolidation path by using the rate of axial stressing of 0.5 kPa per min. The consolidation was successful as the reduced rate was employed. It should be mentioned that the rate of stressing might preferably be less than 1 kPa per min for soft clays.

Figure 3-2 shows the void ratios of the preconsolidated block sample, e_0, and those of the specimen at the end of consolidation, e_c. The e_0 value ranged between 1.5 and 1.6 among most of the samples prepared at different laboratories. It is noteworthy that the change in the void ratio during consolidation was nearly the same among these different laboratories.

3.3 Monotonic loading triaxial tests

Figure 3-3 shows the undrained effective stress paths from the three comparable MTX tests on isotropically consolidated specimens. The stress-strain relationship are compared in Figure 3-4. The results appear to be consistent with each other showing

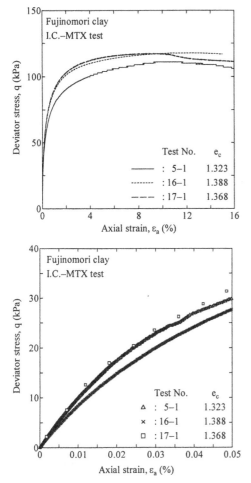

Figure 3-4. Relationship between deviator stress and axial strain in undrained monotonic tests in triaxial compression on Fujinomori clay; a) overall, b) small strain.

insignificant scatter in the undrained shear strength, the pore pressure generation as well as the stress-strain response. Note also that the accuracy of the employed stress/strain measurement methods is high enough to determine the quasi-elastic Young's modulus.

3.4 Maximum Young's modulus at small strains

Figures 3-5 and 3-6 compare the values of E_{max} from the MTX, CTX, CTS and BE tests. A trend may be seen for E_{max} to increase with a decrease in the void ratio. It can also be seen that the E_{max} values are located within a narrow range between 100 and 120 MPa at the same mean effective stress of 130 kPa.

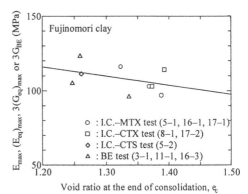

Figure 3-5. Stiffness at strains of about 0.001% plotted against void ratio in tests on Fujinomori clay.

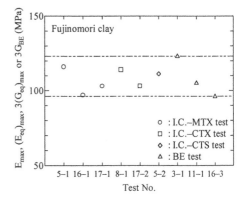

Figure 3-6. Comparison of the stiffness at strains of about 0.001% (Fujinomori clay samples).

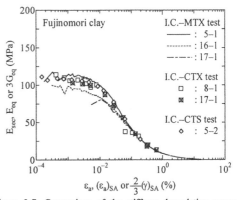

Figure 3-7. Comparison of the stiffness degradation curves from MTX, CTX and CTS tests on Fujinomori clay.

3.5 Comparison between triaxial and torsional shear test data

Figure 3-7 shows the stiffness degradation curves with strain from the MTX, CTX and CTS tests. Note

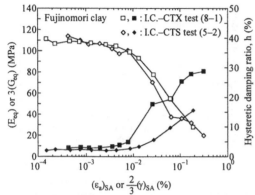

Figure 3-8. Comparison of the results between undrained CTX and CTS tests on Fujinomori clay.

that $3G_{eq}$ $(=E_{eq})$ from the undrained CTS test data are compared with the Young's modulus E_{eq} and E_{sec} values from the triaxial tests. The trend of the stiffness degradation is similar among these different types of tests.

Figure 3-8 shows the results from a comparable set of CTX and CTS tests. The equivalent modulus and damping ratio are similar between these tests for strains up to 0.01 %. As stated earlier in section 2, the larger damping ratio in CTX tests at intermediate strains (larger than about 0.01%) may be attributed to weaker stress-strain response on extension side.

4 RESULTS OF ROUND-ROBIN TESTS ON SOFT ROCK

4.1 Participating laboratories and test conditions

Five laboratories participated into the IRR test programme on sedimentary soft rock. Table 4-1 summarizes the types of tests performed, together with the number of tests reported. Note that the results of fourteen triaxial tests were reported. Table 4-2 summarizes the details of test conditions.

It is now widely recognized that effects of bedding error at the specimen ends could be significant in triaxial tests on stiff geomaterials, such as soft

rocks. The bedding error could bring about a significant underestimate of stiffness at small strains (e.g., Tatsuoka & Shibuya 1992). Capping the specimen ends is also important to improve uniformity in stresses/strains in the specimen subjected to consolidation and shearing. The capping is however not a remedial treatment to get rid of bedding error effects at the specimen ends (Shibuya et al, 1999). All of the participating laboratories employed a load cell installed inside the cell. However, it was reported whether capping is employed in the respective test. In the MTX tests, the rate of axial straining imposed at the rigid boundary was 0.001, 0.018 and 0.01 % per min. It should be pointed out that with a constant external axial strain rate, the rate of local axial strain could not be constant at small strains in tests on soft rocks (Shibuya et al. 1999).

In the CTX tests, the cyclic loading was imposed in a stress-controlled fashion by using sinusoidal wave with a loading frequency ranging between 0.01 and 1 Hz. The axial deformation of the specimen was measured locally by a means of PT or LDT. The nominal diameter of 50 mm with the height/diameter ratio of two was the common feature of the cylindrical specimens.

4.2 Monotonic loading triaxial tests

Figure 4-1 compares the stress-strain relationship from the drained MTX tests. The drained compressive strength ranged between 5.33 and 9.7 MN/m². The scatter may be due to the intrinsic scatter in the mechanical properties of the specimens. The scatter in the axial strains at the peak state is possibly due to the effect of bedding error at the specimen ends.

Similar results from the undrained tests are shown in Figure 4-2. The peak strength from the undrained tests ranged between 7.05 and 7.4 MN/m². The scatter among these different laboratories is much less than that with the drained tests. Also in this case, large differences seen with the axial strains at the peak stress are due likely to the effects of bedding error. The stiffness at small strains (see Fig.4-2d) is also more consistent than it is with the drained tests, except for the result from a laboratory.

Table 4-1. Summary of the test results on Sagamihara Soft Rock.

Lab. No.	Monotonic Test				Cyclic Test						Total
	Triaxial		Torsional		Triaxial		Torsional		Resonant Column		
	Drained	Undrained	Drained	Undrained	Drained	Undrained	Drained	Undrained	Drained	Undrained	
15				1*				1*		1*	1
3	3		2								5
5	2										2
9		4 (2*)				2*					4
13		1				1					2
Total	5	7		1*		3		1*		1*	14

* the same specimen

Table 4-2. Participating research organization, and test conditions on Sagamihara Soft Rock;

(a) Drained Monotonic Triaxial Compression Test (MTX-D Test)

Test No.	σ_c (kPa)	u_b (kPa)	σ_c' (kPa)	σ_v' (kPa)	Axial strain rate (%/min)	t_m (mm)	D_{oc} (mm)	H_c (mm)	Density(g/cm³) Wet ρ_t	Dry ρ_{dc}	Axial strain measurement Local	Internal	External	E_{max} (MN/m²) (LDT)	q_{max} (MN/m²)
3-1	719	247	472	472	0.01	0.3	46.06	104.33	2.016	1.731	LDT		LVDT	15700	6.73
3-2	720	249	471	471	0.001	0.3	47.94	101.53	2.003	1.643	LDT		LVDT	7440	6.69
3-3	920	455	465	465	0.001	0.3	47.79	154.98	1.996	1.635	LDT		LVDT	6540	5.50
5-1	666	196	470	470	0.01	0.3	48.90	120.20	2.025	1.655	LDT		LVDT	3310	8.72
5-2	666	196	470	470	0.01	0.3	47.16	99.54	1.974	1.592	LDT		LVDT	3560	9.69

* Initial diameter and Initial height

LDT: Local Deformation Transducer

(b) Undrained Monotonic Triaxial Compression Test (MTX-U Test)

Test No.	σ_c (kPa)	u_b (kPa)	σ_c' (kPa)	σ_v' (kPa)	Axial strain rate (%/min)	t_m (mm)	D_{oc} (mm)	H_c (mm)	Density (g/cm³) Wet ρ_t	Dry ρ_{dc}	Axial strain measurement Local	Internal	External	E_{max} (MN/m²) LDT	ST	q_{max} (MN/m²)
3-4	670	215	455	455	0.01	0.3	48.40	102.58	1.989	1.595	LDT		LVDT	3080		6.52
3-5	666	194	472	472	0.01	0.3	48.84	102.76	1.998	1.564	LDT		LVDT	10300		7.05
9-1	666	196	470	470	0.01	0.5	49.10	99.90	1.996	1.559	LDT,ST	PT		3370	3250	7.42
9-2	666	196	470	470	0.01	0.5	49.10	99.50	1.993	1.587	LDT,ST	PT		3390	2930	7.40
9-3	666	196	470	470	0.01	0.5	48.60	99.50	1.982	1.538	LDT,ST	PT		3430	3330	7.43
9-4	666	196	470	470	0.01	0.5	48.50	100.00	1.975	1.559	LDT,ST	PT		3330	3240	7.48
13-1	754	284	470	470	0.01	0.3	49.80*	96.60*	2.007	--			LVDT			4.26

ST: Strain gauge, PT: Proximity Transducer

(c) Undrained Monotonic Torsional Shear Test (MTS-U Test) Solid cylindrical specimen

Test No.	σ_c (kPa)	u_b (kPa)	σ_c' (kPa)	σ_v' (kPa)	Shear strain rate (%/min)	t_m (mm)	D_{oc} (mm)	H_c (mm)	Wet ρ_t	Dry ρ_{dc}	Shear strain measurement Local	Internal	External
15-1	670	200	470	470	0.018	0.5	36.20	71.10	1.980	1.588		PT	

(d) Undrained Cyclic Triaxial Compression Test (CTX-U Test)

Test No.	σ_c (kPa)	u_b (kPa)	σ_c' (kPa)	σ_v' (kPa)	f (Hz)	t_m (mm)	D_{oc} (mm)	H_c (mm)	Density (g/cm³) Wet ρ_t	Dry ρ_{dc}	Axial strain measurement Local	Internal	External	E_{max} (MN/m²) LDT	ST
9-3	666	196	470	470	0.2	0.5	48.60	99.60	1.982	1.532	LDT,ST	PT		3140	3180
9-3(2nd)	666	196	470	980	0.2	0.5	48.50	99.50	1.982	1.538	LDT,ST	PT		3180	3220
9-4	666	196	470	470	0.2	0.5	48.50	100.00	1.975	1.559	LDT,ST	PT		3570	3230
9-4(2nd)	666	196	470	980	0.2	0.5	48.50	100.00	1.975	1.559	LDT,ST	PT		3560	3580
13-2	754	284	470	470	0.1	0.3	49.80*	99.00	1.977	--			LVDT		

* Initial diameter and Initial height

(e) Undrained Cyclic Torsional Shear Test and Undrained Resonant Column Test (CTS-U and RC-U Test) Solid cylindrical specimen

Test No.	σ_c (kPa)	u_b (kPa)	σ_c' (kPa)	σ_v' (kPa)	f (Hz)	t_m (mm)	D_{oc} (mm)	H_c (mm)	Wet ρ_t	Dry ρ_{dc}	Shear strain measurement Local	Internal	External
15-1	670	200	470	470	0.01 ,0.1, 1 53 (RC-U)	0.5	36.20	71.10	1.980	1.588		PT	

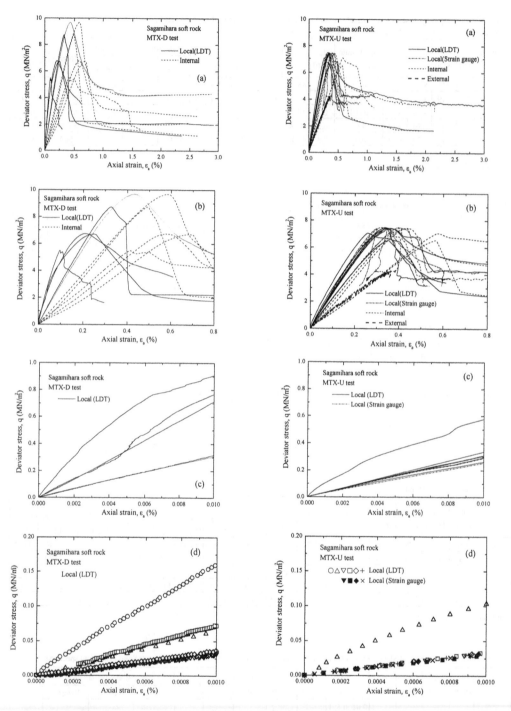

Figure 4-1. Relationship between deviator stress and axial strain in drained monotonic tests in triaxial compression on Sagamihara sedimentary soft rock.

Figure 4-2. Relationship between deviator stress and axial strain in undrained monotonic tests in triaxial compression on Sagamihara sedimentary soft rock.

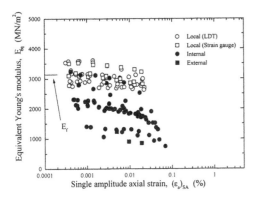

Figure 4-3. Comparison of Equivalent Young's modulus obtained from local axial measurement, internal and external measurement.

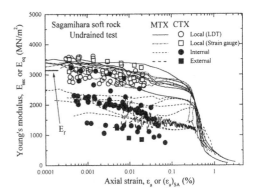

Figure 4-5. Comparison of the stiffness degradation curves between undrained monotonic and cyclic shear tests on Sagamihara sedimentary soft rock.

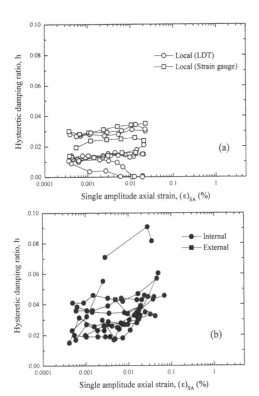

Figure 4-4. Comparison of Hysteretic damping ratio obtained from (a) local axial measurement, (b) internal and external measurement.

4.3 Cyclic triaxial tests

Figure 4-3 shows the relationship between E_{eq} and $(\varepsilon_a)_{SA}$ for the 10th cycle at each stage in the undrained CTX tests. It can be seen that the E_{eq} values obtained by local measurement of axial strain is significantly larger than that obtained by the internal

as well as external measurement. Also, the scatter in the E_{eq} values by the internal and external measurement among the different laboratories is more significant than that of the E_{eq} values by local measurement. This is due likely to the effects of bedding error. Figure 4-4 shows the relationship between h and $(\varepsilon_a)_{SA}$. It can be seen that the scatter of the h values obtained by internal and external measurement is noticeably larger than that of the h values by local measurement. It is also found that the h values by internal and external measurement are generally larger than the h values by local measurement. That is, for the cyclic triaxial test on sedimentary soft rock, it is likely that the damping ratio could be over-estimated when the axial strain is obtained from the displacement of top cap or piston rod.

4.4 Comparison between monotonic and cyclic shear test data

Figure 4-5 shows the stiffness degradation curves with strain from the undrained MTX and CTX tests. The following trends of behaviour may be noted;
i) the local strain measurement yielded a consistent stiffness-strain relationship among the different laboratories;
ii) the stiffness values by the local strain measurement exhibited a similar difference for a strain range below 0.1% between MTX and CTX tests; and
iii) the stiffness at strains of about 0.001 % is consistent with the Young's modulus of 3140 MN/m^2 from in-situ seismic survey (Toki et al. 1995; Tatsuoka and Kohata, 1995).

The results strongly indicate that the monotonic test with local strain measurement could be an alternative to cyclic tests only when the stiffness is of interest.

5 CONCLUDING REMARKS

During a period of four years from 1996 to 2000, a series of international round-robin tests on sand, clay and soft rock was performed by participation of nineteen laboratories in six countries. The programme aimed at providing a valuable data to develop the recommendation for procedures referred to triaxial and torsional shear tests.

The outcomes may be summarized in the followings:

i) The effect of bedding error on the measurement of axial deformation of triaxial specimens was significant in the tests on sedimentary soft rock, whereas it was not so in the tests on fine-grained soils; i.e., Toyoura sand and Fujinomori clay. Local strain measurement should be employed in tests on stiff geomaterials.

ii) The stiffness at strains between 0.001 % and 0.1 % was similar between monotonic and cyclic tests in the tests on sand, clay and soft rock. Monotonic tests may be sufficient in order to evaluate the stiffness-strain relationship of geomaterials at small strains.

iii) In the behaviour of Toyoura sand at intermediate strain levels, the degradation of secant shear moduli in the monotonic torsional shear tests was much more pronounced than those observed in cyclic tests. The degradation in the stiffness with strain in the monotonic triaxial compression tests was masked as the axial strain became about 0.1 % due to the effects of an increase in the axial stress.

iv) The scatter of the test data from the different laboratories was much less in the torsional tests than in the triaxial tests on Toyoura sand.

v) The test results were consistent among the different laboratories only when the test specification was properly fulfilled while using a properly instrumented apparatus.

ACKNOWLEDGEMENTS

On behalf of TC-29, the authors appreciate tremendous co-operation of all the laboratories participating into the IRR tests. They also pay tribute to the late Dr Tej Pradhan, who was deeply involved in the IRR test programme, including preparation of the original draft of specification for test on Fujinomori clay. A review by Professor Tatsuoka, F. is also appreciated.

REFERENCES

Goto, S., Tatsuoka, F., Shibuya, S., Kim, Y-S. & Sato, T. 1991. A simple gauge for local small strain measurements in the laboratory. Soils and Foundations, JGS 31(1): 169-180.

Hardin, B. O. & Richart, F. E. Jr. 1963. Elastic wave velocities in granular soils. *Journal of the Soil Mechanics and Foundations Division*, ASCE 89(1): 33-65.

Hoque, E. & Tatsuoka, F. 1998. Anisotropy in elastic deformation of granular materials. Soils and Foundations, JGS 38(1): 163-179.

Iwasaki, T. & Tatsuoka, F. 1977. Effect of grain size and grading on dynamic shear moduli of sands. *Soils and Foundations*, JGS 17(3): 19-35.

Kim, D.S. & Stokoe, K.H., II 1995. Deformational charavteristics of soils at small to medium strains. Poceedings of the 1st International Conference on Earthquake Geotechnical Engineering, Balkema: 89-94.

Shibuya, S., Mitachi, T., Tanaka, H., Kawaguchi, T. & Lee, I-M. 1999. Measurement and application of quasi-elastic properties of geomaterials in geotechnical site characterisation. Theme Lecture for discussion session 1, Proc. of 11th Asian Regional Conference on SMGE, Seoul, Balkema, Vol.2 (in print).

Tatsuoka, F., Iwasaki, T. & Takagi, Y. 1978. Hysteretic damping of sands under cyclic loading and its relation to shear modulus. Soils and Foundations, JGS 18(2): 25-35.

Tatsuoka, F. & Kohata, Y. 1995. Stiffness of hard soils and soft rocks in engineering applications, Keynote paper, Proceedings of the 1st International Conference on Pre-failure Deformation of Geomaterials, Sapporo, Balkema: 947-1063.

Tatsuoka, F. & Shibuya, S. 1992. Engineering properties of soils and rocks from in-situ and laboratory tests. Keynote paper for discussion session 1, Proc. of 9th Asian Regional Conference on SMFE, Vol.2: 101-170.

Toki, S., Shibuya, S. & Yamashita, S. 1995. Standardization of laboratory test methods to determine the cyclic deformation properties of geomaterials in Japan. Keynote paper, Proceedings of the 1st International Conference on Pre-failure Deformation of Geomaterials, Sapporo, Balkema: 741-784.

Yamashita, S., Toki, S. & Suzuki, S. 1996. Effect of membrane penetration on modulus and Poisson's ratio for undrained cyclic triaxial conditions. Soils and Foundations, JGS 36(4): 127-133.

APPENDIX A: ROUND-ROBIN TESTS ON SAND

1 Outline of the program

The following types of tests to evaluate the pre-failure stress-strain property of sand are planned;
1) Monotonic triaxial compression tests and/or torsional shear tests.
  i) drained monotonic tests on sand specimens with a relative density at the end of consolidation, D_{rc}, equal to 50 and 80 %.
  ii) undrained monotonic tests on sand specimens with D_{rc} = 50 and 80 % (optional).
2) Cyclic triaxial tests and/or torsional shear tests.
  i) drained cyclic tests on sand specimens with a relative density at the end of consolidation, D_{rc}, equal to 50 and 80 %.
  ii) undrained cyclic tests on sand specimens with D_{rc} = 50 and 80 % (optional).

2 Test material

The test material, which will be provided at a later stage, is Toyoura sand (see Fig. A-1). The results obtained from the round-robin tests of cyclic triaxial and torsional tests on the same type of sand that were performed in Japan are described in the paper "Toki, S., Shibuya, S. and Yamashita, S. (1995): Standardization of laboratory test methods to determine the cyclic deformation properties of geomaterials in Japan, *Pre-failure Deformation of Geomaterials (IS-Hokkaido'94)*, Balkema, Vol.2, pp.741-784".

10 kgf of the test material will be distributed to each laboratory. A more amount of the material will be provided on request.

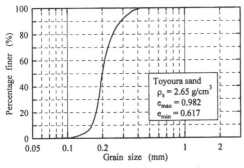

Figure A-1. Grain size distribution of Toyoura sand.

3 Test procedures

In principle, the proposed test procedures follow the test standard of Japanese Geotechnical Society "Method for Cyclic Triaxial Test to Determine Deformation Properties of Geomaterials" (see Appendix D) and "Method for Cyclic Torsional Shear Test on Hollow Cylindrical Specimens to Determine Deformation Properties of Soils" (see Appendix E). The copies are enclosed herein. When radial strains are measured locally, report its details.

4 Sample preparation

Specimens having D_{rc} equal to 50 and 80 % (dry densities ρ_{dc} equal to 1.473 and 1.567 g/cm^3) at the end of isotropic consolidation are prepared by the air-pluviation method; air-dried sand is poured from a copper nozzle with a rectangular inner cross-section of 1.5 mm × 15.0 mm, while maintaining a constant drop-height throughout the preparation.

The final top surface of specimen is made level and smooth by scraping with a thin plate having a straight edge.

These nozzles will be produced by the Japanese domestic committee of TC-29, and will be distributed to the members of TC-29.

Another method can be employed to prepare specimens. In that case, it is requested to describe the details of the method in the report.

5 Specimen set-up

Before the mold is disassembled, a partial vacuum of 5 kPa is applied to the specimen, while the loading piston is unclamped. Subsequently, the mold is dismantled, and then the vacuum is raised to 30 kPa while ensuring that the specimen can deform freely both in the axial and lateral directions. Then, specimen height and diameter are measured, and the results are reported.

6 Consolidation

After the partial vacuum is replaced with a cell pressure of 30 kPa while keeping the effective stress constant throughout the procedure, the specimen is saturated using any of or a combination of appropriate saturation techniques. Thereafter, a back pressure of 100 kPa is applied. Then, the specimen is isotropically consolidated to an effective confining stress σ'_c of 100 kPa.

7 Evaluation of the degree of saturation

The pore pressure parameter B is measured as follows;
  i) the drainage valves are closed.
  ii) the cell pressure is decreased by 50 kPa and the change in the pore pressure is measured.
  iii) the cell pressure is increased by 50 kPa and the change in the pore pressure is measured.
  iv) the B value is calculated by the following equation:

$$B = \frac{\Delta u_u + \Delta u_l}{2\Delta\sigma_3}$$

where

Δu_u = the change in the pore water pressure (absolute value) that occurs as a result of the decrease in the cell pressure;

Δu_l = the change in the pore water pressure (absolute value) that occurs as a result of the increase in the cell pressure; and

$\Delta\sigma_3$ = the change in the cell pressure (absolute value).

v) for an undrained test, the test is continued only when the B value is equal to, or greater than, 0.95.

8 Loading

1) Monotonic loading test

The axial or shear strain rate is set to be equal to 0.05 % per minute.

When feasible, strain is measured for a range from less than about 0.001 % to 15 % (axial strain) or 22.5 % (shear strain) or whatever for a possible range of strain.

2) Cyclic loading test

The specification for cyclic loading tests follows the above-described JGS Standard.

9 Report

1) Monotonic test

i) Outline of the test apparatus used.

ii) Fill up the items listed in form 1 attached.

iii) Values of deviator (shear) stress, q (τ) (kN/m^2), and volumetric strain, ε_v (%) or excess pore water pressure, u (kN/m^2), at appropriated selected axial (shear) strains, ε_a (γ) (%) (i.e. fill up the items listed in form 2 and/or record the data in a 3.5 inch floppy disk in the form of example 2).

iv) Relationship between the secant Young's (shear) modulus, E_{sec} (G_{sec}) (kN/m^2), and the logarithm of axial (shear) strain, ε_a (γ) (%).

v) Any deviations from the procedure outlined in this specification.

2) Cyclic test

i) Outline of the test apparatus used.

ii) Fill up the items listed in form 1 attached.

iii) Values of equivalent Young's (shear) modulus (i.e. peak-to-peak secant modulus), E_{eq} (G_{eq}) (kN/m^2) and hysteretic damping ratio, h (%), at each single amplitude cyclic axial (shear) strains, $(\varepsilon_a)_{SA}$ (γ_{SA}) (%). The values at the second through tenth cycles in each cyclic loading stage are reported (i.e. fill up the items listed in form 3 and/or record the data in a 3.5 inch floppy disk in the form of example 3).

iv) Relationship between the equivalent Young's (shear) modulus, E_{eq} (G_{eq}) (kN/m^2), the hysteretic damping ratio, h (%), and the single amplitude

cyclic axial (shear) strain, $(\varepsilon_a)_{SA}$ (γ_{SA}) (%), at the fifth and tenth cycles in each cyclic loading stage.

v) Any deviations from the procedure outlined in this specification.

APPENDIX B: ROUND-ROBIN TESTS ON CLAY

1 Objective

The main objective of this International Round Robin (IRR) Tests is to investigate the range of scatter in measured pre-failure deformation properties of clay, including stiffness and damping values, obtained by different laboratories in the world. The scatter may be either due to human and or apparatus errors or due to intrinsic properties of soil which exhibit different test results under different test conditions (e.g., different shear modes, different strain rates, different number of loading cycles and so on).

2 Test type

1) Monotonic and/or cyclic triaxial test
2) Monotonic and/or cyclic torsional shear test
3) Bender element test
4) Resonant-column test
5) Others

3 Sample

Samples are distributed in the form of air-dried powder (20 kgf each package).

1) Name: Fujinomori Clay. The index properties are; Liquid limit =62 %, Plasticity index =33, Particle density =2.69 g/cm^3, Clay content (<5µm) =43 % (JIS A 1205).

2) Before producing test specimens from the provided powder, the powder should be first thoroughly mixed with deaired water and passed through a sieve of 0.42 mm or a similar opening (depending on the standard of each country) while crushing gently clay clods.

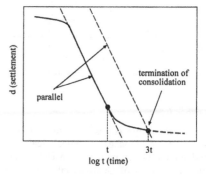

Figure B-2. Settlement (d) – log (time) curve (the 3t method).

4 Pre-consolidation method

An oedometer mold having an internal diameter of about 25 to 30 cm and a height of about 30 cm is prepared.

1) The sieved powder clay is thoroughly mixed with tap water so as to make the water content of the slurry to be about twice of the liquid limit (i.e., $w = 124 \%$).

The slurry is then poured into the mold gently to a thickness (height) of about 25 cm.

2) The slurry is one-dimensionally pre-consolidated with a final vertical pressure of 70 kPa. The consolidation should be terminated based on the 3t method (see Fig. B-1); t is the time at the end of primary consolidation.

For consolidation, stepwise loading is recommended. Drainage is allowed from the both top and bottom ends.

5 Preparation of test specimen

1) The consolidated soil is removed from the oedometer mold and then it is trimmed to the test specimen dimensions in such a way that the vertical direction during consolidation coincides with that in the triaxial or torsional shear apparatus.

2) A disk of porous stone or porous metal together with an incompressible filter material should be placed at both top and bottom drainage ends of the specimen.

3) Side drain of filter paper should not be used.

4) A thin latex rubber membrane with a thickness of 0.2 to 0.3 mm is recommended to use.

6 Saturation of specimen

1) Use a back pressure of either 100 kPa or 200 kPa. Report the value used.

2) Choose any adequate method for saturation. However the Bishop's B-value ($B=\Delta u/\Delta\sigma_c$) should be greater than 0.95. If not, report the value measured.

7 Reconsolidation

1) In the case of isotropic consolidation, the effective confining pressure is 130 kPa. While in the case of anisotropic consolidation, the vertical effective stress is 200 kPa and the horizontal effective stress is about 100 kPa (assuming K_0 value of about 0.5).

2) The rate of mean stress increase during reconsolidation is 5 kPa/min.

3) The reconsolidation time at the final specified stress state is 24 hours.

4) Report the height change and volume change during reconsolidation.

8 Testing procedures

1) The test procedures is basically the same as the test standard of the Japanese Geotechnical Society (JGS) "Method for Cyclic Triaxial Test to Determine Deformation Properties of Geomaterials" (see Appendix D) and "Method for Cyclic Torsional Shear Test on Hollow Cylindrical Specimens to Determine Deformation Properties of Soils" (see Appendix E). The copies of the English version are enclosed herein. When lateral strain is measured locally, report its details.

2) For bender element tests and resonant-column tests, choose your own adequate method and report the method used.

9 Loading

1) Monotonic loading tests

The axial strain rate is set to be equal to 0.05 % per minute. When another strain rate is used, report its value.

It is recommended to measure the axial strain for a range form less than about 0.01 %, or if possible about 0.001 %, to 15 %.

2) Cyclic loading tests

The specification for cyclic loading tests follows the above-mentioned JGS Standard.

10 Report

1) Monotonic loading tests

i) Outline the test apparatus used.

ii) Fill up the items listed in Form 1 attached.

iii) List up the values of deviator stress, q (or shear stress, τ) (kPa), and volumetric strain, ε_v (%) or excess pore water pressure, u (kPa), at appropriately selected axial strains, ε_a (or shear strain, γ) (%) (i.e., fill up the items listed in Form 2 and/or record the data in a 3.5 inch floppy disk in the form of example 2).

iv) Plot the relationship between the secant Young's modulus, E_{sec} (or secant shear modulus, G_{sec}) (kPa), and the axial strain, ε_a (or shear strain, γ) (%).

v) Report any deviations from the procedure outlined in this specification.

2) Cyclic loading tests

i) Outline the test apparatus used.

ii) Fill up the items listed in From 1 attached.

iii) List up the values of equivalent Young's modulus, E_{eq} (or shear modulus, G_{eq}) (kPa) (i.e. peak-to-peak secant modulus), and hysteretic damping ratio, h (%), at each single amplitude cyclic axial strains, $(\varepsilon_a)_{SA}$ (or shear strain, γ_{SA}) (%).

iv) Plot the relationship between the equivalent Young's modulus, E_{eq} (or the shear modulus, G_{eq}) (kPa), the hysteretic damping ratio, h (%), and the single amplitude cyclic axial strain, $(\varepsilon_a)_{SA}$ (or shear strain, γ_{SA}) (%), at the fifth and tenth cycles in each cyclic loading stage.

v) Report any deviations from the procedure outlined in this specification.

3) Bender Element and other tests:

i) Report the shear modulus at the estimated strain level (use your own form).

APPENDIX C: ROUND-ROBIN TESTS ON SOFT ROCK

1 Outline of programme

The following types of tests are planned in order to evaluate the pre-failure stress-strain properties of sedimentary soft rock;

1) Monotonic triaxial compression tests.

consolidated-undrained or consolidated-drained triaxial compression tests subjected to monotonic loading

2) Cyclic triaxial tests;.

consolidated undrained triaxial tests subjected to cyclic loading

3) Any other laboratory testing to determine the prefailure stress-strain properties

In performing the tests of 1) and 2), the common specifications have been prepared, which are described below.

2 Test material

The test material is a sedimentary soft rock (mudstone). The samples were retrieved by block sampling at a depth of 48 m in Kazusa group sediment, having a geological age of about 1.5 million years, at Sagamihara test site, Kanagawa prefecture, Japan. A cyclic triaxial round-robin test using samples of the same soft rock was performed in Japan. The test results are described in "Toki,S., Shibuya,S. and Yamashita,S. (1995); Standardization of laboratory test methods to determine the cyclic deformation properties of geomaterials in Japan, *Pre-failure Deformation of Geomaterials* (*IS-Hokkaido '94*), Balkema, Vol.2, pp.741- 784".

Upon request, the following test specimens will be distributed to each laboratory.

1) A specimen with 5 cm (or 10 cm) in diameter and 12-13 cm (15-18 cm) in height or other dimensions on your request; The specimens will be cored from a large block that has been subjected to in-situ effective vertical stress σ'_v of 470 kPa (4.8 kgf/cm^2). The compressive strength from the previous triaxial compression tests is about 4.9-8.83 MPa (50-90 kgf/cm^2).

2) The number of samples; three for each laboratory; it is preferable to perform two tests under the same conditions, while remaining the other specimen for a spare.

3 Test procedures

The test procedures proposed herein follow in principle the testing standard of the Japanese Geotechnical Society "Method for Cyclic Triaxial Test to Determine Deformation Properties of Geomaterials (see Appendix D)". The copy is enclosed herein.

4 Sample preparation

1) Trim only the top and bottom ends of each specimen and report the trimming method. The lateral surface of each specimen is not trimmed. The height of specimen can be determined at your convenience.

2) Before each specimen is placed on the pedestal of the triaxial cell, it should be confirmed that the surfaces of the top cap and pedestal are reasonably flat and parallel to each other. If a very good parallelism is not assured, the top or bottom ends or both of the specimen should be capped by using an appropriate method, and report the method of capping in detail.

3) Use side drain and report its detail, in particular the method how the side drain is connected to drainage lines. Based on experiences at IIS, the University of Tokyo, side drain is necessary because of a low permeability of the mudstone.

When porous metal or porous stone is used at the top and pedestal, report its detail.

4) Report the material and thickness of rubber membrane.

5) Measure the specimen size (diameter and height) and weight before setting each specimen in the triaxial cell and report these results.

5 Re-consolidation

1) Isotropic consolidation

i) First each specimens is isotropically consolidated under an effective confining stress $\sigma'_c = 30$ kPa (0.3 kgf/cm^2) and then a back pressure of 196 kPa (2.0 kgf/cm^2) is applied.

ii) Then, the specimen is isotropically consolidated to $\sigma'_c = 470$ kPa (4.8 kgf/cm^2), which is equal to the estimated in-situ effective overburden pressure.

iii) The specimen is left at the final consolidation stress state at least for two hours.

iv) For cyclic tests, first consolidate the specimen isotropically to $\sigma'_c = 470$ kPa as described above. Then, perform two-way cyclic loading tests as described below until the single amplitude deviator stress becomes larger than the isotropic consolidation stress σ'_c. If feasible, reconsolidate the specimen at the isotropic stress state $\sigma'_c = 470$ kPa for two hours. Then, increase the effective axial stress σ'_1 to 980 kPa (10 kgf/cm^2) under drained conditions over the duration of about ten minutes and wait for about two hours at the stress condition $\sigma'_1 = 980$ kPa and $\sigma'_c = 470$ kPa. Then, perform again two-way cyclic loading tests as described below.

6 Evaluation of the degree of saturation

The pore pressure parameter B should be measured as follows;
1) Close the specimen drainage valves.
2) Decrease the cell pressure by 49 kPa (0.5 kgf/cm^2) while keeping the isotropic stress state, then measure the change in the pore water pressure.
3) Increase the cell pressure by 49 kPa (0.5 kgf/cm^2) while keeping the isotropic stress state, then measure the change in the pore water pressure.
4) Calculate the B-value using the following equation;

$$B = \frac{\Delta u_u + \Delta u_l}{2\Delta\sigma_3}$$

where

Δu_u = the change in the pore water pressure (absolute value) that occurred as a result of the decrease in the cell pressure, which is measured ten minutes after having changed the cell pressure;

Δu_l = the change in the pore water pressure (absolute value) that occurred as a result of the increase in the cell pressure, which is measured ten minutes after having changed the cell pressure; and

$\Delta\sigma_3$ = the change of cell pressure (absolute value).

Obtain the changes in the height and volume of specimen which occur during re-consolidation (i.e., the start of shearing at a cell pressure at 470 kPa (4.8 kgf/cm^2)).

7 Loading

1) Monotonic loading test
 i) The axial strain rate is set to be equal to 0.01 % per minute. When another strain rate is selected, report its value.
 ii) Select drainage condition (drained or undrained), and report it.
 iii) When feasible, axial strain is measured for a strain range from less than about 0.001 % to 3 % or whatever for a possible range of strain. Report the method of axial strain measurement (local or external, and each specific details). When a local gauge is used, please refer to "A note on the use of local gauge in consolidated undrained triaxial compression tests on sedimentary soft rock" (see Appendix F).
 When radial strain is measured locally, report its details.
 iv) If possible, a monotonic triaxial compression test is carried out after respective cyclic loading test.

In that case, either of the following two methods has to be selected.
 a) After having reconsolidated the specimen to 470 kPa (4.8 kgf/cm^2), a triaxial compression test is carried out under either undrained or drained condition.
 b) The specimen is isotropically unloaded to 0 kPa (0 kgf/cm^2), and then retrieved from the triaxial cell. The specimen is treated as a new sample, and reset the specimen in the triaxial cell.
2) Cyclic loading test;
The specifications for cyclic loading tests are proposed after the above-mentioned JGS Standard (see Appendix D).

8 Report

1) Monotonic loading test
 i) Outline of the test apparatus used
 ii) Fill up the items listed Form 1 attached
 iii) The value of deviator stress, q (MN/m^2), and volumetric strain, ε_v (%) or excess pore water pressure, u (kN/m^2), at appropriately selected axial strains, ε_a (%) (i.e. fill up the items listed in Form 2 and/or record the data in a 3.5 inch floppy disk in the form of example 2).
 iv) Relationship between the deviator stress and the axial strain, and that between the secant Young's modulus, E_{sec} (MN/m^2) and the axial strain ε_a (%).
 v) Any deviations from the procedure described in this specification.
2) Cyclic loading test
 i) Outline of the test apparatus used
 ii) Fill up the items listed Form 1 attached
 iii) Equivalent Young's modulus (i.e. peak-to-peak secant modulus), E_{eq} (MN/m^2) and damping ratio, h (%), at each single amplitude cyclic axial strains, $(\varepsilon_a)_{SA}$ (%). The values at second through tenth cycles in each cyclic loading stage are reported (i.e. fill up the items listed in Form 3 and/or record the data in a 3.5 inch floppy disk in the form of example 3).
 iv) Relationship between the equivalent Young's modulus, E_{eq} (MN/m^2), the damping ratio, h (%), and the single amplitude cyclic axial strain, $(\varepsilon_a)_{SA}$ (%), recorded at the fifth and tenth cycles at each cyclic loading stage.
 v) Any deviations from the procedure described in this specification.

APPENDIX D:

Designation: JGS 0542-2000
Method for Cyclic Triaxial Test to Determine Deformation Properties of Geomaterials

1. GENERAL

1.1 Purpose of Test

This test method covers the determination of the deformation properties of geomaterials in either an isotropic or an anisotropic stress state, when subjected to either drained or undrained cyclic loading in triaxial apparatus.

1.2 Scope of Application

This test method is applicable to various kinds of geomaterials such as sandy soils, cohesive soils, gravely soils, soft rocks, cement-treated soils and so on.

1.3 Definition of Terms

The *drained/undrained cyclic triaxial test used to determine the deformation properties of geomaterials* imposes either uniform, and symmetrical, cyclic axial deviator stress (load control) or cyclic axial deformation (displacement control) on a specimen using a fixed frequency.

The *deformation properties* determined are the equivalent Young's modulus defined from amplitudes of cyclic deviator stress and cyclic axial strain, and the hysteretic damping ratio defined from a hysteresis loop of the relationship between deviator stress and axial strain.

Axial stress is applied in the axial direction of the specimen, while *lateral stress* is applied in the radial direction of the specimen. The values of these stresses and pore-water pressure are defined at the mid-point of the height of the specimen.

Back pressure is the pore-water pressure applied to the specimen by keeping its effective stress constant for the purpose of raising the degree of saturation of the specimen.

Anisotropic consolidation stress ratio is calculated by dividing effective lateral stress by effective axial stress at the end of consolidation.

Consolidation stress is calculated by subtracting back pressure from the outside stress of the specimen during the consolidation process; *axial consolidation stress* is defined for the axial direction, while *lateral consolidation stress* is defined for the radial direction of the specimen.

1.4 Referenced Standards

The specimen shall be prepared and set in accordance with the following JGS standards:

JGS 0520 "Preparation of Soil Specimens for Triaxial Tests", and JGS 0530 "Preparation of Specimens of Coarse Granular Materials for Triaxial Tests".

[Notes]
1. Report details of any procedure(s) employed which are different from those specified in this standard.
1.2a. In principle, this standard is applicable to any kind of geomaterials.
b. Cement-treated soils refer to geomaterials whose properties are modified by chemical(s) such as cement.
c. Saturated, dry and partly saturated specimens are considered.
1.4 The following standards shall be referred to for procedures not described in this standard:
JGS 0522 "Method for Consolidated-Undrained Triaxial Compression Test on Soils",
JGS 0523 "Method for Consolidated-Undrained Triaxial Compression Test on Soils with Pore Water Pressure Measurements",
JGS 0524 "Method for Consolidated-Drained Triaxial Compression Test on Soils", and
JGS 0541 "Method for Cyclic Undrained Triaxial Test on Soils".

2. APPARATUS

The cyclic triaxial apparatus consists of a triaxial cell, cell pressure and back pressure control devices, an axial load or axial deformation control device, and a data acquisition and recording system for the axial load, axial displacement, volume change and pore-water pressure of the specimen.

(1) The triaxial apparatus must be capable of sustaining the desired maximums of the cell and back pressures and compression/extension axial load.

(2) The triaxial apparatus shall be capable of imposing the prescribed cell and back pressures and axial load on a specimen. The specimen is sealed by a top cap, a pedestal and a rubber membrane, and has drainage at the top and bottom. In an undrained cyclic loading test on a saturated specimen, the volume change of all the assembled parts of the pore-water pressure measurement system shall be negligible.

(3) During isotropic or anisotropic consolidation of a specimen in the triaxial cell, the cell pressure, back pressure and axial pressure control devices shall be capable of applying and controlling pressures to within ± 2 kN/m² for pressures less than 200 kN/m² and to within ± 1 % for pressures equal to or greater than 200 kN/m². The axial displacement and volume change of the specimen shall be measured to an accuracy of ± 0.02 % of the height and ± 0.05 % of the volume of the specimen, respectively.

(4) The cyclic loading equipment must be capable of applying specified axial loads or speci-

fied axial displacements to a specimen being held in an isotropic or anisotropic stress state when sheared under drained or undrained conditions. The range of single amplitude cyclic axial strain, $(\varepsilon_a)_{SA}$ (refer to 4.4), that can be applied shall be from 0.001% to 0.1 %. It is recommended that either a sinusoidal or a sawtooth variation of load with time at a frequency within a range of 0.05 and 1.0 Hz be used.

The loading device used in an axial load controlled test must be able to satisfy the following conditions:

i) For the single amplitude cyclic axial loads in compression and in extension, P_C and P_E, as defined from the reference stress state at the start of cyclic loading, the scatter of $(P_C{+}P_E)$ during a stage of cyclic loading shall be less than 10 %.

ii) $0.8 \leqq P_C/P_E \leqq 1.2$

The loading device in an axial displacement controlled test must be able to satisfy the following conditions:

i) For the single amplitude cyclic axial displacements in compression and in extension, ΔL_C and ΔL_E, as defined from the reference state at the start of cyclic loading, the variation of $(\Delta L_C{+}\Delta L_E)$ during a stage of cyclic loading shall be less than 10 %.

ii) $0.8 \leqq \Delta L_C/\Delta L_E \leqq 1.2$

(5) During cyclic loading, the cell pressure control system shall be capable of applying and controlling pressures to within ± 2 kN/m^2 for pressures less than 200 kN/m^2 and to within ± 1 % for pressures equal to or greater than 200 kN/m^2.

(6) During cyclic loading, the cell pressure (and also the pore-water pressure for an undrained cyclic loading test on a saturated specimen) shall be continuously measured to an accuracy of ± 2 kN/m^2 for pressures less than 200 kN/m^2 and to an accuracy of ± 1 % for pressures equal to or greater than 200 kN/m^2.

(7) The axial load cell shall be installed inside the triaxial cell. The axial load is continuously measured to an accuracy of ± 1 % of the magnitude of the double amplitude cyclic load when the single amplitude cyclic axial strain, $(\varepsilon_a)_{SA}$, is greater than 0.01 %.

(8) In cyclic loading stages where the single amplitude cyclic axial strain, $(\varepsilon_a)_{SA}$, is greater than 0.01%, the axial deformation measurement device, which shall have negligible hysteresis, shall be capable of continuously measuring the axial deformation of the specimen to an accuracy of ± 1 % of the magnitude of the double amplitude cyclic axial strain. In cyclic loading stages where $(\varepsilon_a)_{SA}$ is smaller than 0.1 %, the axial deformation of

the specimen may be measured by a displacement measurement device mounted inside the triaxial cell. When either the axial strain or the hysteretic damping ratio measured by means of the above methods is expected to differ by more than 5 % from the relevant average measured on the central portion of the specimen, the axial strain of the specimen is to be measured locally on the lateral surface of the specimen, by which means bedding errors at the specimen ends will be avoided.

(9) During cyclic loading, the data acquisition system shall be capable of monitoring both cyclic axial load and cyclic axial deformation, and also pore-water pressure if necessary, with little time lag between these measurements in each data set.

[Notes]

2.1 Fig. 1 shows a typical cyclic triaxial apparatus.

(1)a. Care shall be taken for the triaxial cell not to move upwards when the maximum axial load is applied in extension.

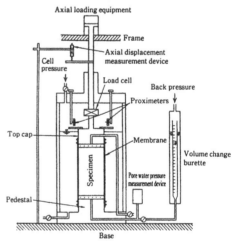

Fig. 1 Typical triaxial apparatus for cyclic loading test

b. A rigid contact must be ensured between the loading piston and the specimen top cap, prior to assembling the triaxial cell (refer to Fig. 2a in JGS 0522 "Method for Consolidated-Undrained Triaxial Compression Test on Soils", and Fig. 1 in JGS 0541 "Method for Cyclic Undrained Triaxial Test on Soils").

(2)a. The diameters of top cap and pedestal shall be equal to that of the specimen, the flat surfaces of the top cap and pedestal shall be parallel to each other, and the angle between these planes and the axis of the loading piston shall be 90 degrees.

b. Porous disk with and without filter paper can be used for drainage. However, care shall be taken with the measurement of axial deformation when filter paper is used.

c. The volume change of all the assembled parts of the pore-water pressure measurement system shall meet the requirement specified in note 2.(2)c. in JGS 0541 "Method for Cyclic Undrained Triaxial Test on Soils".

(3) The volume change of specimens during isotropic or anisotropic consolidation shall be measured using a burette or other device with equivalent accuracy. The burette shall be capable of sustaining the back pressure, and its water level shall preferably stay unchanged when pressurized.

(4)a. As illustrated in Fig. 2, the single amplitudes of axial load in compression and in extension, P_C and P_E, shall be defined from the stress state at the start of cyclic loading, and these are both positive in value. The value of ΔP $(=P_C+P_E)$ refers to the double amplitude of axial load. In Fig. 2, P_0 stands for deviator force, that is the multiplier of the deviator stress and cross section area of the specimen, at the start of cyclic loading.

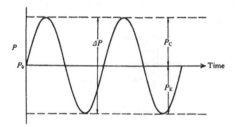

Fig. 2 Definitions of single amplitude of axial load in compression and in extension, P_C and P_E, for sinusoidal application of axial load

b. As can be seen in Fig. 3, single amplitudes of axial displacement in compression and in extension, ΔL_C and ΔL_E, shall be defined from the state at the start of cyclic loading, and these are both positive in value. The value of ΔL $(=\Delta L_C+\Delta L_E)$ refers to double amplitude of the axial displacement.

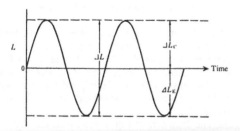

Fig.3 Definitions of single amplitude of axial displacement in compression and in extension, ΔL_C and ΔL_E, for sinusoidal application of axial displacement

c. Loading frequency may be outside of the range between 0.05 and 1.0 Hz, provided that the accuracy

of the stress and deformation measurements is compatible with that specified in this standard.

d. The variation of cyclic axial load or cyclic axial displacement with time shall be either sinusoidal or sawtooth in shape.

(7)a. A strain-gauge type load cell shall be used for the axial load measurement. It shall be installed inside the triaxial cell, and be capable of measuring force both in compression and in extension.

b. The output of the load cell shall not be affected by the change of cell pressure and the eccentricity of the loading. In addition, the drift as well as the change of calibration factor shall be less than 1 % of the axial load as measured with a single amplitude cyclic axial strain, $(\varepsilon_a)_{SA}$, of 0.01 %.

c. The hysteresis of the load cell shall be measured by determining the relationship between the output of the load cell and the axial load, which can be measured by applying slow load-unload cycles (see Fig.4). The apparent hysteresis of the load cell, h_{LC}, shall be less than 5 % of the hysteretic damping ratio of the specimen as measured at $(\varepsilon_a)_{SA}$ equal to 0.01 % for the maximum conceivable axial load.

Definition of apparent hysteretic damping of load cell:

$$h_{LC} = (1/2\pi) \times \Delta X / X$$

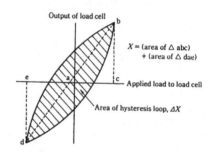

Fig.4 Definition of apparent hysteretic damping of load cell

(8)a. When measuring axial deformation during cyclic loading using a sensor mounted outside the triaxial cell, the system compliance shall be less than 1 % of the axial deformation of the specimen at $(\varepsilon_a)_{SA}$ of 0.1 %.

b. The use of a proximity transducer is recommended for measuring the axial deformation of a specimen from the movement of the top cap.

c. Local axial deformation of the specimen shall be measured when bedding errors at the specimen ends are expected to amount for more than 5 % of the specimen's axial deformation. As can be seen in Fig. 5, the local deformation, ΔL_{local}, shall be measured at two positions, opposite to each other, around the specimen diameter. The gauge length shall be 50 to 80 % of the specimen height. Care shall be taken that the local strain device dose not restrict free deformation of the specimen.

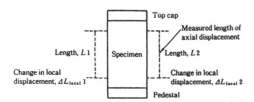

Fig. 5 Example of local axial strain measurement

d. Output of the sensor for the axial deformation measurement shall be unchanged due to the change of cell pressure. In addition, the drift as well as the change of calibration factor shall be less than 1 % of the axial load as measured with a single amplitude cyclic axial strain, $(\varepsilon_a)_{SA}$, of 0.01 % in every cycle.

e. The hysteresis of the sensor for axial deformation measurement shall be measured by determining the relationship between the output of the sensor and the axial deformation which can be measured by applying slow load-unload cycles (see Fig. 6). The apparent hysteresis of the sensor, h_{DT}, shall be less than 5 % of the hysteretic damping ratio of the specimen as measured at $(\varepsilon_a)_{SA}$ equal to 0.01 % for the maximum conceivable axial deformation.

Definition of apparent hysteretic damping of displacement measurement device:

$$h_{DT} = (1/2\pi) \times \Delta Y / Y$$

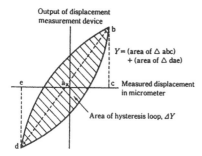

Fig.6 Definition of apparent hysteretic damping of displacement measurement device

(9)a. A digital data logger shall be used for data sampling of the axial load and axial displacement (and cell pressure and pore-water pressure when necessary). The total number of data points for each quantity required to form a single hysteresis loop is more than forty.

b. When the hysteretic damping increases apparently with the increase in the loading frequency, time lag in the measurements between axial load, P, and axial displacement , ΔD, is suspected (see Fig.7). The error must be within 5 % of the hysteretic damping ratio of the specimen as measured at $(\varepsilon_a)_{SA}$ equal to 0.01 %.

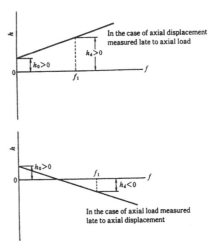

Fig.7 Relationship between apparent damping and loading frequency due to time lag in measurements of axial load and axial displacement (h_0: correct value, h_d: apparent value at a frequency of f_1)

3. TEST PROCEDURE

3.1 Preparation and Set-up of Specimen

Specimen preparation and set-up shall be done in accordance with JGS 0520 "Preparation of Soil Specimens for Triaxial Tests". Specimens should be prepared in any manner that minimizes sample disturbance, especially at the ends. The specimen ends should meet or exceed the flatness and parallelism requirements.

3.2 Evaluation of Degree of Saturation

If necessary, measure the pore pressure coefficient, B (B-value), before and after consolidation of the specimen.

3.3 Consolidation Process

Consolidate the specimen isotropically or anisotropically depending on the objective of the test. During the consolidation process, measure the change in the height of the specimen, ΔH_c (cm), and the volume of water entering and leaving the saturated specimen, ΔV_c (cm³).

(1) Isotropic consolidation process
① Hold the maximum back pressure constant and increase the cell pressure until the difference between the cell pressure and the back pressure equals the desired consolidation pressure.
② Terminate consolidation, at the earliest, at the end of primary consolidation.

(2) Anisotropic consolidation
① In the drained condition, apply axial stress to achieve the desired anisotropic consolidation stress ratio for the value of the effective lateral stress corresponding to the initial isotropic consolidation stress.

② By keeping the desired anisotropic consolidation stress ratio, increase the lateral and axial stresses step by step until they reach the values required at the end of the prescribed anisotropic consolidation.

③ Terminate consolidation, at the earliest, at the end of primary consolidation.

3.4 Cyclic Loading Process

(1) Ensure that the specimen is subjected to the prescribed effective stress state.

(2) Apply either cyclic axial load or cyclic axial displacement to the specimen using the following procedure:

① First stage

i) In an undrained test, close the specimen drainage valve or valves.

ii) Impose a total of eleven cyclic loadings to the specimen using a single amplitude cyclic axial strain, $(\varepsilon_a)_{SA}$, of less than about 0.001 %. The cyclic loading shall be applied in either an axial load controlled manner or an axial displacement controlled manner, for which either sinusoidal or sawtooth load at a constant frequency in the range of 0.05 and 1.0 Hz is to be used. Record the axial load and axial displacement, and the pore-water pressure if the test is undrained during cyclic loading.

iii) Measure the change in the specimen height (and also the volume change of the specimen in a drained test on a saturated specimen). In an undrained cyclic loading test, open the drainage valve or valves after the end of cyclic loading, and measure the changes in the height and volume of the specimen.

② Second stage

i) In an undrained cyclic test, ensure that the rate of axial strain at the end of the consolidation stage is less than 0.01 %/min, and then close the specimen drainage valve or valves.

ii) Impose a similar cyclic loading to the specimen using a value of $(\varepsilon_a)_{SA}$ of about twice that used in the first stage.

iii) Measure the change in the specimen height (and also the volume change of the specimen in a drained test on a saturated specimen). In an undrained cyclic loading test, open the drainage valve or valves after the end of cyclic loadings, and measure the changes in the height and volume of the specimen.

③ Third stage onwards

Repeat the process described for the second stage. Continue the test as long as it is possible to continue to double the amplitude of cyclic loading.

3.5 Observe and sketch the deformation of the specimen.

3.6 Measure the specimen dry mass, m_s (g).

[Notes]

3.1a. Flatness, roughness and parallelism for the specimen ends must be able to satisfy the following values:

i) The maximum deviation of the specimen height must be within 0.5 % of the average specimen height.

ii) As the specimen sits on the pedestal of the triaxial apparatus, the angle of the average specimen top surface relative to the contact surface of top cap shall be less than 0.5 degrees.

iii) The offset between the specimen's center and the center of loading must be within 1 % of the specimen height.

b. The specimen diameter of gravelly soils shall be more than ten times the maximum grain size; however, it can be not less than five times for soils having a wide range in their grain size distribution. The diameter shall be more than 50mm and 35mm for sandy soils and cohesive soils, respectively.

c. The specimen height shall be between 1.5 and 2.5 times the diameter.

3.2 Measure B-value according to the note 3.2 in JGS 0541 "Method for Cyclic Undrained Triaxial Test on Soils". This is done at the end of isotropic consolidation when consolidating the specimen anisotropically.

3.3 The end of primary consolidation shall be judged by following note 3.2(4)a in JGS 0522 "Method for Cyclic Undrained Triaxial Test on Soils".

(2)a. The standard method for anisotropic consolidation is described as follows:

i) Apply axial stress which will result in the desired anisotropic consolidation stress ratio corresponding to the initial isotropic consolidation stress.

ii) The increment of lateral stress, $\Delta\sigma_r$ is determined to not exceed one fifth of the difference in the effective lateral stress between the initial consolidation stress condition and the final consolidation stress condition. Note that the amount of $\Delta\sigma_r$ should not exceed 20 kN/m^2.

iii) Increase the lateral stress by $\Delta\sigma_r$.

iv) Increase the axial stress until the prescribed anisotropic consolidation stress ratio is achieved.

v) Ensure the rate of axial straining is less than 0.1 % per minute.

vi) Repeat procedures iii), iv) and v) until the final stress state is reached.

The following simple method may be used instead of the above method depending on the state of specimens:

i) Consolidate the specimen isotropically until the effective stress, σ'_{rc}, reaches that of the horizontal effective stress state aimed for at the end of anisotropic consolidation.

96

ii) An increment of deviator stress, Δq, is determined as one fifth of the difference between the effective axial stress, σ'_{ac}, and the effective lateral stress, σ'_{rc}, at the end of consolidation.

iii) Apply an increment of axial stress of Δq with drainage valve or valves open. During this process, ensure that the excess pore pressure, Δu, is less than 10 % of the current stress, σ'_{rc}.

iv) Ensure that the rate of axial straining is less than 0.1 % per minute.

v) Repeat procedures iii) and iv) until the final stress state is reached.

3.4(2)

a. Perform the test under either drained or undrained conditions for saturated specimen, and under drained conditions for unsaturated or air-dried specimens.

b. The maximum of cyclic axial load (or axial displacement) amplitude is to be set depending on the state of specimen and the purpose of the test.

c. In drained tests, the axial stress at peak in extension shall be positive. In undrained tests on saturated soil, the total axial stress in extension shall be positive. Back pressure may be increased in order to raise the maximum allowable cyclic amplitude of stress.

d. An example of the results of cyclic triaxial tests is shown in Fig. 8.

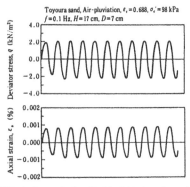

Fig. 8 Typical record of cyclic triaxial test performed under axial load controlled fashion with sinusoidal wave

4. CALCULATION OF TEST RESULTS

4.1 Specimen Properties before Consolidation

Calculate the specimen volume and height before consolidation, V_0 (cm^3) and H_0 (cm), as follows:

$$V_0 = V_i - \Delta V_i$$
$$H_0 = H_i - \Delta H_i$$

where

V_i : initial volume of the specimen (cm^3),

H_i : initial height of the specimen (cm),

ΔV_i : change in the volume of the specimen during the process after measurement of ini-

tial properties and before consolidation (cm^3), and

ΔH_i : change in the height of the specimen during the process after measurement of initial properties and before consolidation (cm).

4.2 Pore Pressure Coefficient B

Calculate the B-value of the specimen as follows:

$$B = \frac{\Delta u}{\Delta \sigma}$$

where

$\Delta \sigma$: increment of isotropic stress (kN/m^2), and

Δu : increase in pore-water pressure caused by $\Delta \sigma$ (kN/m^2).

4.3 Consolidation Process

(1) Calculate the specimen volume after consolidation, V_c (cm^3), as follows:

$$V_c = V_0 - \Delta V_c$$

where

ΔV_c : change in the volume of the specimen during consolidation (cm^3).

(2) Calculate the specimen height after consolidation, H_c (cm), as follows:

$$H_c = H_0 - \Delta H_c$$

where

ΔH_c : change in the height of the specimen during consolidation (cm).

(3) Calculate the cross-sectional area of the specimen after consolidation, A_c (cm^2), as follows:

$$A_c = \frac{V_c}{H_c}$$

(4) Calculate the dry density of the specimen after consolidation, ρ_{dc} (g/cm^3), as follows:

$$\rho_{dc} = \frac{m_s}{V_c}$$

where

m_s : dry mass of the specimen (g).

4.4 Cyclic Loading Process

(1) Calculate the specimen volume, height and cross-sectional area at the start of each cyclic loading stage performed using a uniform and constant axial load or axial displacement amplitude.

① Calculate the specimen volume at the start of each cyclic loading stage, V_n (cm^3), as follows:

$$V_n = V_c - \Delta V_n$$

where

ΔV_n : change in the volume of the specimen measured between the end of consolidation and the start of each cyclic loading stage (cm^3).

② Calculate the specimen height at the start of each cyclic loading stage, H_n (cm), as follows:

$$H_n = H_c - \Delta H_n$$

where

ΔH_n : change in the height of specimen measured between the end of consolidation and the start of each cyclic loading stage (cm).

③ Calculate the cross-sectional area of the specimen at the start of each cyclic loading stage, A_n (cm^2), as follows:

$$A_n = \frac{V_n}{H_n}$$

(2) Calculate the single amplitude cyclic deviator stress and axial strain, equivalent Young's modulus and hysteretic damping ratio at the fifth and tenth cycles in each cyclic loading stage using a uniform and constant axial load or axial displacement.

① Calculate the single amplitude cyclic deviator stress, σ_d (kN/m^2), as follows:

$$\sigma_d = \frac{P_C + P_E}{2A_n} \times 10 \ (\text{kN/m}^2)$$

where

P_C and P_E : single amplitude axial load (in N) on the compression side and the extension side in each cyclic loading stage, respectively (c.f., both are positive in value), and

A_n : cross-sectional area of the specimen at the start of the cyclic loading stage (cm^2).

② Calculate the single amplitude cyclic axial strain, $(\varepsilon_a)_{SA}$ (%), as follows:

$$(\varepsilon_a)_{SA} = \frac{\Delta L}{2H_n} \times 100$$

where

ΔL : double amplitude axial displacement of the specimen (cm), and

H_n : specimen height at the start of each cyclic loading stage (cm).

③ Calculate the equivalent Young's modulus, E_{eq} (MN/m^2), as follows:

$$E_{eq} = \frac{\sigma_d}{(\varepsilon_a)_{SA}} \times \frac{1}{10}$$

④ Calculate the hysteretic damping ratio, h (%), as follows:

$$h = \frac{1}{2\pi} \cdot \frac{\Delta W}{W} \times 100$$

where

ΔW : damping energy in a single loading cycle, which is defined as the area of the hysteresis loop on the deviator load, P, versus axial displacement, ΔH, curve (N·cm), and

W : equivalent elastic energy input in a single cyclic loading, which is defined as:

$$W = \frac{(P_C + P_E)\Delta L}{4} \ (\text{N·cm})$$

[Notes]
4.3(1)

a. For a saturated specimen, the volume change during isotropic consolidation, ΔV_c (cm^3), may be determined from the following:

$$\Delta V_c = \frac{3\Delta H_c}{H_0} V_0$$

where

ΔH_c : axial displacement during consolidation (cm),

V_0 : volume of specimen prior to consolidation (cm^3), and

H_0 : height of specimen prior to consolidation (cm).

b. The volume change of an unsaturated specimen or a specimen subjected to anisotropic consolidation should also be measured or estimated.

(4) If necessary, calculate void ratio, e_c, and relative density, D_{rc} (%), of the specimen after consolidation, using the following equations:

$$e_c = \frac{V_c \rho_s}{m_s} - 1$$

$$D_{rc} = \frac{e_{max} - e_c}{e_{max} - e_{min}} \times 100$$

where

ρ_s : density of the soil skeleton (g/cm^3),

e_{max} : void ratio of the sample evaluated by the minimum density test, and

e_{min} : void ratio of the sample evaluated by the maximum density test.

4.4 When axial deformation is less than 1 %, correction for membrane force on σ_d is in most case not needed.

(1) If necessary, calculate void ratio of the specimen at the start of the cyclic loading stage, e_n, as follows:

$$e_n = \frac{V_n \rho_s}{m_s} - 1$$

N.B.: e_n at the start of the first cyclic stage is equal to e_c at the end of consolidation.

①a. For an unsaturated specimen subjected to isotropic consolidation, the volume change of the specimen associated with cyclic loading may be estimated in the following:

$$\Delta V_n = \frac{3\Delta H_n}{H_c} V_c$$

where

ΔH_n : axial displacement from the end of consolidation to the instant at the start of cyclic loading in each stage (cm), and

V_c and H_c : specimen volume (cm^3) and height (cm) at the end of consolidation, respectively.

b. The volume change of an anisotropically consolidated unsaturated specimen subjected to cyclic loading should also be measured or estimated.

(2)a. In a test with a uniform amplitude of cyclic load, when axial displacement in two continuous cycles, ΔH, is observed as shown in Fig. 9 and the observed deviation of α is larger than 2 % of the measured double amplitude axial deformation, ΔL, should be corrected as follows:

i) if the specimen height shows continuous contraction associated with the increase in number of cycles, N (see Fig. 9(a));

"ΔL corrected at N-th cycle"

= "ΔL measured at N-th cycle" + $\alpha/2$, and

ii) if the specimen height shows continuous extension associated with the increase in number of cycles, N (see Fig.9 (b));

"ΔL corrected at N-th cycle"

= "ΔL measured at N-th cycle" - $\alpha/2$.

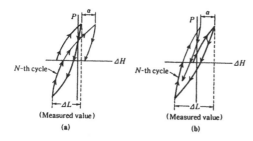

Fig. 9 Deviation of axial displacement α in two continuous cycles

b. In a test with a uniform amplitude of cyclic axial displacement, when the axial load in two continuous cycles, ΔP, is observed as shown in Fig. 10 and the deviation of β is larger than 2 % of the measured double amplitude axial load, ΔP, should be corrected as follows:

i) if the axial load shows continuous increase associated with the increase in number of cycles, N (see Fig. 10(a));

"ΔP corrected at N-th cycle"

= "ΔP measured at N-th cycle" - $\beta/2$, and

ii) if the axial load shows continuous decrease associated with the increase in number of cycles, N (see Fig. 10(a));

"ΔP corrected at N-th cycle"

= "ΔP measured at N-th cycle" + $\beta/2$.

c. Calculate equivalent Young's modulus and hysteretic damping ratio at the Nth cycle other than $N = 5$ and $N = 10$ when necessary.

②a. When using more than 2 proximeters for axial deformation measurement, the averaged value shall be used for determining ΔL.

b. When local axial deformation, $\Delta L_{local}1$ and $\Delta L_{local}2$, is measured in two positions opposite to each other around the specimen's diameter (refer Fig. 5), ΔL shall be determined as follows:

$\Delta L = \{((\Delta L_{local}1)/L1_i + (\Delta L_{local}2)/L2_i))/2\} \times H_i$

where

$L1_i$ and $L2_i$: current gauge length for $\Delta L_{local}1$ and $\Delta L_{local}2$, respectively.

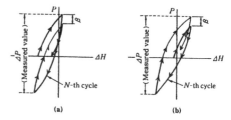

Fig. 10 Deviation of axial load β in two continuous cycles

③ Fig. 11 shows an example of how to determine equivalent Young's modulus E_{eq} (MN/m^2) in the hysteresis loop.

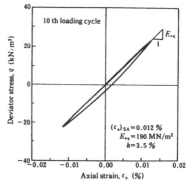

Fig. 11 Typical hysteresis loop

④a. Fig. 12 shows how to determine the stored energy per cycle, W, and the damped energy per cycle, ΔW, to be used to calculate hysteretic damping ratio $h \{= (1/2\pi) \times \Delta W/W\}$.

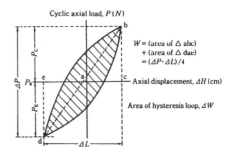

Fig. 12 Hysteretic damping ratio

b. If the hysteresis loop is not closed, ΔW can be the sum of an area gbh on the compression side and an area hdf on the extension side (see Fig. 13). The W-value may be determined as follows:

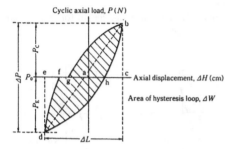

Fig. 13 Hysteretic damping ratio when the hysteretic loop is not closed

i) in a test with uniform amplitude of cyclic axial load:

$$W = \frac{1}{4} \Delta P \cdot \Delta L$$

where

ΔP : the axial load measured, and

ΔL : the axial displacement corrected by the method in note 4.4(2)a. i) or ii).

ii) in a test with uniform amplitude of cyclic axial displacement:

$$W = \frac{1}{4} \Delta P \cdot \Delta L$$

where

ΔL : axial displacement measured, and

ΔP : axial load corrected by the method in note 4.4(2)b.i) or ii).

5. REPORT

The following items shall be reported:

(1) Specimen preparation method used.

(2) Dimensions of the specimen before consolidation.

(3) Back pressure, and B-value together with the method used for an undrained cyclic loading test on a saturated specimen.

(4) Changes in the volume and the height of the specimen during consolidation.

(5) Specimen dry mass and dry density after consolidation.

(6) Effective axial and lateral stresses at the end of consolidation

(7) Loading frequency, wave shape and drainage condition during cyclic loading.

(8) Methods employed for the axial load and axial displacement measurements during cyclic loading.

(9) Dimensions of the specimen at the start of the second and all subsequent cyclic loading stages.

(10) A plot of the variations of axial load and axial displacement with time in every cyclic loading stage, and a plot of the hysteresis loops at the fifth and tenth cycles in each cyclic loading stage.

(11) The values of the equivalent Young's modulus, E_{eq} (MN/m^2), and hysteretic damping ratio, h (%), with respect to the single amplitude cyclic axial strain, $(\varepsilon_a)_{SA}$ (%), at the fifth and tenth cycles in each cyclic loading stage.

(12) Plot of the variations of the equivalent Young's modulus, E_{eq} (MN/m^2), and the hysteretic damping ratio, h (%), against the logarithm of single amplitude cyclic axial strain, $(\varepsilon_a)_{SA}$ (%), at the fifth and tenth cycles in each cyclic loading stage.

(13) Any departures from the procedure outlined in this standard.

(14) Remarks and notations regarding any unusual conditions.

[Notes]

(5) Report, if necessary, void ratio and relative density of the specimen after consolidation.

(8) Report positions of load cell and axial displacement measurement device in the triaxial cell. Report the method used for local strain measurement.

(10) Report, if necessary, the relationship between axial stress and axial strain at the 10th cycle and/or 5th cycle.

(11) Report, if necessary, the variations of E_{eq} and h with $(\varepsilon_a)_{SA}$ in every cycle from 2 to 10.

(12) Fig. 14 shows an example of how to report the test results.

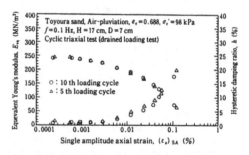

Fig. 14 Example of how to report the test results

(14)a. Give an outline of the testing apparatus used, the method for saturating the specimen, apparent damping of load cell and axial displacement measurement device, time lag in measurements between axial load and axial displacement and the material and thickness of the rubber membrane.

b. Report descriptions of specimen after testing.

Designation: JGS 0543-2000
Method for Cyclic Torsional Shear Test on Hollow Cylindrical Specimens to Determine Deformation Properties of Soils

1. GENERAL

1.1 Purpose of Test

This test method covers the determination of the deformation properties of soils in either an isotropic or an anisotropic stress state, when subjected to either drained or undrained cyclic loading in torsional shear apparatus.

1.2 Scope of Application

This test method is applicable to cohesive soils and sandy soils.

1.3 Definition of Terms

For *cyclic torsional shear test to determine deformation properties*, a hollow cylindrical specimen, which was consolidated in either an isotropic or an anisotropic stress state, is subjected to uniform and symmetrical cyclic torque using a fixed frequency on its horizontal plane under either drained or undrained condition by keeping the external cell pressure and the internal cell pressure constant.

The *deformation properties* determined are the equivalent shear modulus defined from amplitudes of cyclic shear stress and cyclic shear strain, and the hysteretic damping ratio defined from a hysteresis loop of the relationship between shear stress and shear strain.

Axial stress is applied in the axial direction of the hollow cylindrical specimen, while external pressure and internal pressure are applied on its outer and hollow sides, respectively. The external and internal pressure shall be equal to each other, and are denoted as *lateral stress*. The values of these stresses are defined at the mid-point of the height of the specimen.

Back pressure is the pore-water pressure applied to the specimen by keeping its effective stress constant for the purpose of raising the degree of saturation of the specimen.

Anisotropic consolidation stress ratio is calculated by dividing effective lateral stress by effective axial stress at the end of consolidation.

Consolidation stress is calculated by subtracting back pressure from the outside stress of the specimen during the consolidation process; *axial consolidation stress* is defined for the axial direction, while *lateral consolidation stress* is defined for the radial direction of the hollow cylindrical specimen.

1.4 Referenced Standards

The specimen shall be prepared and set in accordance with the following JGS standards:

JGS 0550 "Preparation of Hollow Cylindrical Soil Specimens for Torsional Shear Tests"

[Notes]

1. Report details of any procedure(s) employed which are different from those specified in this standard.

1.2a. In principle, this standard is applied to saturated soils, although it may be used for testing unsaturated soils.

b. This standard may be used for testing other geomaterials.

1.3 Uniform amplitude of cyclic torsional displacement may be used instead of torque.

1.4 The following standards shall be referred to for procedures not described in this standard:

JGS 0541 "Method for Cyclic Undrained Triaxial Test on Soils"

JGS 0542 "Method for Cyclic Triaxial Test to Determine Deformation Properties of Geomaterials"

JGS 0551 "Method for Torsional Shear Test on Hollow Cylindrical Specimens of Soils"

2. APPARATUS

The cyclic torsional shear apparatus consists of a triaxial cell, external, internal and back pressure control devices, torque and axial load control devices, and data acquisition and recording systems for the external and internal cell pressures, torque, axial load, rotation angle, axial displacement, volume change and pore-water pressure of the specimen.

(1) The torsional shear apparatus must be capable of sustaining the desired maximums of the external, internal and back pressures, torque and axial load.

(2) The torsional shear apparatus shall be capable of imposing the prescribed external, internal and back pressures, torque and axial load onto a specimen. The specimen is sealed by a top cap, a pedestal and rubber membranes, and has drainage at the top and bottom. In an undrained cyclic loading test on a saturated specimen, the volume change of all the assembled parts of the pore-water pressure measurement system should be negligible.

(3) During isotropic or anisotropic consolidation of a specimen in the triaxial cell, the external, internal and back pressures and axial pressure control devices shall be capable of applying and controlling pressures to within ± 2 kN/m^2 for pressures less than 200 kN/m^2 and to within ± 1 % for pressures equal to or greater than 200 kN/m^2. The axial displacement and volume change of the specimen shall be measured to an accuracy of ± 0.02 % of the height and ± 0.05 % of the volume of the specimen, respectively.

(4) The cyclic loading equipment must be capable of applying uniform and symmetrical cyclic torque loads to a specimen being held in

an isotropic or anisotropic stress state when sheared under drained or undrained conditions. The range of single amplitude cyclic shear strain, $(\gamma)_{SA}$ (refer to 4.4), that can be applied shall be from 0.001 % to 0.1 %. It is recommended that either a sinusoidal or a sawtooth variation of torque with time at a frequency within a range of 0.05 and 1.0 Hz be used.

The loading device must be able to satisfy the following conditions:

i) For the single amplitude cyclic torque in the clockwise and in counterclockwise direction, T_R and T_L, as defined from the reference stress state at the start of cyclic loading, the scatter of (T_R+T_L) during one stage of cyclic loading should be less than 10 %.

ii) $0.8 \leqq T_R/T_L \leqq 1.2$

(5) During cyclic loading, the external and internal cell pressure control systems shall be capable of applying and controlling pressures to within ± 2 kN/m^2 for pressures less than 200 kN/m^2 and to within ± 1 % for pressures equal to or greater than 200 kN/m^2.

(6) During cyclic loading, the external and internal cell pressures (and also the pore-water pressure for an undrained cyclic loading test) shall be continuously measured to an accuracy of ± 2 kN/m^2 for pressures less than 200 kN/m^2 and to an accuracy of ± 1 % for pressures equal to or greater than 200 kN/m^2.

(7) The torque load cell should be installed inside the triaxial cell. The torque shall be continuously measured to an accuracy of ± 1 % of the magnitude of the double amplitude cyclic torque when the single amplitude cyclic shear strain, $(\gamma)_{SA}$, is greater than 0.01 %.

(8) In cyclic loading stages where the single amplitude cyclic shear strain, $(\gamma)_{SA}$, is greater than 0.01 %, the rotational angle measurement device, which should have negligible hysteresis, shall be capable of continuously measuring the rotational angle of the specimen to an accuracy of ± 1 % of the magnitude of the double amplitude cyclic shear strain. In cyclic loading stages where $(\gamma)_{SA}$ is smaller than 0.1 %, the rotational angle of the specimen may be measured by a rotational angle measurement device mounted inside the triaxial cell.

(9) During cyclic loading, the data acquisition system shall be capable of monitoring both cyclic torque and cyclic rotational angle, and also pore-water pressure if necessary, with little time lag between these measurements in each data set.

[Notes]

2.1 Fig. 1 shows a typical cyclic torsional shear apparatus.

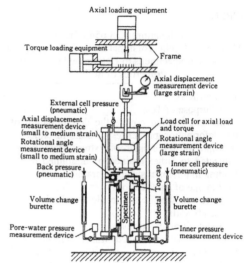

Fig. 1 Typical cyclic torsional shear apparatus

(1) Care should be taken for the triaxial cell not to rotate when the torque is applied.

(2)a. The diameters of top cap and pedestal should be equal to that of the specimen. The flat surfaces of the top cap and pedestal shall be parallel to each other, and the angle between these planes and the axis of loading piston should be 90 degrees.

b. Porous plate with metal ribs shall be used for drainage. Filter paper may be used when necessary. Other types of material may also be used if no slippage occurs between the top cap/pedestal and the specimen and if the specimen is not disturbed extensively.

c. The volume change of all the assembled parts of the pore-water pressure measurement system shall meet the requirement specified in note 2.(2)c. in JGS 0541 "Method for Cyclic Undrained Triaxial Test on Soils".

(3) The volume change of specimens during isotropic or anisotropic consolidation shall be measured using a burette or other device with equivalent accuracy. The burette should be capable of sustaining the back pressure, and its water level should preferably stay unchanged when pressurized.

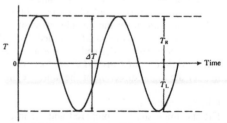

Fig. 2 Definitions of single amplitude of torque in clockwise and counterclockwise directions, T_R and T_L, for sinusoidal application of torque

(4)a. As illustrated in Fig. 2, the single amplitudes of torque in clockwise and counterclockwise directions, T_R and T_L, should be defined from the stress state at the start of cyclic loading, and these are both positive in value. The value of ΔT $(=T_R+T_L)$ refers to the double amplitude of torque applied.

b. Loading frequency may be out of the range between 0.05 and 1.0 Hz, provided that the accuracy of the stress and deformation measurements is compatible with that specified in this standard.

(7)a. A strain-gauge type torque meter should be used for the torque measurement. It should be installed inside the triaxial cell.

b. The output of the torque meter should not be affected by the changes of cell pressure and the axial load. In addition, the drift as well as the change of calibration factor shall be less than 1 % of the torque as measured with single amplitude cyclic shear strain $(\gamma)_{SA}$ of 0.01 %.

c. The hysteresis of the torque meter should be measured by knowing the relationship between the output of the torque meter and the torque, which can be measured by applying slow load-unload cycles (see Fig. 3). The apparent hysteresis of the torque meter, h_{LC}, shall be less than 5 % of the hysteretic damping ratio of specimen as measured at $(\gamma)_{SA}$ equal to 0.01 % for conceivably the maximum torque.

Definition of apparent hysteretic damping of torque meter:

$$h_{LC} = (1/2\pi) \times \Delta X / X$$

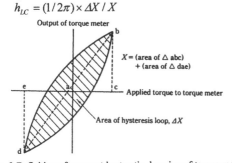

Fig. 3 Definition of apparent hysteretic damping of torque meter

(8)a. When measuring the rotational angle of a specimen during cyclic loading using a sensor mounted outside the triaxial cell, the system compliance should be less than 1 % of the rotational angle of specimen at $(\gamma)_{SA}$ of 0.1 %.

b. Output of the sensor for the rotational angle measurement shall be unchanged due to the change of cell pressure. In addition, the drift as well as the change of calibration factor shall be less than 1 % of the torque as measured with single amplitude cyclic shear strain, $(\gamma)_{SA}$, of 0.01 % in every cycle.

c. The hysteresis of the sensor for rotational angle measurement should be measured by knowing the

relationship between the output of the sensor and the rotational angle which can be measured by applying slow load-unload cycles (see Fig. 4). The apparent hysteresis of the sensor, h_{DT}, shall be less than 5 % of the hysteretic damping ratio of the specimen as measured at $(\gamma)_{SA}$ equal to 0.01 % for conceivably the maximum torsional deformation.

Definition of apparent hysteretic damping of rotational angle measurement device:

$$h_{DT} = (1/2\pi) \times \Delta Y / Y$$

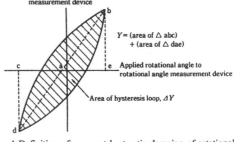

Fig. 4 Definition of apparent hysteretic damping of rotational angle measurement device

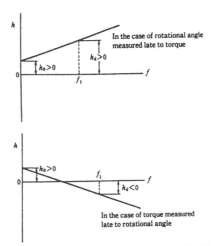

Fig. 5 Relationship between apparent damping and loading frequency due to time lag in measurements of torque and rotational angle (h_0: correct value, h_d: apparent value at a frequency of f_1)

(9)a. A digital data logger is used for data sampling of torque and rotational angle (and external, internal and pore-water pressure when necessary). The total number of data points for each quantity required to form a single hysteresis loop shall be more than forty.

b. When the hysteretic damping increases apparently with the increase in the loading frequency, time lag in the measurements between torque, T, and rotational angle, $\Delta\theta$, is suspected (see Fig. 5). The error shall be within 5 % of the hysteretic damping ratio of the specimen as measured at $(\gamma)_{SA}$ equal to 0.01 %.

3. TEST PROCEDURE

3.1 Preparation and Set-up of Specimen
Specimen preparation and set-up shall be done in accordance with JGS 0550 "Preparation of Hollow Cylindrical Soil Specimens for Torsional Shear Tests".

3.2 Evaluation of Degree of Saturation
The pore pressure coefficient B (B-value) shall be measured before consolidation of the specimen, and after consolidation if necessary.

3.3 Consolidation Process
Consolidate the specimen isotropically or anisotropically depending on the objective of the test. During the consolidation process, measure the change in the height of the specimen, ΔH_c (cm), and the volume change of the specimen, ΔV_c (cm^3). Hold the back pressure constant during consolidation.

(1) Isotropic consolidation process
① Keeping the isotropic stress state in the drained condition, increase the lateral stress to the prescribed value at the end of consolidation.
② Terminate consolidation, at the earliest, at the end of primary consolidation.

(2) Anisotropic consolidation
① In the drained condition, apply axial stress to achieve the desired anisotropic consolidation stress ratio for the value of the effective lateral stress corresponding to the initial isotropic consolidation stress.
② By keeping the desired anisotropic consolidation stress ratio, increase the lateral and axial stresses step by step until they reach the values required at the end of the prescribed anisotropic consolidation.
③ Terminate consolidation, at the earliest, at the end of primary consolidation.

3.4 Cyclic Loading Process
(1) Ensure that the specimen is subjected to the prescribed effective stress state.
(2) Apply cyclic torque to the specimen using the following procedure:
① First stage
i) In an undrained test, close the specimen drainage valve or valves.
ii) Impose a total of eleven cyclic loadings to the specimen using a single amplitude cyclic shear strain, $(\gamma)_{SA}$, of less than about 0.001 %. A cyclic loading shall be applied, for which either sinusoidal or sawtooth torque in at a constant frequency in the range of 0.05 and 1.0 Hz is to be used. Record the torque and rotational angle of the specimen, and the pore-water pressure if the test is undrained during cy-

clic loading.
iii) Measure the changes in the height and volume of the specimen. In an undrained cyclic loading test, open the drainage valve or valves after the end of cyclic loading, and measure the changes in the height and volume of the specimen.
② Second stage
i) In an undrained cyclic test, close the specimen drainage valve or valves.
ii) Impose a similar cyclic loading to the specimen using a value of $(\gamma)_{SA}$ of about twice that used in the first stage.
iii) Measure the changes in the height and volume of the specimen. In an undrained cyclic loading test, open the drainage valve or valves after the end of cyclic loading, and measure the changes in the height and volume of the specimen.
③ Third stage onwards
Repeat the process described for the second stage. Continue the test as long as it is possible to continue to double the amplitude of cyclic loading.

3.5 Observe and sketch the deformation of the specimen.

3.6 Measure the specimen dry mass, m_s (g).

[Notes]
3.2 Refer to note 3.2 in JGS 0541 "Method for Cyclic Undrained Triaxial Test on Soils". Measure the B value at the final isotropic stress state in the case of anisotropic consolidation.
3.3 Refer to note 3.3 in JGS 0551 "Method for Torsional Shear Test on Hollow Cylindrical Specimens of Soils".
3.4(2)
a. The maximum amplitude of torque is to be set depending on the state of specimen and the purpose of the test.
b. An example of the results of a cyclic torsional shear test is shown in Fig. 6.

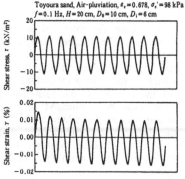

Fig. 6 Typical record of cyclic torsional shear test performed using sinusoidal wave

4. CALCULATION OF TEST RESULTS

4.1 Specimen Properties before Consolidation

(1) Calculate the specimen volume before consolidation, V_0 (cm^3), as follows:

$$V_0 = V_i - \Delta V_i$$

where
V_i : initial volume of the specimen (cm^3), and
ΔV_i : change in the volume of the specimen during the process after measurement of initial properties and before consolidation (cm^3).

(2) Calculate the specimen height before consolidation, H_0 (cm), as follows:

$$H_0 = H_i - \Delta H_i$$

where
H_i : initial height of the specimen (cm), and
ΔH_i : change in the height of the specimen during the process after measurement of initial properties and before consolidation (cm).

(3) Calculate the outer and inner diameters of the specimen, D_{o0} (cm) and D_{i0} (cm), using the following equations:

$$D_{o0} = D_{oi} \times \sqrt{(1 - \varepsilon_{vi})/(1 - \varepsilon_{ai})}$$

$$D_{i0} = D_{ii} \times \sqrt{(1 - \varepsilon_{vi})/(1 - \varepsilon_{ai})}$$

where
D_{oi} : initial outer diameter of the specimen (cm),
D_{ii} : initial inner diameter of the specimen (cm),
ε_{vi} : volumetric strain during the process after measurement of initial properties and before consolidation (= $\Delta V_i/V_i$), and
ε_{ai} : axial strain during the process after measurement of initial properties and before consolidation (= $\Delta H_i/H_i$).

4.2 Pore Pressure Coefficient B

Calculate the B-value of the specimen as follows:

$$B = \frac{\Delta u}{\Delta \sigma}$$

where
$\Delta \sigma$: increment of isotropic stress (kN/m^2), and
Δu : increase in pore-water pressure caused by $\Delta \sigma$ (kN/m^2).

4.3 Consolidation Process

(1) Calculate the specimen volume after consolidation, V_c (cm^3), as follows:

$$V_c = V_0 - \Delta V_c$$

where
ΔV_c : change in the volume of the specimen during consolidation (cm^3).

(2) Calculate the specimen height after consolidation, H_c (cm), as follows:

$$H_c = H_0 - \Delta H_c$$

where
ΔH_c : change in the height of the specimen during consolidation (cm).

(3) Calculate the cross-sectional area of the specimen after consolidation, A_c (cm^2), as follows:

$$A_c = \frac{V_c}{H_c}$$

(4) Calculate the outer and inner diameters of the specimen, D_{oc} (cm) and D_{ic} (cm), using the following equations:

$$D_{oc} = D_{o0} \times \sqrt{(1 - \varepsilon_{vc})/(1 - \varepsilon_{ac})}$$

$$D_{ic} = D_{i0} \times \sqrt{(1 - \varepsilon_{vc})/(1 - \varepsilon_{ac})}$$

where
ε_{vc} : volumetric strain during consolidation (= $\Delta V_c/V_0$), and
ε_{ac} : axial strain during consolidation (= $\Delta H_c/H_0$).

(5) Calculate the dry density of the specimen after consolidation, ρ_{dc} (g/cm^3), as follows:

$$\rho_{dc} = \frac{m_s}{V_c}$$

where
m_s : dry mass of the specimen (g).

4.4 Cyclic Loading Process

(1) Calculate the specimen volume, height, cross-sectional area, and outer and inner diameters at the start of each cyclic loading stage performed using a uniform and constant torque amplitude.

① Calculate the specimen volume at the start of each cyclic loading stage, V_n (cm^3), as follows:

$$V_n = V_c - \Delta V_n$$

where
ΔV_n : change in the volume of specimen measured between the end of consolidation and the start of each cyclic loading stage (cm^3).

② Calculate the specimen height at the start of each cyclic loading stage, H_n (cm), as follows:

$$H_n = H_c - \Delta H_n$$

where
ΔH_n : change in the height of specimen measured between the end of consolidation and the start of each cyclic loading stage (cm).

③ Calculate the cross-sectional area of the specimen at the start of each cyclic loading stage, A_n (cm^2), as follows:

$$A_n = \frac{V_n}{H_n}$$

④ Calculate the outer and inner diameters of the specimen at the start of each cyclic loading stage, D_{on} (cm) and D_{in} (cm), using the following equations:

$$D_{on} = D_{on} \times \sqrt{(1 - \varepsilon_{vn})/(1 - \varepsilon_{an})}$$

$$D_{in} = D_{in} \times \sqrt{(1 - \varepsilon_{vn})/(1 - \varepsilon_{an})}$$

where

ε_{vn} : volumetric strain measured between the end of consolidation and the start of each cyclic loading stage $(= \Delta V_n/V_c)$, and

ε_{an} : axial strain measured between the end of consolidation and the start of each cyclic loading stage $(= \Delta H_n/H_c)$.

(2) Calculate the single amplitude cyclic shear stress and shear strain, equivalent shear modulus and hysteretic damping ratio at the fifth and tenth cycles in each cyclic loading stage using a uniform and constant torque amplitude.

① Calculate the single amplitude cyclic shear stress, τ_d (kN/m²), as follows:

$$\tau_d = \frac{T_R + T_L}{2\pi(r_{on}^2 + r_{in}^2)(r_{on} - r_{in})} \times 10$$

where

T_R and T_L : single amplitude torque (in N·cm) on the clockwise side and the counterclockwise side in each cyclic loading stage, respectively (c.f., both are positive in value),

r_{on} : outer radius of the specimen at the start of the cyclic loading stage (cm) $(= D_{on}/2)$, and

r_{in} : inner radius of the specimen at the start of the cyclic loading stage (cm) $(= D_{in}/2)$.

② Calculate the single amplitude cyclic shear strain, $(\gamma)_{SA}$ (%), as follows:

$$(\gamma)_{SA} = \frac{\Delta\theta(r_{on} + r_{in})}{4H_n} \times 100$$

where

$\Delta\theta$: double amplitude rotational angle of the specimen (rad), and

H_n : specimen height at the start of each cyclic loading stage (cm).

③ Calculate the equivalent shear modulus, G_{eq} (MN/m²), as follows:

$$G_{eq} = \frac{\tau_d}{(\gamma)_{SA}} \times \frac{1}{10}$$

④ Calculate the hysteretic damping ratio, h (%), as follows:

$$h = \frac{1}{2\pi} \cdot \frac{\Delta W}{W} \times 100$$

where

ΔW : damping energy in a single loading cycle, which is defined as the area of the hysteresis loop on the torque, T, versus rotational angle, $\Delta\theta$, curve (N·cm), and

W : equivalent elastic energy input in a single cyclic loading, which is defined as:

$$W = \frac{(T_R + T_L)\Delta\theta}{4} \text{ (N·cm).}$$

[Notes]

4.3(5) If necessary, calculate the void ratio, e_c, and relative density, D_{rc} (%), of the specimen after consolidation, using the following equations:

$$e_c = \frac{V_c \rho_s}{m_s} - 1$$

$$D_{rc} = \frac{e_{max} - e_c}{e_{max} - e_{min}} \times 100$$

where:

ρ_s = density of the soil skeleton (g/cm³),

e_{max} = void ratio of the sample evaluated by the minimum density test, and

e_{min} = void ratio of the sample evaluated by the maximum density test.

4.4(1) If necessary, calculate the void ratio of the specimen at the start of the cyclic loading stage, e_n, as follows:

$$e_n = \frac{V_n \rho_s}{m_s} - 1$$

N.B.: e_n at the start of the first cyclic stage is equal to e_c at the end of consolidation.

(2) If necessary, calculate the equivalent shear modulus and hysteretic damping ratio at the fifth and tenth cycles in each cyclic loading stage.

③ Fig. 7 shows an example of how to determine equivalent shear modulus G_{eq} (MN/m²) in the hysteresis loop.

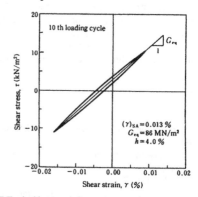

Fig. 7 Typical hysteresis loop

④a. Fig. 8 shows how to determine the stored energy per cycle W and the damped energy ΔW per cycle to be used to calculate hysteretic damping ratio h $\{= (1/2\pi) \times \Delta W/W\}$.

b. When the hysteresis loop is not closed, ΔW can be the sum of area gbh and area hdf (see Fig. 9).

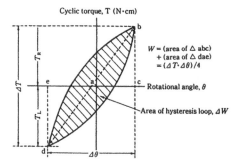

$$W = \text{(area of } \triangle \text{ abc)} + \text{(area of } \triangle \text{ dae)} = (\Delta T \cdot \Delta \theta)/4$$

Area of hysteresis loop, ΔW

Fig. 8 Hysteretic damping ratio

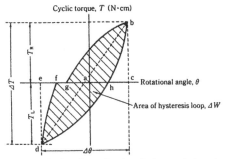

Area of hysteresis loop, ΔW

Fig. 9 Hysteretic damping ratio when the hysteretic loop is not closed

5. REPORT

The following items shall be reported:

(1) Specimen preparation method used.

(2) Dimensions of the specimen before consolidation.

(3) B-value together with the method used for determination.

(4) Changes in the volume and the height of the specimen during consolidation.

(5) Specimen dry mass and dry density after consolidation.

(6) The values of axial stress, external, internal and back pressures at the end of consolidation.

(7) Effective axial and lateral consolidation stresses at the end of consolidation, σ'_{ac} and σ'_{rc}; and their ratio, $\sigma'_{ac}/\sigma'_{rc}$ $(=K)$, if necessary.

(8) Loading frequency, wave shape and drainage condition during cyclic loading.

(9) Methods employed for the torque and rotational angle measurements during cyclic loading.

(10) Dimensions of the specimen at the start of the second and all subsequent cyclic loading stages.

(11) A plot of the variations of torque and rotational angle with time in every cyclic loading stage, and plot of the hysteresis loops at the fifth and tenth cycles in each cyclic loading

stage.

(12) The values of the equivalent shear modulus, G_{eq} (MN/m^2), and hysteretic damping ratio, h (%), with respect to the single amplitude cyclic shear strain, $(\gamma)_{SA}$ (%), at the fifth and tenth cycles in each cyclic loading stage.

(13) Plot of the variations of the equivalent shear modulus, G_{eq} (MN/m^2), and the hysteretic damping ratio, h (%), against the logarithm of single amplitude cyclic shear strain, $(\gamma)_{SA}$ (%), at the fifth and tenth cycles in each cyclic loading stage.

(14) Any departures from the procedure outlined in this standard.

(15) Remarks and notations regarding any unusual conditions.

[Notes]

(4) If it takes a long time to consolidate the specimen, as in the case with cohesive soils, report the consolidation period, axial displacement and volume change of the specimen.

(5) Report, if necessary, void ratio and relative density of the specimen after consolidation.

(9) Report positions of the torque meter and the rotational angle measurement device in the triaxial cell.

(10) Report, if necessary, void ratios of the specimen at the start of each cyclic loading stage.

(11) Report, if necessary, the relationship between shear stress and shear strain at the 10th cycle and/or 5th cycle.

(12) Report, if necessary, the variations of G_{eq} and h with $(\gamma)_{SA}$ in every cycle from 2 to 10.

(13) Fig. 10 shows an example of how to report the test results.

(15)a. Report the outline of testing apparatus used, the method for saturating the specimen, the apparent damping of the torque meter and the rotational angle measurement device, and the material and thickness of the rubber membrane.

b. Report the dimension, size and number of metal ribs mounted in the porous plate, or any alternative device to avoid slippage.

c. Report descriptions of specimen after testing.

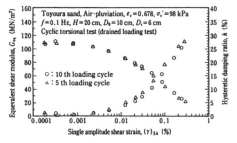

Fig. 10 Example of how to report the test results

APPENDIX F:

A note on the use of a local gauge in consolidated undrained triaxial compression tests on sedimentary soft rock

prepared by Tatsuoka, F., July 1996

Capping the top or bottom ends, or both, of specimen is essential to ensure sufficient parallelism and contact between the specimen ends and the bottom and top surfaces of the cap and pedestal of the triaxial cell. In that case, some precautions are required, which include the following.

1) Possible problems associated with re-setting of the cap on the specimen top end after the capping material has hardened.

By this resetting, sufficient parallelism and contact between the specimen ends and surfaces of the cap and pedestal may be lost. Even this loss is very small, the effects on stress-strain measurements at small strain could be tremendous.

One example (Test B30) is shown in Fig. 1. A core sample of sedimentary softrock (mudstone) obtained by a rotary core tube sampling from Sagamihara Test Site was used. The diameter and height of the sample were 5.5 cm and 14 cm. First, the speci-

men was placed on the pedestal which was fixed to the base platen of the triaxial cell. The bottom end of the specimen was not capped. Then, the top end was capped with gypsum. While the gypsum was still in slurry conditions, the top cap was placed on the sample. As the cap had been fixed to the guided loading piton, the cap could not rotate about any horizontal axis.

After the gypsum had hardened, un-intentionally, the top cap was separated from the top end of the specimen and was rotated about the vertical axis of the loading piston by an unknown amount of rotation. By this procedure, good contact and parallelism between the sample ends and the cap and pedestal was lost. Subsequently, pair of LDTs were set on each end of the specimen diameter as shown in Fig. 2.

Then the sample was re-consolidated to the in-situ overburden pressure equal to 3.6 kgf/cm2, followed by an undrained triaxial compression test.

The result is shown in Fig. 1. It may be seen that readings of the pair of LDTs are very different, particular at small strains (Fig. 1b); one reading showed extension immediately after the start of loading. This peculiar behaviour is most likely due to stress concentration to part of the specimen end which was in contact with the cap and pedestal while no contact in the opposite end of the specimen diameter. This should have resulted into bending of the specimen. In such case, the average of the two readings of the pair of LDTs may not provide a reasonable result as seen from Fig. 3b. The result of a similar test (Test C14), for which good contact and parallelism between the specimen ends and the cap and pedestal was not lost, is also shown for comparison in Fig. 3b. Fig. 3a shows the whole stress-strain curves from this second test. The initial stress-strain relation from

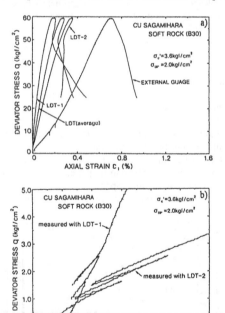

Fig. 1 Result of a CU TC test (Test B30) in which good parallelism and contact between the specimen ends and the cap and pedestal was lost, a) he whole stress-strain relation and b) initial part; sedimentary soft mudstone from Sagamihara Test Site

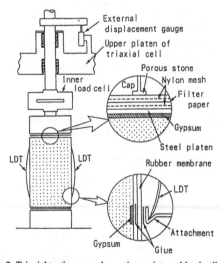

Fig. 2 Triaxial testing procedure using an internal load cell and a local gauge (LDT)

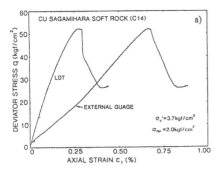

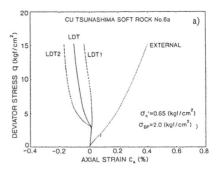

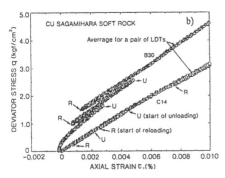

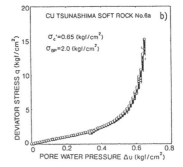

Fig. 3 a) Result of a CU TC test (Test C14) in which good parallelism and contact between the specimen ends and the cap and pedestal was ensured, and b) comparison of initial stress-strain relations between two similar CU TC tests with and without good parallelism and contact between the specimen ends and the cap and pedestal; sedimentary soft mudstone from Sagamihara Test Site

Fig. 4 Result of a CU TC test (Test 6a) in which slipping between the outer surface of the specimen and the inner surface of the membrane occurred; a) relationships between deviator stress and axial strain and b) relationship between deviator stress and excess pore pressure; sedimentary soft mudstone from Sagamihara Test Site

the second test is very smooth.

Note that axial strains obtained from external reading of the axial compression of the specimen were much larger than those obtained by local measurements, due mainly to large effects of bedding error and partly to some effects of system compliance.

2) Possible problems associated with slip between the membrane and the surface of the specimen

For LDTS, the top and bottom ends of a thin strip made of phosphor bronze are fixed to a pari of very small attachment, which are glued on the surface of the membrane. Similar to other types of local gauges, LDTs are measuring the deformation at the outer surface of the membrane. Therefore, if the membrane deforms differently from the specimen by slipping between the lateral surface of the specimen and the inner surface of the membrane, false results will be obtained.

In general, this type of slipping can take place more easily at lower effective confining pressures. In a CU triaxial compression test on a sedimentary soft mudstone, the effective confining pressure could drop to a very low value due to its high contractancy.

A typical example is shown in Fig. 4. In this case, by undrained shearing, the effective lateral stress $\sigma_3'= \sigma_c (= 0.65$ kgf/cm^2) $-\triangle u$ dropped to a value less than 0.05 kgf/cm^2. In this test, differently from the arrangement illustrated in Fig. 1, the inner membrane and the outer surface of the specimen was not fixed by gluing at and near the place of the attachments. Namely, the same deformation between the specimen and the membrane is ensured only when friction between them is sufficient. As seen from Fig. 4a, at a deviator stress q equal to about 3 kgf/cm^2, the readings of the LDTs started becoming very unnatural, while the reading of the external gauge shows a smooth result (although it includes a large amount of bedding error). This peculiar behaviour of the LDT reading should be due to slipping between the inner membrane and the outer surface of the specimen resulting from very low effective stresses. Despite the spring force in each LDT is very small (of the order of 10 kgf), this force can cause the slipping at the pair of attachment into the opposite directions (i.e., upward and downward), which resulted into an apparent extensional behaviour upon triaxial compression.

This slipping can be effectively prevented by a simple procedure as illustrated in Fig. 2 and Plate 1;

i.e., the inner membrane and the outer surface of the specimen are glued to each other. To this end, very thin groove was made on the surface of the specimen and filled with gypsum slurry. Glue was smeared on the surface of hardened gypsum. The result of a CU TC test (Test 6) with the above-mentioned procedure is shown in Fig. 5. The result is very smooth.

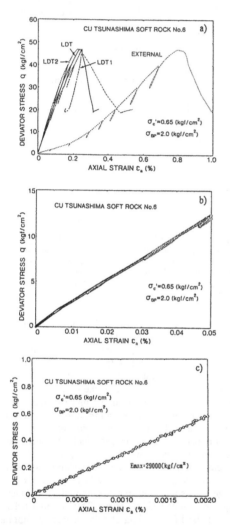

Fig. 5 Result of a CU TC test (Test 6) in which slipping between the outer surface of the specimen and the inner surface of the membrane did not occur; a) whole relationships between deviator stress and axial strain, b) relationship at axial strains of 0.05 % or less, and c) relationship at axial strains of 0.002 % or less; sedimentary soft mudstone from Sagamihara Test Site

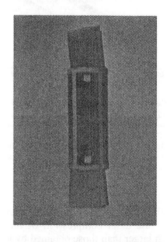

Plate 2 A method for fixing the membrane on the specimen surface

Advanced Laboratory Stress-Strain Testing of Geomaterials, Tatsuoka, Shibuya & Kuwano (eds),
© *2001 Taylor & Francis, ISBN 90 2651 843 9*

Report on applications of laboratory stress-strain test results of geomaterials to geotechnical practice in Japan

J. Koseki & F. Tatsuoka
University of Tokyo, Japan

M. Yoshimine
Tokyo Metropolitan *University*, Japan

M. Hatanaka
Takenaka Research And *Development* Institute, Japan

K. Uchida
Kobe University, *Japan*

N. Yasufuku
Kyusyu University, *Japan*

I. FURUTA
Oyo Corporation, *Japan*

ABSTRACT: As a part of the activities of the Technical Committee TC-29, typical case histories in Japan of the applications of advanced laboratory stress-strain test results of geomaterials to geotechnical practice are summarized. They consist of eighteen cases: eight buildings, two bridges, two LNG tanks, one nuclear power plant, one rock-fill dam, one road embankment, one tunnel, one subway station and one test site for vertical shaft and tunnels. Advanced triaxial tests were conducted in twelve cases, among which axial strains were measured locally in eight cases. In six cases, the laboratory test results were employed before construction to predict full-scale behavior. In the other cases, they were used during/after construction to simulate full-scale behavior and to analyze field test results.

1 INTRODUCTION

Recent developments in laboratory testing techniques have made it possible to evaluate the stress-strain relationships of geomaterials much more addurately than it was before, as reviewed by Tatsuoka et al. (1997). At the same time, field measurement techniques to evaluate in-situ soil properties, often performed prior to construction, and to observe the full-scale behavior of actual structures during and after construction have been improved remarkably. Comparisons between results of relevant laboratory stress-strain tests and field measurements have been made by several researchers, and some Japanese case histories were reported by Tatsuoka and Kohata (1995).

In the above-mentioned circumstances, Technical Committee TC-29 of the International Society for Soil Mechanics and Foundation Engineering was established in 1994 in order; a) to promote co-operation and exchange of information about recent developments in laboratory stress-strain testing; b) to develop recommendations for procedures referred to triaxial and torsional shear tests; and c) to work out a uniform frame for the comparison of stiffness measured by means of different laboratory techniques, with emphasis on their relevance for the solution of problems of practical interest. In relation to term c), an inquiry was made among overseas members of TC-29 into case histories on the applications of laboratory stress-strain test results of geomaterials to geotechnical practice, as has been summarized by Higgins and Jardine (1997) for their experiences in the United Kingdom.

This report summarizes eighteen Japanese case histories collected by Japanese members of TC 29.committee's preliminary study.

2 JAPANESE CASE HISTORIES

Table 1 summarizes the collected Japanese case histories. Further information on them, along with references, is described in the Appendices, and includes the following: a) construction period, b) location, c) type of project, d) ground conditions, e) laboratory tests, f) field tests, and g) agreement between prediction/simulation and measurement. These are eighteen cases in total, consisting of eight excavation works, six building foundations, two bridge foundations, and two embankment works for a rock-fill dam and a road embankment (Fig. 1). The excavation works include one nuclear power plant, two LNG tanks, one tunnel, one subway station built by an open-cut method, and one test site for a vertical shaft and tunnels. The location of each site is indicated in Fig. 2.

Table 1　Summary of case histories in Japan

No.	Type of project	Ground conditions	Lab./ field tests	Application of lab. test results
JP-1	Bridge foundation (Akashi Kaikyo Bridge)	Gravels and sedimentary soft rocks	Triax.*, unconf./ seismic survey, PLT, PMT	Pre/post-analysis of seismic behavior, post-analysis of full-scale behavior during construction and PLT results
JP-2	Bridge foundation (Rainbow Bridge)	Sedimentary soft rocks	Triax.*, oedometer/ seismic survey, PLT, PMT	Post-analysis of full-scale behavior during construction
JP-3	Excavation (Sagamihara test site)	Sedimentary soft rocks	Triax.*, unconf., ultrasonic log, oedometer/ seismic survey, PLT, PMT	Post-analysis of full-scale behavior during excavation and PLT/PMT results
JP-4	Excavation (LNG tanks at Negishi)	Sedimentary soft rocks	Triax., unconf., ultrasonic log, sprit/ seismic survey, PMT	Pre/post analysis of full-scale behavior during excavation
JP-5	Excavation (Nuclear power plant)	Sedimentary soft rocks	Triax.*/ seismic survey, PLT, PMT	Post-analysis of full-scale behavior during excavation
JP-6	High-rise building	Sedimentary soft rocks	- / Seismic survey	(Pre-analysis of full-scale behavior during construction, based on lab. test results on similar soils)
JP-7	Building	Pleistocene sands, gravels and clays	- /Seismic survey, PLT, PMT	(Comparison between field test results and full-scale behavior)
JP-8	High-rise building with pile foundation (OAP Towers)	Pleistocene clay	Triax.* /seismic survey, vertical loading of test pile	Post-analysis of behavior of test pile, pre-analysis of full-scale behavior during construction
JP-9	High-rise building with pile foundation (Kagoshima Prefectural Office)	Shirasu	- /Seismic survey, vertical loading of test pile	(Post-analysis of behavior of test pile and pre-analysis of full-scale behavior, based on seismic survey results and empirical formulations)
JP-10	High-rise building complex with spread and pile foundation	Pleistocene sands and sedimentary soft rocks	Triax.*/ seismic survey	Post-analysis of full-scale behavior during excavation
JP-11	Tunneling	Holocene sands and clays	Triax./ seismic survey	Pre-analysis of full-scale behavior during tunneling
JP-12	Excavation (LNG tank at Chita)	Sedimentary soft rocks (Neogene sands and silts)	Triax.*, ultrasonic log, BE/ seismic survey, BE	Post-analysis of full-scale behavior during excavation
JP-13	Rock-fill dam	Tuffs	Model tests using artificial soil/ seismic survey	Post-analysis of full-scale behavior during dam construction
JP-14	Road embankment	Holocene sands and clays	Oedometer, unconf./ -	Pre-analysis of full-scale behavior during replacement of embankment with expanded polystyrene
JP-15	Excavation (Building construction)	Holocene and Pleistocene sands and clays	Triax., oedometer, unconf./ seismic survey, permeability	Post-analysis of full-scale behavior during excavation
JP-16	Excavation with ground improvement (subway station)	Holocene sands and Pleistocene sands and clays	Triax., oedometer/ PMT	Pre-analysis of full-scale behavior during excavation
JP-17	High-rise buildings with mat foundations	Holocene and Pleistocene sands and clays	Triax.*/ seismic survey	Pre-analysis of full-scale behavior during excavation and building construction
JP-18	Excavation (Building construction)	Volcanic ash soils, gravels, sands and silts	- / seismic survey	(Pre-analysis of full-scale behavior, based on seismic survey results and lab. test results on similar soils)

*: with local strain measurements, BE: bender element tests, PLT: plate loading tests, PMT: pressuremeter tests

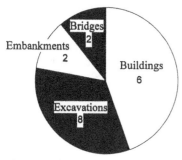

Note: no laboratory tests conducted for
three buildings and one excavation.

Fig. 1   Types of projects

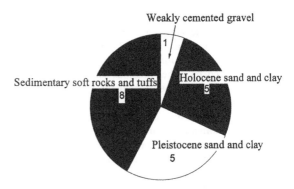

Note: at sites where tests were conducted on different
types of soils, they were independently counted.

Fig. 4   Soil types of samples for laboratory tests

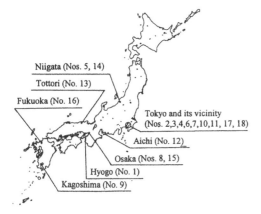

Fig. 2   Location of sites listed in Table 1

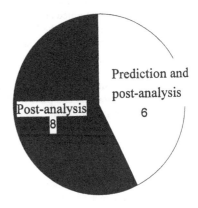

Fig. 5   Analysis type using laboratory test results

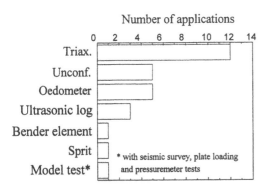

Fig. 3   Application of laboratory tests

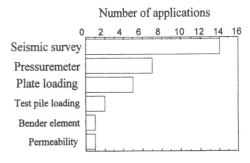

Fig. 6   Application of field tests

It is to be noted that in case Nos. 6, 7, 9 and 18, no laboratory test relevant to the estimation of pre-failure deformation was conducted, while the strain dependency of the deformation characteristics were estimated based on either previous laboratory test results on similar soil types or analyses of field tests and full-scale behavior. In the other fourteen cases, triaxial tests were mainly employed to evaluate the pre-failure deformation characteristics on a variety of soil types, including sedimentary soft rock, tuff, volcanic ash, gravel, sand and clay (Figs. 3 and 4). In particular, in eight cases out of twelve for which the triaxial tests were employed, axial deformation was measured locally with LDTs.

Based on the results of laboratory tests, numerical simulations of the full-scale behavior and the field tests results were conducted in the above-mentioned fourteen cases. In six of these cases, prediction of the full-scale behavior was performed at the design stage, while in the others, the results of the laboratory tests were used during/after construction to simulate full-scale behavior and to analyze field test results (Fig. 5).

For field tests, a seismic survey was performed in all cases, while pressuremeter and plate loading tests were employed in about a half of the cases (Fig. 6).

3 CONCLUDING REMARKS

In Japan, the number of construction works for large scale structures has increased recently, and stability-based designs are being replaced with displacement-based designs. Therefore, further accumulation of case histories will enhance the importance of relevant laboratory stress-strain testing in geotechnical practice.

ACKNOWLEDGEMENTS

The authors would like to thank the individuals who kindly provided information on their respective case histories referred to in the present report.

REFERENCES

Higgins,K.G. and Jardine,R.J. 1997. Experience on the use of non-linear pre-failure constitutive soil models, *Géotechnique* 47, No. 3, pp.409-411.

Tatsuoka,F. and Kohata,Y. 1995. Stiffness of hard soils and soft rocks in engineering applications, *Prefailure Deformation of Geomaterials*, Shibuya, Mitachi and Miura (eds.), Balkema, Vol.2, pp.947-1063.

Tatsuoka, F., Jardine, R.J., Lo Presti, D., Di Benedetto, H. and Kodaka, T. 1997. Theme lecture: Characterising the pre-failure deformation properties of geomaterials, *Proc. of 14th ICSMFE*, Vol.4, pp.2129-2164.

APPENDIX 1. CASE NO.1 (AKASHI KAIKYO BRIDGE)

A1.1 *Construction period*

1986-1998.

A1.2 *Location*

Hyogo, Japan.

A1.3 Type of project

A suspension-type highway bridge having a central span of 1990 m and side spans of 960 m (3910 m in total length, Fig. A1.1), which crosses over the Akashi strait, was constructed.

A1.4 Ground conditions

Pier 2P is founded on a weakly cemented gravel deposit (Akashi Formation), having a largest thickness of about 50 m, while pier 3P and anchorage 1A are founded on a sedimentary softrock (Kobe Group) of Early Neogene Period (Fig. A1.2). Anchorage 4A is founded on granites. Two piers were constructed by the laying-down caisson method (Fig. A1.3).

A1.5 *Laboratory tests*

At the design stage, undrained cyclic triaxial tests were conducted on undisturbed gravel samples with a diameter of 30 cm retrieved from a depth of about 40 m below the sea bottom (Fig. A1.4).

At the design stage and during construction, a very large amount of unconfined or triaxial compression tests were conducted on undisturbed core samples of softrock and granite with external measurements. A series of triaxial compression tests with local measurements were also conducted on silt-to-sandstone samples that were block-sampled at the bottom of excavation for anchorage 1A (Fig. A1.5). Values of the initial Young's moduli E_{max} obtained at strains of less than about 0.001 % based on local measurements agreed with the E_f values measured by field seismic surveys (Fig. A1.6). On the other hand, values of E_{50} as obtained from unconfined compression tests with external measurements scattered very largely, and their average values were considerably smaller than the values of E_{max} and E_f (Fig. A1.6).

A1.6 *Field tests*

At the design stage and during construction, seismic surveys, pressuremeter tests (PMTs) and plate loading tests (PLTs) were conducted. Values of the tangent Young's moduli E_{PLT} obtained from primary loading curve in PLTs and those of E_{BHLT} obtained from primary loading curves in PMTs were much smaller than the values of E_{max} and E_f (Fig. A1.6), due to the strain levels that were much larger than the elastic limit strain (about 0.001 %) and bedding errors between softrock surface and a loading platen.

A1.7 *Comparison of tests results with full-scale behavior during construction*

After the construction, axi-symmetric elasto-plastic FEM analyses were conducted on piers 2P and 3P (Fig. A1.7), after validating them by simulating successfully the results of PLTs performed at the bottom of excavation at anchorage 1A (Fig. A1.8). The material properties were evaluated based on the E_f values obtained by the field seismic survey while taking into account the dependency of stiffness on shear stress and pressure levels obtained from triaxial compression tests. Note that in Fig. A1.8, the linear isotropic elastic FEM solutions by using the average values of E_{50} and E_{BHLT} are also shown, which exhibited responses much softer than those measured.

The simulated pressure-settlement relationships during construction of pier 3P agreed well with the measured ones (Fig. A1.9b), while those of pier 2P showed some deviation from the measured relationship (Fig. A1.9a). The latter deviation is likely to be caused by insufficient amount of data in modeling the deformation characteristics of the ground supporting pier 2P. Note that the agreement between the measured and simulated relationships between in-ground settlement and the depth along centerline of pier 3P (Fig. A1.10) was not as good as those as shown in Fig. A1.9b. Investigations are in progress on this issue.

Based on the vertical strain distribution in the ground along center line of piers 2P and 3P (Fig. A1.11), which were measured with a sliding micrometer (Fig. A1.12), the equivalent Young's moduli E_{DBA} at each depth were back-calculated using linear elastic FEM analysis. The ratio of E_{DBA} to E_{BHLT} was much larger than unity, and it decreased with the increase in the measured vertical strain ε_1 (Fig. A1.13a). Fig. A1.13b shows a similar plot, in which the ratio of E_{DBA} to E_f and that of E_{BHLT} to E_f are plotted against the log of in-ground strain ε_1. The ratio E_{DBA}/E_f tends to approach the unity as ε_1 becomes smaller.

A1.8 *Comparison of tests results with full-scale behavior during earthquake*

At the design stage, the maximum displacement of pier 2P under the design seismic condition was estimated by a two-dimensional pseudo-static FEM analysis (Figs. A1.14 and A1.15) using the deteriorated stiffness values of the gravel evaluated based on the undrained cyclic triaxial test results. The analysis was conducted in combination with an equivalent linear dynamic response analysis.

The 1995 Hyogoken-Nanbu earthquake had an epicenter located very close to the construction site of the bridge and induced a seismic condition that was much severer than the designed one. At the time of this earthquake, the piers with towers and the anchorages had been already completed, and squeezing works of the main cables after installation had been in progress. Inspections and surveys after the earthquake revealed that there was no structural damage and that the piers and anchorages suffered only minimal residual displacements (Fig. A1.16). Note that the residual settlement of pier 2P relative to the granite layer at a depth of about 200 m from the pier bottom was 20 mm (Fig. A1.17), which was measured with the sliding micrometer.

After the earthquake, a static FEM analysis was conducted in order to evaluate the residual settlement of pier 2P, following the procedures of the previous FEM analysis, while using a relevant strong motion record and considering the effects of dissipation of excess pore pressures generated in the gravel deposit based on the results of the undrained cyclic triaxial tests. Effects of sample disturbance on the test results were considered in the analysis by reducing the strain potential of the gravel deposit with a factor of 1/4, which was determined from the ratio of the initial Young's modulus $E_{initial}$ evaluated in the triaxial tests with external measurements to the E_f values measured by the field seismic survey. The computed distribution of residual settlements were qualitatively consistent with the observed distribution, whereas the computed settlement at the bottom of the pier was about 3.5 times larger than the observed one (Fig. A1.17).

A similar analysis on the residual settlement of pier 3P founded on a softrock is in progress.

A1.9 *References for case No.1*

Kohata,Y., Tatsuoka,F., Dong,J., Teachavorasinskun,S. and Mizumoto,K. 1994. Stress states affecting elastic deformation moduli of geomaterials, *Pre-failure Deformation of Geomaterials* (Shibuya et al., eds.), Balkema, Vol.1, pp.3-9.

Saeki,M., Kurihara,T., Koseki,J., Manabe,S. and Tatsuoka,F. 1998. Investigation on earthquake-induced residual settlements of large scale bridge foundations, *Proc. of the 53rd Annual Conf. of the Japan Society of Civil Engineers*, 1-B, pp.618-619 (in Japanese).

Siddiquee,M.S.A., Tatsuoka,F., Hoque,E., Tsubouchi,T., Yoshida,O., Yamamoto,S. and Tanaka,T. 1994. FEM simulation of footing settlement for stiff geomaterials, *Pre-failure Deformation of Geomaterials* (Shibuya et al., eds.), Balkema, pp.531-537.

Siddiquee,M.S.A., Tatsuoka,F., Inoue,A., Kohata,Y., Yoshida,O., Yamamoto,Y. and Tanaka,T. 1995. Settlement of a pier foundation for Akashi-Kaikyo Bridge and its numerical analysis, *Rock Foundation* (Yoshinaka and Kikuchi, eds.), Balkema, pp.413-420.

Tatsuoka,F., Yamada,K., Yasuda,M., Yamada,S. and Manabe,S. 1991. Cyclic undrained behavior of an undisturbed gravel for aseismic design of a bridge foundation, *Proc. of 2nd Int. Conf. on Recent Advances in Geotechnical Earthquake Engineering and Soil Dynamics*, St. Louis, Vol.1, pp.141-148.

Tatsuoka,F. and Kohata,Y. 1995. Stiffness of hard soils and soft rocks in engineering applications, *Pre-failure Deformation of Geomaterials* (Shibuya et al., eds.), Balkema, Vol. 2, pp.947-1063.

Tatsuoka,F. et al. 1997. Non-linearity of geomaterials at small strain and its application to deformation issues -part 6, case histories, *Tsuchi-to-Kiso*, Vol. 45, No. 9, pp.43-48 (in Japanese).

Yamada,K., Manabe, S and Tatsuoka,F. 1990. Prediction of earthauake-induced deformation of large scale bridge foundations, *Proc. of 25th Japan National Conf. on Soil Mechanics and Foundation Engineering*, pp.951-954 (in Japanese).

Yamagata,M., Nitta,A. and Yamamoto,S. 1995. Design and its evaluation through displacement measurement for the Akashi Kaikyo Bridge foundation, *Rock Foundation* (Yoshinaka and Kikuchi, eds.), Balkema, pp.35-46.

Yamagata,M., Yasuda,M., Nitta,A. and Yamamoto,S. 1996. Effects on the Akashi Kaikyo Bridge, *Special Issue of Soils and Foundations on Geotechnical Aspects of the January 17 1995 Hyogoken-Nambu Earthquake*, pp.179-187.

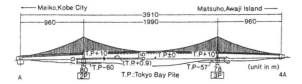

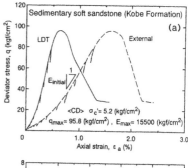

Fig. A1.1 Side view of the Akashi Kaikyo Bridge (Tatsuoka et al., 1991)

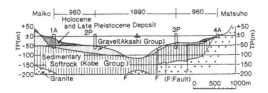

Fig. A1.2 Geological profile (Tatsuoka et al., 1991)

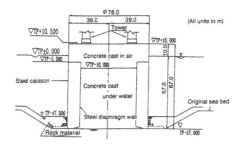

Fig. A1.3 Pier 3P (Siddiquee et al., 1995)

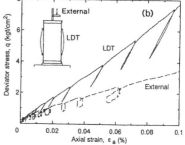

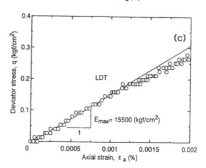

Fig. A1.5 Typical stress-strain relations using different scales from drained triaxial compression test on sandstone from Kobe Group (Kohata et al., 1994)

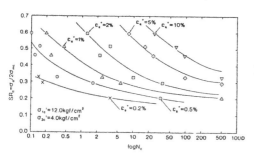

Fig. A1.4 Results from undrained cyclic triaxial tests on undisturbed gravel from Akashi Formation (Tatsuoka et al., 1991)

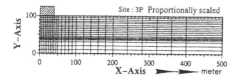

Fig. A1.7 Axi-symmetric model of pier 3P and its supporting ground (Siddiquee et al., 1995)

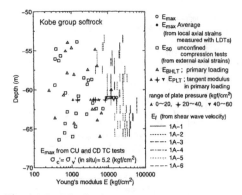

Fig. A1.6 Distribution of Young's moduli with depth (Siddiquee et al., 1994)

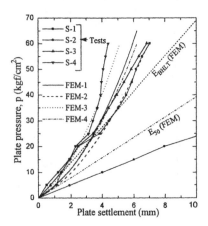

Fig. A1.8 Comparison of measured and analyzed pressure-settlement relations from PLTs at anchorage 1A (Siddiquee et al., 1994)

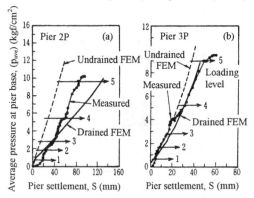

Fig. A1.9 Comparison of measured and analyzed pressure-settlement relationships during the construction of piers: (a) 2P and (b) 3P (Siddiquee et al., 1995, Tatsuoka et al., 1997)

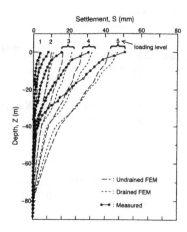

Fig. A1.10 Comparison of measured and analyzed relationships between in-ground settlement and the depth along centerline of pier 3P (Siddiquee et al., 1995)

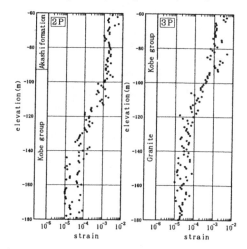

Fig. A1.11 Vertical strain distributions measured with sliding micrometer (Yamagata et al., 1995)

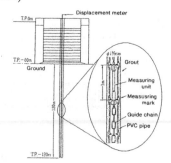

Fig. A1.12 Setting layout of sliding micrometer (Yamagata et al., 1995)

118

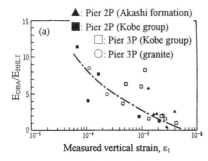

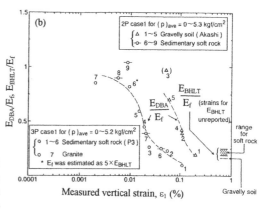

Fig. A1.13 Relationships between the measured vertical strain and (a) the ratio of E_{DBA} to E_{BHLT} (Tatsuoka et al., 1997) and (b) the ratio of E_{DBA} to E_f and the ratio of E_{BHLT} to E_f (Tatsuoka and Kohata, 1995)

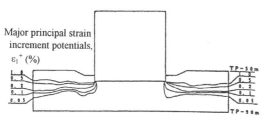

Fig. A1.14 Distribution of major principal strain increment potentials under design seismic condition estimated for Akashi Formation beneath pier 2P (Yamada et al., 1990)

Maximum displacements (x, y) at each node

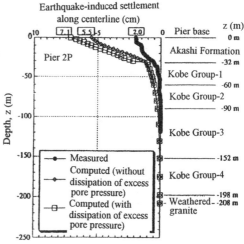

Fig. A1.15 Computed maximum displacement of pier 2P under design seismic condition (Yamada et al., 1990)

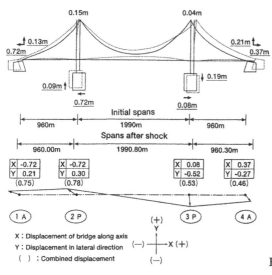

Fig. A1.16 Vertical and horizontal residual displacements induced by the 1995 Hyogoken-Nanbu earthquake (Yamagata et al., 1996)

Fig. A1.17 Comparison of measured and analyzed relationships between in-ground settlement and the depth along centerline of pier 2P induced by the 1995 Hyogoken-Nanbu earthquake (Saeki et al., 1998)

119

APPENDIX 2. CASE NO.2 (RAINBOW BRIDGE)

A2.1 *Construction period*

1988-1991.

A2.2 *Location*

Tokyo, Japan.

A2.3 *Type of project*

A suspension-type highway bridge having a central span of 570 m and side spans of 114 m (798 m in total length, Fig. A2.1) was constructed.

A2.4 *Ground conditions*

The anchorages and the piers of the bridge are supported by a sedimentary softrock (Kazusa Group), consisting mostly of mudstones of Neogene to Early Pleistocene Period, which is overlain by a Holocene deposit, consisting mostly of soft clayey soil (Fig. A2.2).

A2.5 *Laboratory tests*

Two series of triaxial compression tests were conducted on softrock core samples retrieved from a depth of about 40 to 80 m (Kc layer in Fig. A2.2). In the first series that was performed at the design stage, axial strains were obtained by external measurements, and the Young's moduli $E_{initial}$ were obtained from the apparently linear portion of each stress-strain curve. On the other hand, in the second series, which was performed after the construction, the initial Young's moduli E_0 were obtained at a strain level of 0.0005 % based on local measurements, while obtaining the $E_{initial}$ values based on external measurements as well (Fig. A2.3). The E_0 values were much larger than the $E_{initial}$ values. The values of shear moduli G_0 that converted from the E_0 values by using Poisson ratios of 0.5 under undrained condition and of 0.12 under drained condition agreed with the results of field seismic surveys, which were performed after the construction of the foundations (Fig. A2.5).

At the design stage, oedometer tests were also conducted on mudstone samples to evaluate the tangential coefficient of volume compressibility m_v, considering that the deformation of the mudstone layer below the bridge foundation is closely one-dimensional. Note that values of $1/m_v$ as shown in Fig. A2.4 are very close to the drained Young's

modulus, since the drained Poisson's ratio is about 0.12 as obtained from triaxial tests.

The triaxial tests in the second series were conducted because the stiffness values of the soft rock obtained at the design stage was found to be too small to explain the full-scale behavior of the ground observed during construction.

A2.6 *Field tests*

At the design stage and during construction, pressuremeter tests (PMTs) were conducted in a predrilled borehole and plate loading tests (PLTs) on the excavated ground surface (Figs. A2.5 and A2.6). Only primary loading was applied in the PMTs. As seen from Fig. A2.6, The Young's moduli obtained from the PMTs were smaller than those obtained from the PLTs, due probably to the larger strain levels involved in the former tests.

A2.7 *Comparison of test results with full-scale behavior*

The anchorage is composed of a caisson 70 m wide by 45 m deep and an anchorage body which supports the cables. The weight of the anchorage upon completion of concrete filling into the caisson was about 280,000 tonf, with a further 140,000 tonf for the anchorage body.

Three-dimensional linear or non-linear elastic finite element analyses (Fig. A2.7) were conducted to simulate the behavior of the anchorage during construction of the anchorage body. In the linear analysis (method 2), the Young's modulus determined from PLTs and PMTs (Fig. A2.6) was employed. In the non-linear analysis (method 3), average values of G_0 obtained from triaxial tests with local measurements (Fig. A2.5) were used, and the non-linearlity was modeled as a function of strain level based on the measured relations from drained triaxial compression tests (Fig. A2.8). For comparison, two-dimensional plane strain linear elastic analysis (method 1) using the reloading Young's modulus obtained from the oedometer tests (Fig. A2.4) and conventional consolidation calculation using deformation characteristics at primary loading from the oedometer tests were also conducted.

Methods 1 and 2 overestimated the observed settlements of the anchorage by about three and two times, respectively, while the conventional calculation yielded much larger settlements (Fig. A2.9). Possible reasons are the effects of bedding errors in the oedometer tests and strain levels involved in the field loading tests that were larger than the those actually operating in the ground (about 0.01%). On the other hand, method 3 simulated very well the observed behavior (Figs. A2.9 and A2.10). Note that the results of method 1 were converted into the equivalent values for the 3D model and that creep effects were considered in the results of methods 1 through 3 by introducing a reduction factor of 0.87 for stiffness values, which was determined based on the results of triaxial creep tests.

A2.8 *References for case No.2*

Izumi,K., Ogihara,M. and Kameya,H. 1997. Displacements of bridge foundations on sedimentary soft rock: a case study on small-strain stiffness, *Géotechnique* 47, No. 3, pp.619-632.

Izumi,K., Ogihara,M. and Kameya,H. (1996): Deformation behavior of sedimentary soft rock beneath foundation of the Rainbow Bridge, *Tsuchi-to-Kiso*, Vol. 44, No. 11, pp.5-8 (in Japanese).

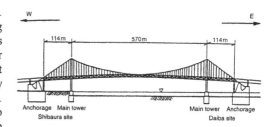

Fig. A2.1   Side view of the Rainbow Bridge (Izumi et al., 1997)

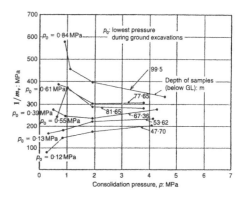

Fig. A2.4   Results of oedometer tests during reloading (Izumi et al., 1997)

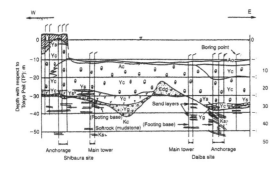

Fig. A2.2   Geological profile (Izumi et al., 1997)

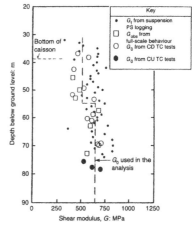

Fig. A2.5   Results of field seismic surveys and triaxial tests (Izumi et al., 1997)

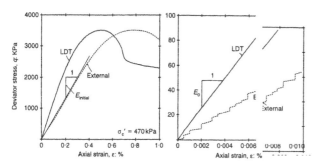

Fig. A2.3   Typical stress-strain relations using different scales from drained triaxial compression test (Izumi et al., 1997)

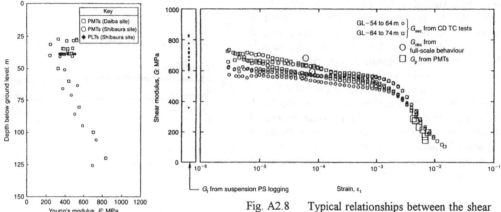

Fig. A2.6 Results of pressuremeter tests and plate loading tests (Izumi et al., 1997)

Fig. A2.8 Typical relationships between the shear moduli and strain levels (Izumi et al., 1997)

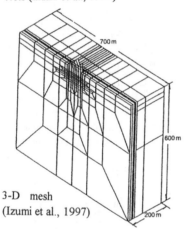

Fig. A2.7 3-D mesh (Izumi et al., 1997)

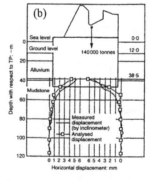

Fig. A2.9 Comparison of measured and analyzed settlements during the construction of the anchorage body at Daiba site (Izumi et al., 1997)

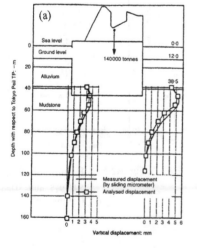

Fig. A2.10 Comparison of measured and analyzed ground displacements by method 3 at Daiba site: (a) vertical displacements ; (b) horizontal displacements (Izumi et al., 1997)

A3.1 Construction period

1989-1992.

A3.2 Location

Kanagawa, Japan.

A3.3 Type of project

A 50 m-deep shaft and a series of short tunnels were excavated to investigate the rational construction method in a soft rock deposit and the deformation characteristics of soft rock (Fig. A3.1). The structures of the vertical shaft and tunnels are shown in Fig. A3.2. Generally, the support system is very light when compared to those employed in actual construction projects under similar conditions. Yet, the lateral displacement of the vertical shaft and the axial force of the rockbolts were very small as shown later.

A3.4 Ground conditions

Subsurface layers of loam and gravel to a depth of 21 m from the ground surface is underlain by a very thick soft rock deposit (Fig. A3.1). It is a sedimentary soft mudstone (Kazusa Group) of a geological age of about 1.5 million years, which has not been noticeably weathered or distorted by the tectonic force.

Above a depth of about 35 m, about 15 cm thick sandy layers and about 5 cm thick scoria layers were occasionally found. Below that level, the mudstone is mostly massive. The stratification formed a simple slope structure and aligned along a NW-SE axis with a dip of about 10 degrees toward the NE.

A3.5 Laboratory tests

Before the excavation of the shaft, core samples were obtained from a bore hole drilled from the ground surface. When the excavation of the shaft reached a depth of 35 m, core samples were also retrieved from another bore hole drilled from the excavated surface. Triaxial compression tests with local measurements were conducted on these samples under a confining stress equal to the in-situ effective overburden pressure, and a maximum deviator stress q_{max} of 5.5 MPa was obtained, on the average, for depths 25 to 50 m (Kim et al., 1994).

Unconfined compression tests and oedometer tests with one-dimensional cyclic loading were also conducted on the cored samples.

Distribution of Young's moduli obtained from these different laboratory tests are compared in Fig. A3.3. The average value of E_{max} obtained from the triaxial compression tests, based on the initial part, at strain levels smaller than about 0.001 %, of the relationship between the deviator stress and the locally measured axial strain, was 32,000 kgf/cm^2. On the other hand, that of E_i when based on the apparently linear part of the relationship between the deviator stress and the externally measured axial strain was 8,000 kgf/cm^2. Further, that of E_{50} obtained from the unconfined compression tests was 3,000 kgf/cm^2, which was almost similar to the range of E_{OED} obtained from the oedometer tests. For reference, results from a down-hole seismic survey are shown as E_f in the figure, which were consistent with the E_{max} values.

A3.6 Field tests

Before the excavation of the shaft, a down-hole seismic survey and pre-boring type pressuremeter tests (PMTs) were performed. After the excavation of the shaft to a depth of 35 m or 50 m, pre-boring type or self-boring type PMTs were also conducted, respectively. In the tunnel E at a depth of 50 m (Fig. A3.1), plate loading tests (PLTs) using two rigid plates with a diameter of 30 cm and 60 cm were performed, while measuring strains in the ground (Fig. A3.4). In a test adit at a depth of 35 m (Fig. A3.1), similar PLTs were also conducted using a rigid plate with a diameter of 60 cm.

Distribution of Young's moduli obtained from these different field tests are compared in Fig. A3.5. Results from the pre-boring type PMTs, based on the apparently linear part of the relationship between the cavity pressure and the cavity strain during primary loading, are shown as $E_{BHLT(1)}$ and $E_{BHLT(2)}$. Note that a Menard type pressuremeter which measures the volume change of the cavity was employed for $E_{BHLT(1)}$, while a different type pressuremeter which directly measures the change in the cavity diameter was employed for $E_{BHLT(2)}$. Results from the self-boring type PMTs, based on the tangential modulus during unloading at a cavity strain level of about 0.1 %, referring to the procedures proposed by Jardine (1992), are shown as $E_{BHLT(3)}$, which were similar to the values of $E_{PLT(D)}$ obtained from PLTs conducted at a depth of 50 m. On the other hand, the values of $E_{PLT(D)}$ obtained from PLTs conducted at a depth of 35 m were larger than these values, and similar to the E_f values obtained from the down-hole seismic survey. Note also that a suspension-type seismic survey was also performed at the site, which showed similar E_f values to those from the down-hole survey.

For reference, results from ultrasonic wave velocity measurements of cored samples under an isotropic confining stress equal to the in-situ effective overburden pressure are shown in Fig. A3.5 as E_d. Since the values of E_d and E_f were almost similar to those of E_{max} as shown in Fig. A3.3, it is suggested that essentially the same elastic deformation characteristics can be obtained from static tests and dynamic tests (Tatsuoka and Kohata, 1995).

A3.7 Comparison of tests results with full-scale behavior during construction

Distribution of measured inward lateral displacements immediately behind the shaft wall is shown in Fig. A3.6. Based on these full-scale behaviors, the Young's moduli were back-calculated as summarized in Table A3.1 by assuming the soft rock as an isotropic elastic body with a Poisson's ratio of 0.33, which was determined to be the average of Poisson's ratios of 0.2 and 0.45 measured during drained and undrained triaxial tests, respectively. The back-calculated values of major principal strain increment ε_{BA} were also shown in Table A3.1.

Relationships between the Young's moduli and the associated major strain increments were compared in Fig. A3.7. Note that the Young's moduli obtained from the full-scale behavior and the in-situ test results were normalized by E_f, while those from the laboratory test results were normalized by $E_{max}(CU)$ measured with consolidated undrained triaxial tests, since the values of E_f and $E_{max}(CU)$ were almost identical (Fig. A3.3). It is seen from Fig. A3.7 that the relationships obtained from the full-scale behavior and the in-situ test results were well in accordance with those obtained from the laboratory test results. It is, therefore, estimated that the effects of joints and cracks in the tested soft rock on its deformation behavior were small.

Based on Fig. A3.7, modeling of nonlinear deformation characteristics was made on the tested soft rock, and FE analyses were conducted to simulate the plate loading tests (Fig. A3.8), pressuremeter tests (Fig. A3.9) and full scale behavior during the excavation of tunnel B (Fig. A3.10). Good agreement between the measured and simulated results were obtained.

A3.8 References for case No.3

Jardine,R.J. 1992. Non-linear stiffness parameters from undrained pressuremeter tests, *Canadian Geotechnical Journal*, Vol. 29, pp.436-447.

Kim,Y.-S., Tatsuoka,F. and Ochi,K. 1994. Deformation characteristics at small strains of sedimentary soft rocks by triaxial compression tests, *Geotechnique*, Vol.44, No.3, pp.461-487.

Ochi,K., Tsubouchi,T. and Tatsuoka,F. 1994. Deformation characteristics of sedimentary soft rock evaluated by full-scale deformation, *Pre-failure Deformation of Geomaterials* (Shibuya et al., eds.), Balkema, Vol.1, pp.601-607.

Ochi,K., Tsubouchi,T. Nakashita,K., Ito,R. and Amano,S. 1997. Lecture on non-linearity of geo-materials at small strain and its application to the ground deformation problems, No. 6 case histories (2), *Tuchi-to-Kiso*, Journal of JGS, Vol.45, No.10, pp.53-58 (in Japanese).

Siddiquee,M.S.A., Tatsuoka,F., Hoque,E., Tsubouchi,T., Yoshida,O., Yamamoto,S. and Tanaka,T. 1994. FEM simulation of footing settlement for stiff geomaterials, *Pre-failure Deformation of Geomaterials* (Shibuya et al., eds.), Balkema, Vol.1, pp.531-537.

Tatsuoka,F. and Kohata,Y. 1995. Stiffness of hard soils and soft rocks in engineering applications, *Pre-failure Deformation of Geomaterials* (Shibuya et al., eds.), Balkema, Vol.2, pp.947-1063.

Tsubouchi,T., Ochi,K. and Tatsuoka,F. 1994. Non-linear FEM analyses of pressuremeter tests in a sedimentary soft rock, *Pre-failure Deformation of Geomaterials* (Shibuya et al., eds.), Balkema, Vol.1, pp.539-544.

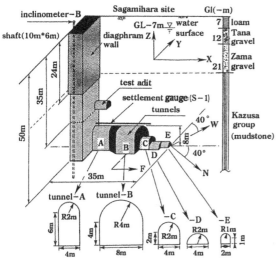

Fig. A3.1 Overall view of shaft and tunnels at Sagamihara site (Ochi et al., 1994)

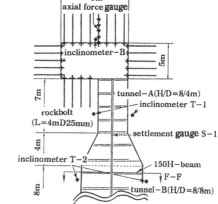

(b) Plan of shaft and tunnels A and B at the depth of 50 m (E-E)

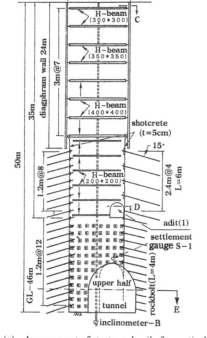

(a) Arrangement of struts and nails for vertical shaft

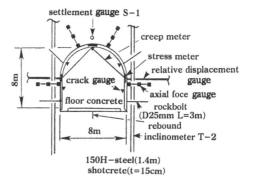

(c) Cross-section of tunnel B (F-F)

Fig. A3.2 Structures of shaft and tunnels (Ochi et al., 1994)

E_{max}: Drained (CD)/ undrained(CU)
triaxial compression tests
E_{50}: Unconfined compression tests
E_{OED}: Oedometer tests
(E_f: In-situ seismic surveys)

E_{BHLT}: Pressuremeter tests
E_{PLT}: Plate loading tests
E_f: In-situ seismic surveys
(E_d: Ultra sonic wave measurements
in laboratory triaxial tests)

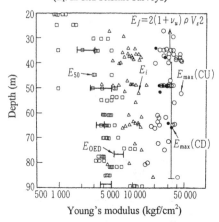

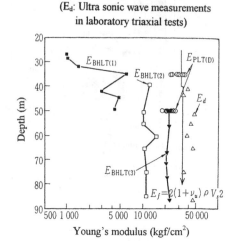

Fig. A3.3 Distribution of Young's moduli obtained from laboratory tests (Ochi et al., 1997)

Fig. A3.5 Distribution of Young's moduli obtained from field tests (Ochi et al., 1997)

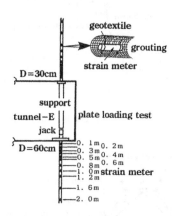

Fig. A3.4 Plate loading tests in tunnel E measuring in-ground strains (Ochi et al., 1994)

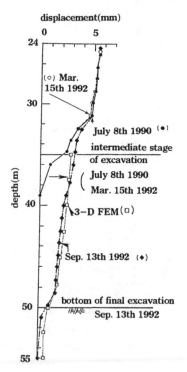

Fig. A3.6 Distribution of measured inward lateral displacements immediately behind the shaft wall (Ochi et al., 1994)

126

Table A3.1 Back-analyzed Young's moduli from full-scale behavior (Ochi et al., 1997)

No.	E_{DBA} (kgf/cm^2), ε_{BA}, σ_m (kgf/cm^2) and Ko values
	Observed seismic response at GL-45 m:
1	E_{DBA}/E_f=1.09, γ_{ref}=0.0000029 (E_f=32000 kgf/cm^2)
2	E_{DBA}/E_f=0.98, γ_{ref}=0.0000029 (E_f=32000 kgf/cm^2)
	Horizontal deformation of shaft and vertical deformation of adit at GL-35 m:
3	E_{DBA}=12,600, ε_{BA}=0.002, σ_m=25.2, Ko=1.5
	Convergent strain of adit at GL-35 m:
4	E_{DBA}=24,000, ε_{BA}=0.0004, σ_m=9.6, Ko=1.5
	Horizontal deformation of shaft and axial stress of strut at GL-20~-35 m:
5	E_{DBA}=28,000, ε_{BA}=0.0002, σ_m=5.6, Ko=1.5
6	E_{DBA}=20,000, ε_{BA}=0.0010, σ_m=20.0, Ko=1.5
7	E_{DBA}=17,000, ε_{BA}=0.0012, σ_m=20.4, Ko=1.5
8	E_{DBA}=14,000, ε_{BA}=0.0019, σ_m=26.9, Ko=1.5
9	E_{DBA}=12,000, ε_{BA}=0.002, σ_m=24.0, Ko=1.5
	Horizontal deformation of shaft at GL-35~-50 m:
10	E_{DBA}=14,000, ε_{BA}=0.0012, σ_m=16.8, Ko=1.5
	Vertical and horizontal cavity strain of tunnel at GL-50 m:
11	E_{DBA}=14,000, ε_{BA}=0.0018, σ_m=25.2, Ko=1.3

E_{DBA}: Back-analyzed values
E_f: Results from seismic surveys
γ_f: Effective shear strain increment
ε_{BA}: Major principal strain increment
σ_m: Stress increment (=$E_{DBA} \cdot \varepsilon_{BA}$)
Ko: Earth pressure coefficient (=σ_h/σ_v)

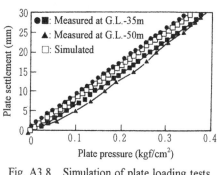

Fig. A3.8 Simulation of plate loading tests (Siddiquee et al., 1994)

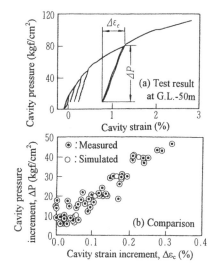

Fig. A3.9 Simulation of pressuremeter tests (Ochi et al., 1997)

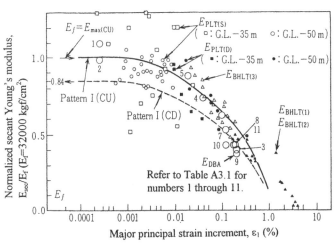

Fig. A3.7 Comparison of Young's modulus from various methods (Ochi et al., 1997)

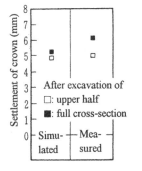

Fig. A3.10 Comparison of settlement of crown of tunnel B (Ochi et al., 1994)

APPENDIX 4. CASE NO. 4 (LNG TANKS AT NEGISHI)

A4.1 *Construction period*

1991-1992.

A4.2 *Location*

Kanagawa, Japan.

A4.3 *Type of project*

Large-scale cylindrical excavation works for in-ground LNG tanks were carried out by vertical NATM. The largest excavation is 76m in diameter and 57m in depth (Fig. A4.1).

A4.4 *Ground conditions*

LNG tanks are located in a reclaimed land mainly consisting of a silt deposit with a thickness between 10 to 20 m. It is underlain by sedimentary soft rock layers named Kazusa Group (Fig.A4.2). The soft rock layers are intact and non-permeable.

A4.5 *Laboratory tests*

At the design stage of tanks with a volume of 85,000 kl, physical property tests, unconfined compression tests, sprit tests, triaxial compression tests (CU), ultra-sonic log and slaking tests were conducted on the soft rock. Typical properties are shown in Table A4.1. At this stage, samples were retrieved by using the tube sampler. The unconfined compression strength, q_u, of the soft rock was about 20 to 30 kgf/cm^2, and the coefficient of permeability, k, was 1×10^{-6} cm/sec $\sim 1 \times 10^{-7}$ cm/sec.

During the excavations for the 85,000 kl tanks, block sampling was conducted at the depth of 17m from the ground surface, and a series of laboratory tests which are similar to those as mentioned above was carried out.

A4.6 *Field tests*

Seismic surveys and pressuremeter tests were conducted at the design stage (Table A4.2).

For the case of 200,000 kl tanks, seismic surveys

were conducted at the bottom of excavation in order to examine decrease in stiffness of the soft rock due to excavation.

A4.7 *Comparison of numerical prediction with full-scale behavior during construction*

At the design stage of the 85,000 kl tanks, linear elastic finite element analysis under plain strain or axi-symmetric condition was conducted to estimate the ground behavior during excavation (Fig. A4.3a.). After excavation, back-analysis was performed by using the same procedure with modified parameter values to simulate the observed behavior (Fig. A4.3b).

In designing 200,000 kl tanks, nonlinear elastic finite element analysis was employed. Its applicability was verified at the design stage through simulating the observed behavior of the 85,000 kl tanks, which was also reconfirmed through comparison with the field measurements and investigations on the 200,000 kl tanks (Fig. A4.4).

The above comparisons are briefly summarized in the following.

1) Linear elastic analyses

In the linear elastic analysis conducted at the design stage of 85,000 kl tanks, Young's modulus of the soft rock was set equal to 2800 kgf/cm^2 based on E_{50} values obtained from unconfined compression tests. The calculated horizontal displacements of the excavated face were much lager than the measured values, and their distribution in depth was different from the measured one (Fig. A4.3a). These discrepancies were considered to be due to the strain levels for defining E_{50} which was much larger than the actual order of 10^{-4}, and to the difference in the loading/unloading directions between the unconfined compression tests and the actual excavation works.

In the back-analysis based on linear elasticity, conducted after excavation for the 85,000 kl tanks, the Young's modulus of the soft rock was parametrically changed. As a result, when it was increased up to 10000 kgf/cm^2, calculated horizontal displacements in the lower part of the excavated ground agreed with the measured values (Fig. A4.3b), while those in the upper part underestimated the measured values. The possible reasons for such discrepancy would be subgrade reaction of sheet piles embedded in the upper part to support earth pressures from the overlying reclaimed soil layers, reduction of arching effects due to non-axi-symmetric behavior, and/or dependency of soil stiffness on the confining stress, which were not considered in the axi-symmetric linear elastic analysis.

2) Nonlinear elastic analyses

In the analyses based on axi-symmetric nonlinear elasticity, the constitutive model proposed by Motojima et al. (1978) was adopted (Table A4.3). By setting parameters based on results from triaxial compression tests and sprit tests (Figs. A4.5 and A4.6) as listed in Table A4.3, the pre-failure stress-strain relationship of the cored specimen could be reasonably simulated as typically shown in Fig. A4.7. In the back-analyses simulating the full-scale behavior, the initial Young's moduli, E_0, were modified to be 6000 kgf/cm^2 for the upper part of the soft rock denoted as Kac_1 and 12000 kgf/cm^2 for the lower part denoted as Kac_2 (see Fig. A4.4), referring to results of pressuremeter tests under unloading conditions. Based on the results of parametric studies, the coefficients of earth pressure at rest, K_0, were set to be 1.0 and 0.75 for Kac_1 and Kac_2 layers, respectively.

As seen from Fig. A4.4, the calculated horizontal displacements of the 85,000 kl and 200,000 kl tanks agreed well with the observed behavior. It is also seen from Fig. A4.8 that for the case with the 200,000 kl tank, the horizontal displacement of the soft rock layer was rather independent of the circular directions, suggesting the applicability of the axi-symmetric analysis.

It should be noted that the initial Young's moduli E_0 employed in the above analyses were twice as large as E_{50}, while they were about half of the values of E_f obtained from shear wave velocities (Table.A4.2). It is estimated that reduction of confining pressure and loosening of ground due to

excavation, which were not considered in the analyses, may have caused the decrease in the stiffness. Note also that the value of E_d obtained from the ultrasonic log on cored samples was almost the same as E_f, demonstrating the consistency between in-situ tests and laboratory tests. On the other hand, the value of E_0 obtained from triaxial compression tests (=4100 kgf/cm^2, Table A4.3) was considerably smaller than those of E_d and E_f. This discrepancy is due possibly to bedding errors in the triaxial tests with external measurements.

In Fig. A4.4, the loosened zones induced by excavation are also compared for the case of the 200,000 kl tank. Loosened zones estimated from the shear wave velocities measured after the excavation were slightly smaller than those predicted by the nonlinear analysis, due possibly to the fact that the stiffness of the soft rock layer below G.L.-35m level is larger than the design value.

A4.8 *References for case No.4*

Ito, R., Watanabe, K., Ueno, M. and Nakano, M. 1994. Analytical and observed results during cylindrical excavation of mudstone layer, Proc. 9th Japan Symposium on Rock Mechanics, pp. 593-598 (in Japanese).

Ito, R., Watanabe, K., Takagi, A., Ueno, M. and Nakasita, K. 1995. Deep excavation of soft rock with NATM, *Rock Foundation* (Yoshinaka and Kikuchi, eds.), Balkema, pp. 293-298.

Komatsubara, T., Aoki, H. and Amano, S. 1992a. A study on the behavior of the ground surrounding a large-scaled cylindrical cavern excavation in sedimentary soft rock, Proc. 27th Japan National Conf. on SMFE, pp. 2037-2038 (in Japanese).

Komatsubara, T., Aoki, H., Nishihara, Y. and Tanaka, Y. 1992b. The study on the observed results during the large-scale cylindrical excavation with shotcrete and rock bolts, Proc 24th Symposium on Rock Mechanics, Committee of Rock Mechanics, JSCE, pp. 331-335 (In Japanese).

Motojima, M., Hibino, S. and Hayashi, M. 1978. Development of the computer program for the stability analysis of rock ground during excavation, Technical Report of Central Research Institute of Electric Power Industry, No. 377012 (in Japanese).

Ochi,K., Tsubouchi,T. Nakashita,K., Ito,R. and Amano,S.

1997. Lecture on non-linearity of geo-materials at small strain and its application to the ground deformation problems, No. 6 case histories (2), *Tuchi-to-Kiso*, Journal of JGS, Vol.45, No.10, pp.53-58 (in Japanese).

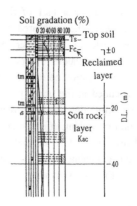

Fig. A4.2 Soil profile at Negishi
(Ochi et al., 1997)

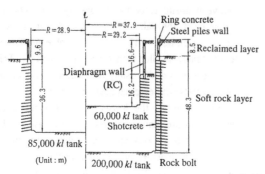

Fig. A4.1 Cross section of vertical NATM at Negishi (Ito et al., 1995)

Table A4.1 Properties of soft rock
(Ito et al., 1994)

Unit weight	$\gamma_t = 1.9$ tf/m³
Unconfined compression Strength	$q_u = 20 - 30$ kgf/cm²
Natural water content	$w_n = 30\%$
Void ratio	$e = 0.7 - 0.9$

Table A4.2 Stiffness of soft rock (Ito et al., 1994)

Young's moduli		Kac₁	Kac₂
Initial values for nonlinear elastic analysis	E_0	6,000	12,000
Shear wave survey (in-situ)	E_f	9,900	20,000 - 23,000
		$E_0 \doteqdot 0.5 - 0.6\ E_f$	
Ultrasonic log (in lab.)	E_d	-	19,700
		$E_0 \doteqdot 0.5 E_d$	
Pressuremeter test	D_b	-	9,600 - 12,300
		$E_0 \doteqdot 0.5 - 0.6\ D_b$	
Unconfined compression test	E_{50}	3,000	5,000 - 6,000
		$E_0 \doteqdot 0.5 E_{50}$	

(Unit : kgf/cm²)

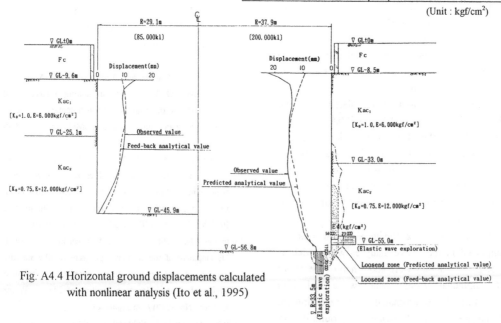

Fig. A4.4 Horizontal ground displacements calculated with nonlinear analysis (Ito et al., 1995)

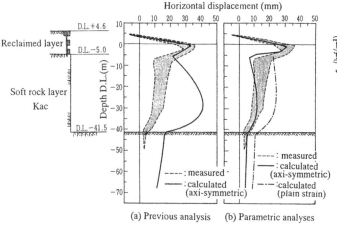

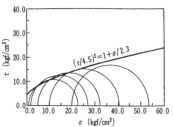

Fig. A4.5 Failure envelope with parabolic modeling (Ochi et al., 1997)

(a) Previous analysis   (b) Parametric analyses

Fig. A4.3 Distribution of horizontal ground displacement for 85,000 kl tank (Komatsubara et al., 1992b)

Table A4.3 Nonlinear elastic model of soft rock (Ito et al., 1995)

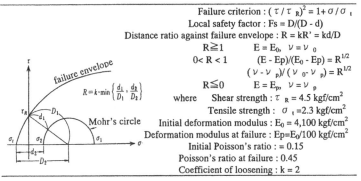

Failure criterion : $(\tau/\tau_R)^2 = 1 + \sigma/\sigma_t$

Local safety factor : $Fs = D/(D - d)$

Distance ratio against failure envelope : $R = kR' = kd/D$

$R \geqq 1$	$E = E_0,\ \nu = \nu_0$
$0 < R < 1$	$(E - Ep)/(E_0 - Ep) = R^{1/2}$
	$(\nu - \nu_p)/(\nu_0 - \nu_p) = R^{1/2}$
$R \leqq 0$	$E = E_p,\ \nu = \nu_p$

where Shear strength : $\tau_R = 4.5$ kgf/cm²

Tensile strength : $\sigma_t = 2.3$ kgf/cm²

Initial deformation modulus : $E_0 = 4,100$ kgf/cm²

Deformation modulus at failure : $Ep = E_0/100$ kgf/cm²

Initial Poisson's ratio : $= 0.15$

Poisson's ratio at failure : 0.45

Coefficient of loosening : $k = 2$

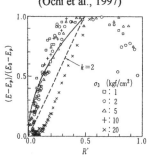

Fig. A4.6 Young's modulus (E) vs. Distance ratio (R') (Ochi et al., 1997)

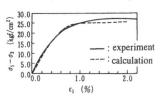

Fig. A4.7 Typical stress - strain curve during triaxial test (Ochi et al., 1997)

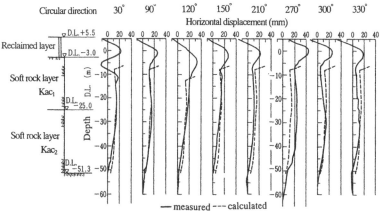

— measured --- calculated

Fig. A4.8 Distribution of horizontal ground displacement for 200,000 kl tank (Ochi et al., 1997)

APPENDIX 5. CASE NO.5 (A NUCLEAR POWER PLANT)

A5.1 Construction period

Dec. 1978 : Construction start of unit 1.
Jul. 1997 : Commercial operation start of unit 7 (the last unit of the site).

A5.2 Location

Kashiwazaki-Kariwa Nuclear Power Station of Tokyo Electric Power Company : Kashiwazaki City and Kariwa Village, Niigata Pref., Japan.

A5.3 Type of project

To construct a nuclear power plant, the original ground at an elevation of +45m was excavated by an open-cut method down to the elevation of –10m. Then a deep excavation with vertical sides was conducted for the power plant that has a bottom elevation of about –40m (Fig. A5.4). The vertical sides were supported by deep diaphragm walls and ground anchors (Fig. A5.6). The deformation of the vertical walls and the retained ground due to the excavation was analyzed in detail.

A5.4 Ground conditions

The ground conditions at the site comprise Deluvial clay (Yasuda Layer) and soft rock (Nishiyama Layer, sedimentary mudstone of Neogene period) that had a unit weight of $17.1 kN/m^3$ and a compressive strength of around 3.0MPa. (Fig. A5.4 and A5.6)

A5.5 Laboratory tests

Triaxial tests in monotonic loading, simple shear tests in monotonic and cyclic loading conditions were conducted on the material obtained from the site. The specimens were consolidated under the stress equal to the overburden stress at the original position (isotropic consolidation in TC tests), and then sheared undrained.

A5.6 Field tests

Plate loading tests, borehole pressuremeter tests, borehole PS logging tests, monotonic and cyclic plate loading tests, and elastic wave tests in tests pits were conducted to examine the elastic properties of the mudstone at various elevations.

A5.7 Elastic modulus from laboratory and field tests

Young's moduli of the mudstone were measured from various types of field and laboratory tests listed in previous sections and plotted in Fig. A5.1. In case of monotonic pressuremeter tests and monotonic plate loading tests, the moduli were obtained from the linear part of the stress-strain curve. The secant modulus, E_{50}, was adapted for triaxial test results. Young's moduli for small strain level of 10^{-5} to 10^{-6} was evaluated from the cyclic plate loading tests and the cyclic simple shear tests using Hardin-Drnevich method. Young's moduli from PS logging tests and elastic wave tests were calculated from the observed shear wave velocity. Fig. A5.1 shows that the evaluated elastic moduli were different depending on the types of the tests, i.e. monotonic loading tests exhibited lower Young's modulus.

Fig. A5.2 shows the relationship between elastic shear modulus and shear strain level obtained from pressuremeter tests, plate loading tests and simple shear tests. It may be seen that elastic shear modulus was largely affected by strain level. To examine this aspect in more detail, additional triaxial compression tests with small strain measurements using LDT (local displacement transducer) was conducted as shown in Fig. A5.3. The elastic modulus at the strain level of 10^{-6} measured by LDT in monotonic triaxial tests was consistent with the elastic moduli obtained from cyclic and dynamic tests including cyclic plate loading tests, cyclic simple shear tests, elastic wave tests and PS logging tests as shown in Fig. A5.5. This fact confirmed that the apparent scatter of measured elastic moduli plotted in Fig. A5.1 was due to the strain level effect.

A5.8 Analysis of the ground deformation

The deformation of the vertical walls and the retained ground due to the deep excavation was analyzed by two-dimensional FEM in plane strain conditions (Fig. A5.4). Linear elastic model was adapted and the Young's moduli obtained from the pressuremeter tests in unloading process (Table A5.1) were used in the analysis. The result was compared with the measurements in Fig. A5.6. A Poisson's ratio of $\nu = 0.46$ could simulate the horizontal earth pressure after the open-cut before the deep excavation (Fig. A5.7). When this Poisson's ratio was used in the analysis, the deformation of the retained ground due to the deep excavation was precisely predicted as shown in Fig. A5.6. On the contrary, if Poisson's ratio of $\nu = 0.34$ obtained from K_0 consolidation tests was used, the calculated horizontal earth pressure after the open-cut was larger than the reality (Fig. A5.7), and the ground deformation during the successive deep excavation was overestimated (Fig. A5.6).

A5.9 References for case No.5

Terada, K. and S. Fukui. 1997. Ground performance
during a construction of a nuclear power plant.
Tuchi-to-Kiso, Vol. 45, No.11, pp.49-51
(in Japanese).

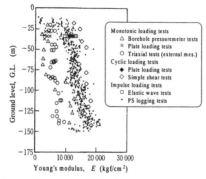

Fig. A5.1 Young's moduli from various types of
tests (Terada and Fukui, 1997)

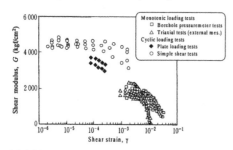

Fig. A5.2 Shear modulus versus shear strain
(Depth=G.L.-10m ~ -35m) (Terada and
Fukui, 1997)

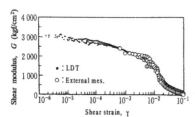

Fig. A5.3 Shear modulus from triaxial tests (Terada
and Fukui, 1997)

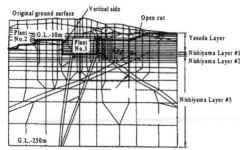

Fig. A5.4 Ground model for FEM analysis (Terada
and Fukui, 1997)

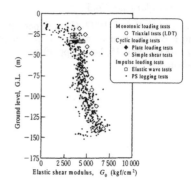

Fig. A5.5 Measured elastic shear modulus at small
strain level (Terada and Fukui, 1997)

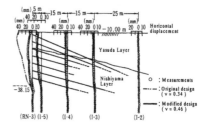

Fig. A5.6 Deformation of the diaphragm wall and
retained ground (Terada and Fukui, 1997)

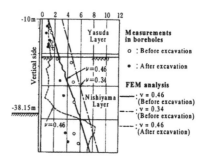

Fig. A5.7 Lateral earth pressure in the retained
ground (Terada and Fukui, 1997)

	Original design parameters			Modified parameters obtained after the open cut		
	Poisson's ratio		Young's modulus (kgf/cm²)	Poisson's ratio		Young's modulus (kgf/cm²)
	Before open cut	After open cut		Before open cut	After open cut	
Yasuda Layer			1400~2300			1400~2300
Nishiyama Layer #1	0.49	0.34	2300	0.49	0.46	2300
Nishiyama Layer #2			4800			4650
Nishiyama Layer #3			7300			7000

Table A5.1 Properties of soil layers (Terada and
Fukui, 1997)

APPENDIX 6. CASE NO.6 (HIGH-RISE BUILDING)

A6.1 *Construction period*

1992-1994.

A6.2 *Location*

Kanagawa, Japan.

A6.3 *Type of project*

A 152 m high high-rise building having 34 stories above the ground with 2-story basement was constructed (Figs. A6.1 and A6.2). It had a base area of about 6,000 m². Rebound of the ground during excavation and instantaneous settlement during construction of the building were evaluated based on elastic wave velocities, in order to compare with measured results.

A6.4 *Ground conditions*

The building is supported by mat slabs of 4m thick constructed on alternative soft rock layers consisting of cemented silt (T_m) and fine sand (T_s) deposits (Kami-Hoshikawa Formation of Kazusa Group) of Pleistocene Period (Fig. A6.3). Above the Kami-Hishikawa Formation, soft fill (F) and alluvial sand (A_s) layers are deposited at a depth of 6 to 9 m.

As summarized in Table A6.1, the T_m layer had unconfined compression strength exceeding 20 kgf/cm² and consolidation yield stress exceeding 65 kgf/cm², thus it was estimated that there would be no problem with respect to bearing capacity and consolidation settlement.

A6.5 *Laboratory tests*

No laboratory test that is relevant to the deformation properties of the foundation soils was conducted. They were estimated based on empirical proposals for the G/G_f γ relationships of clayey and sandy soils (Ishihara, 1982 and Zen et al., 1985). Some other laboratory tests such as consistency test, unconfined compression test and consolidation test were performed.

A6.6 *Field Tests and Simulation Procedures*

Borehole PS logging and density logging tests were performed as shown in Fig. A6.3. A small strain

Young's modulus E_f was obtained from the shear wave velocity V_s , Poisson's ratio ν and wet density ρ_t, employing the following equation.

$$E_f = 2(1+\nu) \cdot \rho_t \cdot V_s^{2} \qquad (1)$$

Then, the value of E_f was corrected for the effects of confining pressure, except for the clayey soils. The vertical stress increment $\Delta\sigma_v$ due to an average load per unit area Δp was obtained by employing the procedures for the stress calculation proposed by Steinbrenner (1936), with the assumption that the foundation plan is rectangular. Coefficients depending on confining pressure at the center and the corner of the foundation, α_1 and α_3, respectively, were obtained from the following equation:

$$\alpha_{1,3} = \sqrt{(\sigma_{v0}'+\Delta\sigma_{v1,3})/\sigma_{v0}'} \qquad (2)$$

where σ_{v0}' is the effective overburden pressure. The corrected Young's modulus E_{fc} was calculated from the following equation:

$$E_{fc} = E_f(1+\alpha_{ave})/2, \quad \alpha_{ave} = (\alpha_1+\alpha_3)/2 \qquad (3)$$

In addition, these Young's moduli (i.e., $E_0=E_{fc}$ for non-clayey soils and $E_0=E_f$ for clayey soils) were corrected for the mobilized strain level. Settlements of each layer at the center and the corner of the foundation, δ_1 and δ_3, respectively, due to Δp were calculated based on the elastic deformation calculation proposed by Steinbrenner (1936), as explained in A6.8. Then, the average shear strain of each layer was estimated as:

$$\gamma_{ave} = (1+\nu)\varepsilon_{ave} \quad \varepsilon_{ave} = (\delta_1+\delta_3)/2L \qquad (4)$$

where L is the thickness of the concerned layer. A reduction factor for the stiffness λ was obtained from the $G/G_f - \gamma$ curve, and the Young's modulus to be used in the simulation was determined by $E_1 = E_0 \cdot \lambda$.

A6.7 *Comparison of simulation results with full-scale rebound behavior during excavation*

The simulation results were compared with the measured ones on the rebound of the ground during excavation. Table A6.2 shows the scale of the excavation, and Fig. A6.4 shows the comparison between measured and simulated rebound, along with the parameters used in the simulation. As shown in Table A6.2, the maximum difference between measured and simulated rebound was about 25%, while it can be seen from Fig.A6.4 that the simulated results were considerably consistent with the measured ones.

A6.8 Comparison between results from elastic settlement analysis and in-situ measurements

The model used in the elastic settlement analysis consisted of 3-dimensional beam elements for the foundation and elastic spring elements for the ground. Concentrated loads that were evaluated as a sum of the design axial loads and the self weight of the foundation were applied to the nodal points. Their values normalized with the concerned area were 50 to 60 tf/m^2 for the high-rise parts and 15 to 30 tf/m^2 for the lower parts. The mat slabs and the foundation beams were converted into equivalent beams having the same bending stiffness.

Table A6.3 shows the elastic parameters of each soil layer used in the analysis, where E_f is the Young's modulus evaluated from the elastic wave velocity; and E_l is the corrected Young's modulus as explained in A6.6. To determine the E_l values, an average load of 35 tf/m^2 were applied on the foundation with a plan of 100$\times$70m. The Poisson's ratios ν for sandy and clayey soils were assumed to be 0.33, considering that drained condition would be maintained for a long term periods.

Fig. A6.5 shows time histories of measured relative settlement. Each result denotes the settlement of the foundation basement relative to the concerned depth. The final settlement relative to the depth of G.L.-100m was about 12 mm.

Fig. A6.6 shows the comparison between measured and simulated distributions of the final settlement. Simulated settlements relative to the depths of G.L.-200 m and G.L.-100 m, respectively, are denoted as cases 1 and 2. In case 2, the simulated settlement at the foundation basement was about 2 mm larger than the measured one. This is possibly caused by an underestimation of the Young's modulus of the layer #3 at a depth of G.L.-64~-86 m (Table A6.3), because the simulated results deviated from the measured ones within this layer. From the difference between the simulated results of cases 1 and 2, it was estimated that the settlement of the layer at a depth of G.L.-100~-200 m was about 4 mm, which corresponds to about 22 % of the total settlement.

Fig. A6.7 compares the horizontal distribution of foundation settlement at the central section. The maximum values of the measured absolute and differential settlements of the foundation were about 16 mm and 8 mm, respectively. The simulated results were consistent with the measured ones in terms of the maximum absolute settlement, while the simulated maximum differential settlement was larger by 50% than the measured value. The latter deviation was possibly caused by an underestimation of the stiffness of beam elements for mat slabs and foundation beams.

A6.9 References for case No.6

Ishihara, K. 1982. Fundamentals of soil dynamics, *Kajima Publication*, pp.196-202 (in Japanese).

Majima, M., Nagao, T. and Seno, H. 1993. Estimation of ground rebound due to excavation, *Proceedings of Annual Meeting of AIJ*, pp.1413-1414 (in Japanese).

Steinbrenner, W. 1936. Bodenmechanik und neizeitlicher strass enbau symposium by 24 authors, Volk und Reich Verlag, Berlin.

Terada, K., Fukui, S., Majima, M. and Tamaoki, K. 1997. Lecture on non-linearity of geo-materials at small strain and its application to the ground deformation problems, No.6 case histories (3), *Tsuchi-to-Kiso*, Journal of JGS, Vol.45, No.11, pp.49-54 (in Japanese).

Zen, K. and Umehara, Y. 1985. Lecture on Dynamic properties of soils for earthquake response analysis, No.2 dynamic properties of soils, *Tsuchi-to-Kiso*, Journal of JSSMFE, Vol.33, No.12, pp.63-70 (in Japanese).

Table A6.1 Soil properties of Kami-Hoshikawa Formation (Terada et al., 1997)

Layer	Cemented Silt T_{m1} - T_{m6}	Sandy Soil T_{s1} - T_{s4}
Fines Content	60-90%	10-90(Ave.20)%
Water Content	20-35%	20-30%
Wet Density	1.35g/cm^3	2.15g/cm^3
Unconfined Comp. Strength	20-40kgf/cm^2	--
Cohesion	c_u=8-20kgf/cm^2	c_d=0.6-20kgf/cm^2
Friction Angle	ϕ_u=0-30°	ϕ_d=38°-40°
Consolidation Yield Stress	65-80kgf/cm^2	--

Table A6.2 Scale of excavation and maximum difference between measured and simulated rebound (Terada et al., 1997)

| No. | Excavation Scale | | | Max. |
	Width L*B(m)	Depth D(m)	Load $\triangle$p (tf/m²)	Difference*
1	110*70	22.5	40	18%
2	100*50	20.0	35	22%
3	100*75	18.0	32	25%
4	105*30	23.0	45	14%

Table A6.3 Soil parameters employed in elastic settlement analysis (Terada et al., 1997)

Layer #	Depth	E_f (tf/m²)	E_1 (tf/m²)	ν
#1	G.L.−19〜−49m	140,000	102,700	0.33
#2	G.L.−49〜−64m	200,000	155,600	0.33
#3	G.L.−64〜−86m	140,000	105,200	0.33
#4	G.L.−86〜−100m	180,000	151,200	0.33

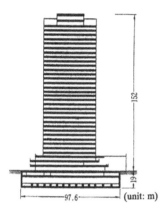

Fig. A6.1 Cross-section of building (Terada et al.,1997)

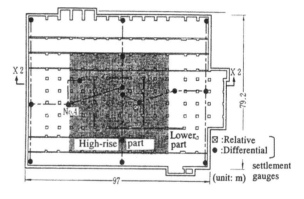

Fig. A6.2 Plan of building (Terada et al., 1997)

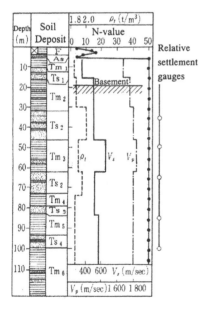

Fig. A6.3 Soil profile and results from PS logging (Terada et al., 1997)

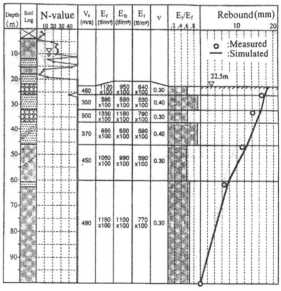

Fig. A6.4 Soil parameters employed in elastic rebound analysis and comparison of measured and simulated rebound values (Terada et al., 1997)

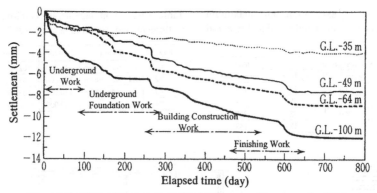

Fig. A6.5   Time histories of relative settlements (Terada et al., 1997)

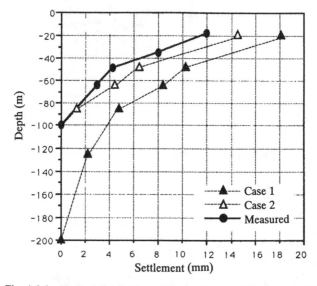

Fig. A6.6   Vertical distribution of final settlement (Terada et al., 1997)

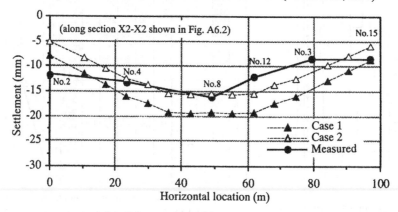

Fig. A6.7   Horizontal distribution of final foundation settlement at central section (Terada et al., 1997)

APPENDIX 7. CASE NO. 7 (K BUILDING)

A7.1 *Construction period*

1988-1990.

A7.2 *Location*

Tokyo, Japan.

A7.3 *Type of project*

A 9-story building with 3 basement floors (Fig. A7.1) was constructed.

A7.4 *Ground conditions*

A spread foundation is supported by a Pleistocene clay layer at a depth of 14.3 m from the ground surface (Fig. A7.2).

A7.5 *Laboratory tests*

No laboratory test that is relevant to the estimation of pre-failure deformation of soil layers affecting the settlement of the building was conducted.

A7.6 *Field tests*

Seismic surveys (Fig. A7.2) were conducted before the construction. Monitoring of ground settlements and measurement of shear wave velocities by using the same bore hole were performed at two locations during the construction (Figs. A7.1 and A7.2).

A7.7 *Secant Young's modulus based on ground settlements*

Vertical strains were evaluated from the measured distribution of ground settlements (Fig. A7.3). Vertical stresses were estimated from the excavated soil depth and the building weight by using the approximated solution by Steinbrenner assuming elasticity (Terzaghi, 1943), where the Poisson's ratios of sand, gravelly sand and clay layers were assumed to be 0.35, 0.3 and 0.45, respectively. Based on these values, secant Young's moduli, E_{sec}, were estimated and plotted versus the vertical strains, ε_v, which were evaluated for excavation and construction processes, separately (Fig. A7.4). E_{sec} values decreased with ε_v, and their relationship curves were upward-convex during excavation and downward-convex during construction of the building. These behaviors were similar to those during unloading and loading processes in plate loading tests, which may be affected by the difference in the confining stress and that of plastic deformation mode.

A7.8 *Comparison with other field test results*

The secant Young's moduli based on the ground settlements and results of seismic surveys are plotted versus SPT N-values in Fig. A7.5. Results obtained at other construction sites, including those of plate loading tests (PLTs) and pressuremeter tests (PMTs), are also shown in the figure.

It should be noted that, when based on the ground settlements, the values of E_{sec} at the final stage of excavation/construction were plotted. These E_{sec} values were generally larger than or equal to those obtained by an empirical formulation employed in the design practice in Japan (AIJ, 1988) as:

$$E=28 \cdot N$$

where N is the SPT N-value; and E is the Young's modulus in kgf/cm^2. On the other hand, results of PMTs were considerably lower than the E_{sec} values for sand and gravelly sand layers with SPT N-values larger than 20, while they were on the lower bound of the range of the E_{sec} values for Pleistcene clay layers with SPT N-values smaller than 10. It is suggested that effects of stress release due to drilling on the Young's modulus may be different depending on the soil type.

It should be also noted that PLTs referred in Fig. A7.5 were conducted at the surface of the excavated ground. The Young's moduli evaluated from reloading behaviors measured in these PLTs were similar to the E_{sec} values at the final stage of construction, suggesting the applicability of PLTs in evaluating the Young's modulus of subsoils near the excavated ground surface.

The range of the E_{sec} values shown in Fig. A7.5 were almost comparable to that of $0.2E_f$ and $0.5E_f$, where E_f denotes the Young's modulus based on seismic survey results. Such deviation of the field test results from the E_{sec} values may be due to the differences in the involved strain level, the confining stress and the previous stress histories.

The E_{sec} values shown in Fig. A7.4 are normalized by the corresponding E_f values and are shown in Fig. A7.6. Note that results obtained at other construction sites are also shown and that the E_f values during construction (i.e., after excavation) are corrected for the change in the confining stress by using

$$V_s = \alpha \cdot (\sigma_v')^{0.2}$$

where V_s is the shear wave velocity; σ_v' is the current effective overburden pressure; and α is a coefficient that is determined from the relationship between V_s and σ_v' values measured before excavation. Since rather unique relationships are found in Fig A7.6, they may be employed in evaluating the Young's modulus of subsoils at various strain levels by using V_s values measured before excavation. Note that the above relationship between V_s and σ_v' was verified based on the data measured during excavation work for K building (Fig. A7.7).

A7.9 *References for case No. 7*

Architectural Institute of Japan 1988. Recommendations for Design of Building Foundations, pp.146-154 (in Japanese).

Tamaoki,K., Katsura,Y., and Kishida, S. 1992. Young's moduli of bearing strata based on vertical deformation measured during construction work, *Research Report of Shimizu Corp.*, No. 55, pp.11-20 (in Japanese).

Tamaoki,K., Katsura,Y., and Kishida, S. 1993a. Young's moduli of bearing strata estimated from vertical deformation during excavation and construction, *Journal of Struct. Constr. Engng, AIJ*, No. 446, pp.73-80 (in Japanese).

Tamaoki,K., Katsura,Y., Nishio, S. and Kishida, S. 1993b. Estimation of Young's moduli of bearing soil strata, *Excavation in Urban Areas*, KIGForum '93 (Adachi ed.), pp.23-33.

Terzaghi, K. 1943. Theoretical Soil Mechanics, Wiley, pp.423-425.

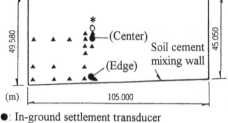

●: In-ground settlement transducer
○: Pore water pressure gauge
▲: Earth pressure cell to measure contact pressure
*: Bore hole for investigation before excavation

Fig. A7.1   Plan of K building (Tamaoki et al., 1993a,b)

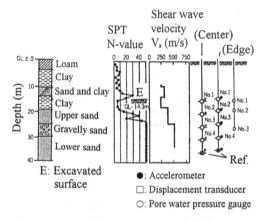

E: Excavated surface

●: Accelerometer
□: Displacement transducer
○: Pore water pressure gauge

Fig. A7.2   Soil profile, instrumentation and depth of excavation for spread foundation (Tamaoki et al., 1993a,b)

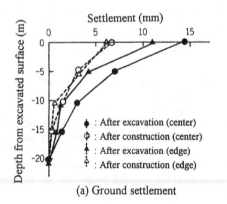

♦: After excavation (center)
φ: After construction (center)
▲: After excavation (edge)
△: After construction (edge)

(a) Ground settlement

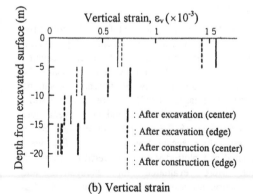

: After excavation (center)
: After excavation (edge)
: After construction (center)
: After construction (edge)

(b) Vertical strain

Fig. A7.3   Measured ground settlement and vertical strain (Tamaoki et al., 1993a)

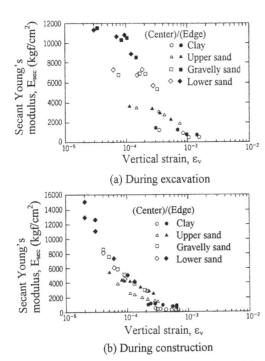

(a) During excavation

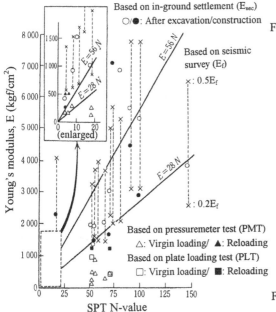

(b) During construction

Fig. A7.4 Relationships between Young's modulus and vertical strain (Tamaoki et al., 1993a,b)

E_{sec}: Based on in-ground settlement
E_f: Based on seismic survey
E_f': Based on seismic survey and corrected for change in confining stress

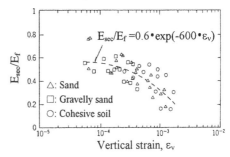

$$E_{sec}/E_f = 0.6 \cdot \exp(-600 \cdot \varepsilon_v)$$

△: Sand
□: Gravelly sand
○: Cohesive soil

(a) During excavation

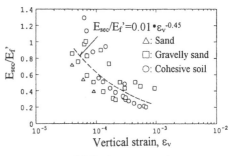

$$E_{sec}/E_f' = 0.01 \cdot \varepsilon_v^{-0.45}$$

△: Sand
□: Gravelly sand
○: Cohesive soil

(b) During construction

Fig. A7.6 Relationships between normalized Young's modulus and vertical strain (Tamaoki et al., 1993a,b)

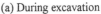

Based on in-ground settlement (E_{sec})
○/●: After excavation/construction

Based on seismic survey (E_f)
×: 0.5E_f

×: 0.2E_f

Based on pressuremeter test (PMT)
△: Virgin loading/ ▲: Reloading
Based on plate loading test (PLT)
□: Virgin loading/ ■: Reloading

Fig. A7.5 Relationships between Young's modulus and SPT N-value (Tamaoki et al., 1992)

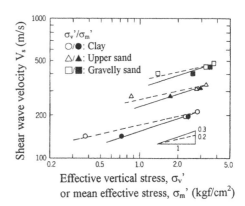

σ_v'/σ_m'
○/●: Clay
△/▲: Upper sand
□/■: Gravelly sand

Fig. A7.7 Change of shear wave velocity during excavation work for K building (Tamaoki et al., 1993a,b)

APPENDIX 8. CASE NO. 8 (OSAKA AMENITY PARK TOWERS)

A8.1 *Construction period*

1992-1996.

A8.2 *Location*

Osaka, Japan.

A8.3 *Type of project*

High-rise buildings (39 stories for the office building with 3 basement floors and 24 stories for the hotel building with 2 basement floors) supported by pile and wall foundations (Fig. A8.1) were constructed.

A8.4 *Ground conditions*

Cast-in-place concrete piles having a diameter of 1.6 to 2.4 m with enlarged tips having a diameter of 4.0 m at maximum were constructed in Holocene/Pleistocene clay, sand and gravel layers (Fig. A8.2). Wall foundations having a maximum plan dimension of 2.0 by 4.4 m were also constructed to support the office building.

A8.5 *Laboratory tests*

Triaxial compression tests were conducted on Pleistocene clays (Oc1 and Oc3 layers in Fig. A8.2), measuring locally axial strains from less than 0.001%. The measured elastic Young's moduli (E_{LDT}) agreed with the results of field seismic surveys (E_{PS}) (Fig. A8.3).

It was one of the most critical geotechnical judgements at the design stage to decide the depth of the bottom ends of the pile foundations to locate at the present depth. The depth was eventually 37 m, with the piles not penetrating the underlying clay layer, existing between about 40 m to about 60 m. The judgement was made based on the sufficiently large small strain stiffness obtained from the seismic survey, which was confirmed by results from triaxial tests using undisturbed samples from the site (Mukabi et al., 1994; Tatsuoka and Kohata, 1995). When based on the results from the conventional geotechnical survey, the stiffness of the clay layer be not sufficiently high to support the piles.

A8.6 *Field tests*

Seismic surveys (Fig. A8.3) and vertical loading tests of a test pile (Fig. A8.4) were conducted before the construction. Monitoring of ground settlements and pile loads were performed during and after the construction (Figs. A8.5 and A8.6).

A8.7 *Prediction*

Three-dimensional elasto-plastic finite element analysis (Fig. A8.7) was conducted to simulate the results of the vertical loading tests of the test pile. The non-linear Young's modulus E_C of the subsoil was evaluated from an empirical relationship, as shown in Fig. A8.8, based on the estimated operating strain level ε and the Young's modulus E_{PS} obtained by the seismic surveys (as confirmed by the triaxial tests). This formulation, shown in Fig. A8.8, was made based on the results of back-analyses of previous case histories. Elasto-plastic frictional-spring elements were used at the interface between the pile and the subsoil.

Because of the good performance of the simulation (Fig. A8.4), similar analytical procedures were employed to predict the settlement of the buildings and the axial loads of the pile foundations.

A8.8 *Agreement between prediction and measurement*

The predicted settlements and pile loads of the office building were compared with the measured values in Figs. A8.5 and A8.6. A good agreement was also observed for the behaviours of the hotel building and its foundation piles.

It should be noted that the effect of the temperature, which was high during the hardening process of the cast-in-place concrete, on the measured pile loads was found to be significant. The measured values in Fig. A8.6 had been corrected for this effect.

Heaving of the subsoil at the bottom of the excavation was observed to have been caused by the upward seepage force due to the lowered free ground water table and the existence of confined ground water under an artesian condition. The predicted values in Figs. A8.5 and A8.6 were obtained by considering this upward seepage force. During construction, the prediction was updated by considering this effect, which was not considered in the original prediction conducted before the construction.

A8.9 *References for case No. 8*

Suzuki,T., Ogaki,S., Kawamura,H., Akino,N., Hokazono,T. and Kishida,H. 1996. FEM analysis and field measurement of foundation of OAP (Osaka Amenity Park) towers (Parts 1 through 7), *Proc. of Annual Conf. of Architectural Institute of Japan*, pp.529-542 (in Japanese).

Akino,N. 1990. Estimation of rigitity of ground and prediction of settlement of building -prediction of immediate settlement of building (part 1)-, *Journal of Struct. Constr. Engng.*, AIJ, No. 412, pp.109-119 (in Japanese).

Akino,N. 1992. Elasto-plastic analysses of settlement of pile foundations -prediction of immediate settlement of building (part 2)-, *Journal of Struct. Constr. Engng.*, AIJ, No. 442, pp.79-89 (in Japanese).

Akino,N. and Sahara,M. 1994. Strain-dependency of ground stiffness based on measured ground settlement, *Pre-failure Deformation of Geomaterials* (Shibuya et al., eds.), Balkema, Vol. 1, pp.181-187.

Mukabi,J.N., Tatsuoka,F., Kohata,Y., Tsuchida,T. and Akino,N. 1994. Small strain stiffness of pleistocene clays in triaxial compression, *Pre-failure Deformation of Geomaterials* (Shibuya et al., eds.), Balkema, Vol. 1, pp.188-195.

Tatsuoka,F. and Kohata,Y. 1995. Stiffness of hard soils and soft rocks in engineering applications, *Pre-failure Deformation of Geomaterials* (Shibuya et al., eds.), Balkema, Vol. 2, pp.947-1063.

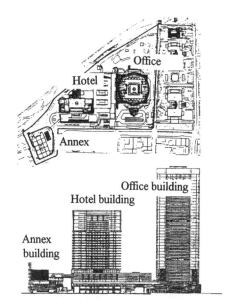

Fig. A8.1 Plan and cross-section of Osaka Amenity Park Towers (Suzuki et al., 1996)

E_{PS}: Young's modulus by seismic surveys
E_{LDT}: Young's modulus by triaxial comp. tests

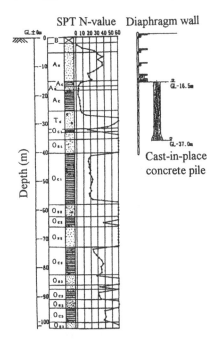

Fig. A8.2 Soil profile and depth of foundations (Suzuki et al., 1996)

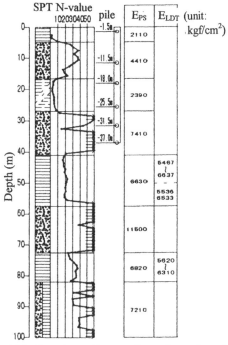

Fig. A8.3 Location of test pile and results of seismic surveys and small strain triaxial compression tests (Suzuki et al., 1996)

143

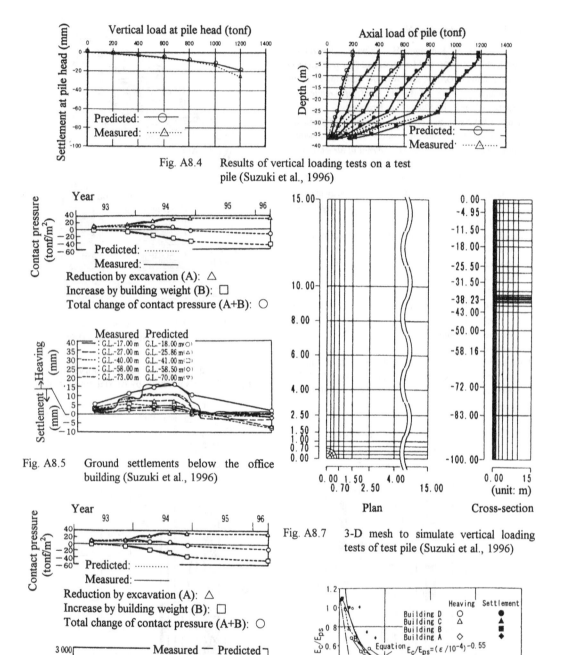

Fig. A8.4   Results of vertical loading tests on a test pile (Suzuki et al., 1996)

Fig. A8.5   Ground settlements below the office building (Suzuki et al., 1996)

Fig. A8.6   Axial loads of piles supporting the office building (Suzuki et al., 1996)

Fig. A8.7   3-D mesh to simulate vertical loading tests of test pile (Suzuki et al., 1996)

Fig. A8.8   Strain level dependency of Young's modulus obtained by back-analyses (Akino and Sahara, 1994)

APPENDIX 9. CASE NO. 9 (KAGOSHIMA PREFECTURAL OFFICE BUILDING COMPLEX)

A9.1 *Construction period*

1993-1996.

A9.2 *Location*

Kagoshima, Japan.

A9.3 *Type of project*

A building complex was constructed, consisting of a prefectural congress building with 7 stories, an administrative office building with 18 stories and a police office building with 9 stories, connected by a common basement floor (Fig. A9.1).

A9.4 *Ground conditions*

Friction-type cast-in-place concrete piles having a diameter of 1.2 to 1.9 m with a tip depth of 63.5 m from the ground surface level (G.L.) were constructed in Holocene/Pleistocene "Sirasu" (local soil with volcanic origin) layers (Fig. A9.2) by means of the reverse circulation drill method.

A9.5 *Laboratory tests*

No laboratory test that is relevant to the estimation of pre-failure deformation of soil layers affecting the settlement of the building was conducted.

A9.6 *Field tests*

Seismic surveys and vertical loading tests of a test pile (Fig. A9.3) were conducted before the construction. Monitoring of ground settlements and pile loads was performed during and after the construction (Figs. A9.8 and A9.9).

A9.7 *Prediction*

Three-dimensional elasto-plastic finite element analysis (Fig. A9.4) was conducted to simulate the results of the vertical loading tests of the test pile. The non-linear Young's modulus E_C of the subsoil was evaluated from an empirical relationship, as previously shown in Fig. A8.8, based on the estimated operating strain level ε and the Young's modulus E_{PS} obtained by the seismic surveys (refer to Appendix No. 8 for details). Values of parameters for the elasto-plastic frictional-spring elements used at

the interface between the pile and the subsoil were determined based on this back-analysis (Figs. 9.5 and 9.6).

A similar analysis as above was conducted (Fig. 9.7) to predict the settlement of the buildings and the axial loads of the pile foundations. Based on the results, it was determined when to connect the construction joint which would be installed between the administrative office building and the police office building for absorbing the differential settlement.

A9.8 *Agreement between prediction and measurement*

The predicted settlements and pile loads of the administrative office building were compared with the measured values in Figs. A9.8 and A9.9. A good agreement was observed for the general behaviours of the building and its foundation piles.

After connecting the construction joint, pumping up of the ground water was stopped. As a consequence, recovery of the ground water table from G.L.-9.5 m to G.L.-3.0 m was observed. During this stage, the behaviours of the building and its foundation piles could be also simulated reasonably, as seen from Figs. A9.8 and A9.9, by considering the effects of additional buoyancy on the piles and those of additional vertical loads on the underlying impermeable soil layers in the analysis.

A9.9 *References for case No. 9*

Kuwabara,K., Enami, A., Akino, N., Kojima, M., Miyazaki,Y., Morita, K., Yamamoto, K., Ooyama, H., Moriwaki, M., Yotsumoto, H., Hamada, K., Kawakami, S. and Suematsu, S. 1996. Non-linear FEM analysis and field measurement of foundation of Kagoshima prefectural office building (Parts 1 through 4), *Proc. of 31st Japan National Conf. of Geotechnical Engineering*, pp.1519-1526 (in Japanese).

Akino,N., Aoki, M., Hanami, K. and Saskajo, S. 1997. Lecture on non-linearity of geo-materials at small strain and its application to the ground deformation problems, No. 6 case histories (4), *Tuchi-to-Kiso*, Journal of JGS, Vol.45, No.12, pp.59-65 (in Japanese).

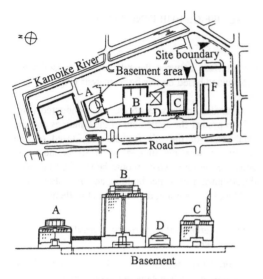

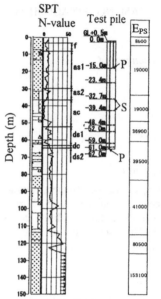

P: Pile displacement transducer
S: Strain gauge for steel bar
E$_{PS}$: Young's modulus by
seismic surveys (tonf/m^2)

A: Prefectural congress building
B: Administrative office building
C: Police office building
D: Auditorium
E: North parking lot building
F: South parking lot building

Fig. A9.1 Plan and cross-section of Kagoshima prefectural office building complex (Kuwabara et al., 1996)

Fig. A9.3 Instrumentation for test pile and results of seismic surveys (Kuwabara et al., 1996)

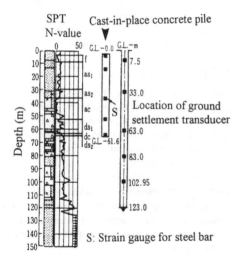

S: Strain gauge for steel bar

Fig. A9.2 Soil profile and depth of foundations (Akino et al., 1997)

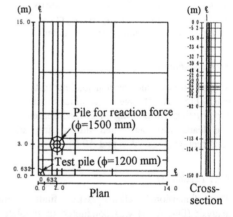

Fig. A9.4 3-D mesh to simulate vertical loading tests of test pile (Kuwabara et al., 1996)

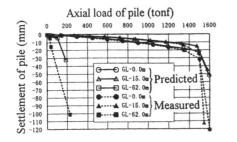

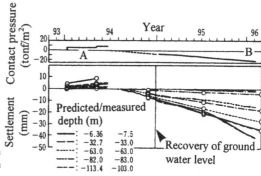

Fig. A9.5 Load and settlement relationships from vertical loading tests on a test pile (Kuwabara et al., 1996)

A: Reduction by excavation
B: Increase by building weight

Predicted values indicated by "○"

Fig. A9.8 Ground settlements below administrative office building (Akino et al., 1997, refer to Fig. A9.7 for location of ground settlement transducer)

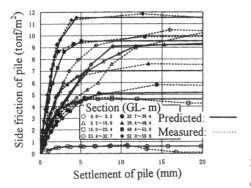

Fig. A9.6 Side friction and settlement relationships from vertical loading tests on a test pile (Kuwabara et al., 1996)

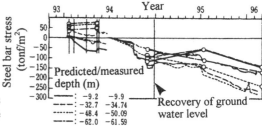

Predicted values indicated by "○".

Fig. A9.9 Vertical loads at pile A below administrative office building (Akino et al., 1997, refer to Fig. A9.7 for location of pile)

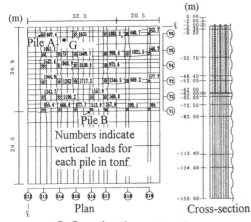

G: Ground settlement transducer
Piles A and B: Instrumented piles (refer to Fig. A9.2)

Fig. A9.7 3-D mesh to simulate subsoil behavior during construction of administrative office building (Kuwabara et al., 1996)

APPENDIX 10. CASE NO.10 (HIGH-RISE BUILDING COMPLEX WITH SPREAD AND PILED FOUNDATION)

A10.1 Construction period

1993-1995

A10.2 Location

Tokyo, Japan

A10.3 Type of project

High-rise building (234 m in height, Fig.A10.1) complex with spread and piled foundation was constructed.

A10.4 Ground conditions

The foundation soil of the construction site consists of Kanto loam (to a depth of about 10 m), Quaternary formation (to a depth of about 90 m) and Tertiary formation. Quaternary formation consists of Edogawa layer, Tokyo gravel layer and Tokyo sand layer in order of depth. Three buildings (T, A and C) are founded on Tokyo gravel layer (Pleistocene) at a depth of 29 m below the ground surface (Fig. A10.1). T-building has a spread foundation, while A and C-buildings are supported by pile foundations (the tip of piles also stands on the Tokyo gravel). These buildings have a common foundation but their super structures are connected by expansion joints. At first, buildings T and C were constructed simultaneously. After that, the building A was constructed.

A10.5 Laboratory test

At the design stage, drained and undrained triaxial compression tests and undrained cyclic triaxial tests were performed on undisturbed samples (5 cm in diameter, 10 to 12 cm high) obtained by rotary-type triple tube method from five depths (Fig.A10-2). The axial strain was measured by using Local Displacement Transducer. The initial shear modulus (G_0) at a strain level of 2 to $5×10^{-5}$ was measured at four levels of confining stress smaller than the in-situ overburden stress. The stress dependency of G_0 was shown in Fig.A10.3. The G_0 obtained in laboratory tests was 40 to 80 % of that calculated from shear wave velocity observed at the same site by performing P-S wave logging tests (Fig.A10-2). The difference of G_0 between them are mainly due to the sample disturbance and the difference of the confining stress between laboratory tests and field tests.

The confining stress used in undrained cyclic triaxial test for obtaining $G \sim \gamma$ relation was set equal to the effective overburden pressure at the sampling depth.

In the present study, the G_0 was determined from V_s and the strain dependency of G was based on the undrained cyclic triaxial test results (Fig. A10.4).

A10.6 Field tests

At the design stage, P-S wave logging tests were performed to measure the shear wave velocity (V_s) (Fig. A10-2). The rebound and settlement behavior were monitored at 5 points by using double-tube type settlement transducers. The location of these measurements were also shown in Figs. A10.1 and A10.3 as plan and vertical section, respectively.

In order to take into account of the effect of the change of the ground water table at each point on settlement measurement, ground water table during construction was also monitored.

A10.7 Comparison of test results with full-scale behavior during construction

(a) Outline of the method for estimation of rebound and settlement

The procedure for estimating the rebound and settlement is shown in Fig.A10.5 associated with the diagrams indicating the methods to determine soil constants for estimation. The Steinbrenner's multi-layers elastic method was used as a numerical method.

The effective stress changes at the middle depth of each layer due to the excavation, building construction and change of the ground water table were first calculated. Then the decrease of G_0 due to stress reduction was estimated. The Poison's ratios of cohesive soil and sandy soil are postulated to be 0.4 and 0.3, respectively.

The distribution of the ground water table was determined based on the observation results.

(b) Comparison of the vertical displacement of the ground during excavation between estimation and the observation

Fig. A10.6 shows the distribution of vertical displacement in depth during the excavation. Fig.A10.7 indicates the plan of the vertical displacement at the depth of 42 meters below the ground surface. It can be seen that the rebound of the building-C is observed at first, after that the building-T, having larger depth of excavation, shows the increase of the rebound. The estimation of the vertical movement of the ground by the present method during the construction procedures well corresponds to the field performance. These results indicate that the settlement and rebound during excavation and building construction can be estimated by the present study.

A10.8 References for case No.10

Aoki, M., Kakurai, M., Ishii, O. and Ishihara K.1997. Field measurements and predictive estimates of ground heave and settlement of a bearing stratum supporting the spread foundation of a skyscraper, AIJ. J. Technol. Des. No.5, pp.80-84.

Goto, S., Tatsuoka, F., Shibuya, S., Kim, Y-S. and Sato, T. 1991. A simple gauge for local small strain measurements in the laboratory, Soils and Foundations, Vol.31, No.1, pp.169-180.

Iwasaki, T. and Tatsuoka, F. 1977. Effects of grain size and grading on dynamic shear moduli of sands, Soils and Foundations, Vol.17, No.3, pp.19-35.

Aoki, M., Shibata, Y., Maruoka, M. and Tanimura, K. 1990. Heave and rebound of foundation soil during the underground excavation, AIJ. J. Annual Meeting, pp.1649-1652 (in Japanese).

Ishihara, K. 1976. Fundamentals of soil dynamics, Kashima Publication, pp.23-25 (in Japanese).

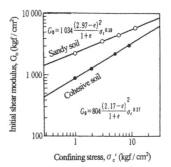

Fig. A10.3 Correlation between G_0 and σ_c' (Aoki et al, 1997)

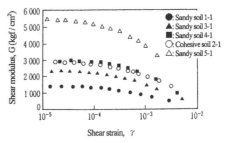

Fig. A10.4 Strain dependency of shear modulus (Aoki et al, 1997)

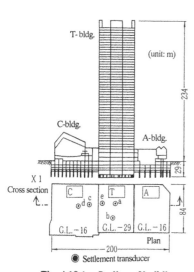

Fig. A10.1 Outline of buildings (Aoki et al, 1997)

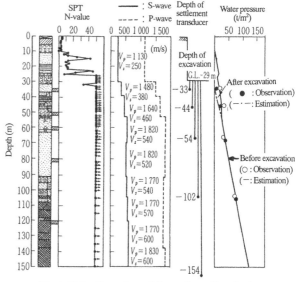

Fig. A10.2 Ground condition and sampling depth (Aoki et al, 1997)

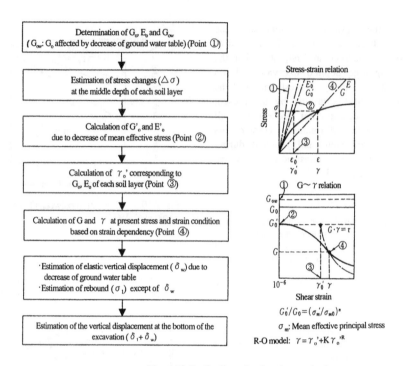

Fig. A10.5   Outline of estimation method
(Aoki et al, 1997)

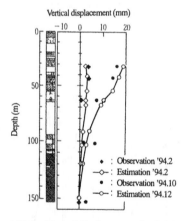

Fig. A10.6   Vertical displacement distribution
at point (a) (Aoki et al, 1997)

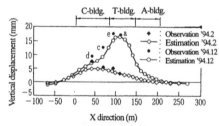

Fig. A10.7   Plan of vertical displacement
(cf. Fig. A10.1: X1 Cross section, G.L.–42m)
(Aoki et al, 1997)

APPENDIX 11. CASE NO.11 (A SHIELD TUNNEL)

A11.1 *Construction period*

1990-1993.

A11.2 *Location*

Tokyo, Japan

A11.3 *Type of project*

A shield tunnel was constructed adjacent to an existing subway in Koutou-district in Tokyo, which is a famous low land. The diameter and length of the new tunnel are 5.55m and 1,342m respectively.

A11.4 *Ground condition*

30m thick alluvial layers named the upper (A_{s1}) and the lower (A_{s2} and A_{c2}) Yuraku-cho layers are deposited beneath the subsoil layer (B). The lower clay layer (A_{c2}) is very soft with the N values of 0 to 4. The lower sand layer (A_{s2}) is with N values of 20 on average. Beneath that, Nana-go layers (A_{s3} and A_{c3}) are deposited. The A_{s3} layer is a rather hard sand layer with N values of 20 to 35 (Fig.A11.1).

A11.5 *Laboratory tests*

In order to model the consolidation settlement for clay layers after the cyclic loading at the tail void, volumetric strains of K_0-consolidated clay specimen were measured under drained condition after un-drained cyclic loading using tri-axial compression test apparatus. These tests results were used to define the relationship between the volumetric strain, or the shrinkage ratio due to consolidation and the un-drained shear strain due to soil disturbance by tail void closure.

A11.6 *Field tests*

The characteristics of deformation of sand layers were investigated using PS wave exploration tests and standard penetration tests. The non-linear relations were estimated for sand layers on the basis of the non-linear modelling proposed by Tatsuoka

and Shibuya (1992; Fig.A11.2). The V_s values for the A_{s1}, A_{s2} and A_{s3} are 140, 205 and 300(m/sec) respectively.

A11.7 Numerical model

A numerical procedure proposed by Mori and Akagi (1980, 1983) was improved to be used for the shield tunnel construction, taking account of an immediate settlement due to tail void closure and the consolidation settlement due to the dissipation of excess pore water pressure (Fig.A11.3).

A11.8 Comparison of measurements and numerical results

The prediction analysis was carried out at the section where the shield tunneling is mostly close to the existing subways and another tunnel. The FEM mesh is shown in Fig.A11.4, and the soil parameters are summarized in Table 1. The predicted results with considering the non-linearity were obtained, together with the results assuming the linear elasticity. The deformation was measured by the settlement gauges and inclinometers. The settlement becomes constant and 18mm at point 1, located 1m above the machine clown, when 15 days passed after the tail void closure (Fig.A11.5). The comparison between measurements and computations is shown in Fig.A11.6. The prediction for the case without considering the non-linearity of G-γ relation underestimated the settlement except for the point just above the tunnel clown. On the other hand, the case with considering the nonlinearity of G-γ could simulate the observed results better than the above case assuming linear elasticity.

A11.9 *References for case No.11*

Akiba, Y., Okadome, K. and Sakajo, S. 1999. Application of non-linearity of ground at small strain to a shield tunneling, Proc. of 2nd Int. Conf. on Pre-Failure Deformation Characteristics of Geomaterials, pp.827-831.

Akino, N., Aoki, M., Hanami, K. and Sakajo,S. 1997. Lecture on application of non-linear at small strain of soil materials to the ground deformation problems, No.6 case histories (4), Tuchi-to-Kiso, Journal of JGS, Vol.45, No.12, pp.59-65 (in Japanese)

Mori, A and Akagi, H. 1980. Consolidation settlement due to the soil disturbance in shield tunneling, The Journal of Tunnel and Underground, Vol.11, No.6, pp.563-567 (in Japanese).

Mori, A and Akagi, H. 1983. Consolidation phenomena due to the soil disturbance in un-drained shear, Proc. of Japan Society of Civil Engineers, No.335, pp.117-125 (in Japanese).

Okazaki, K., Iwasaki, M. and Okadome, K. 1992.Prediction of ground deformation considering consolidation settlement of clay induced by a shield construction (2), Proceedings of the 27th annual conference of Japanese Geotechnical Society, pp.2085-2086 (in Japanese).

Sakajo, S., Yoshimura, T. and Kamimura, M. 1996. Analytical and geotechnical consideration on ground settlement induced by tail void closure of shield tunnel construction, Proc. of International Symposium on Geotechnical Aspects of Underground Construction in Soft Ground, London, UK, pp.585-590.

Tatsuoka, F. and Shibuya, S. 1992. Deformation characteristics of soils and rocks from field and laboratory tests, Report of Institute of Industrial Science, The University of Tokyo, Vol.37, No.1, 1992.

Uto, K. 1967. Investigation of foundations for underground structures, Kanto Branch of Society of Japan Civil Enginees (in Japanese).

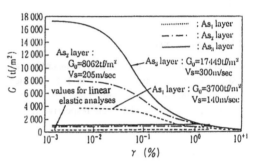

Fig.A11.2 Stress-strain relationship estimated for the sand layers of A_{s1}, A_{s2} and A_{s3} (Akiba et al., 1999)

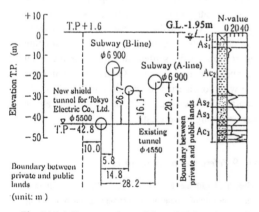

Fig.A11.1 Cross section of the existing subways and a new shield tunnel (Akiba et al., 1999)

1.Initial stress analysis

2.Immediate settlement due to tail void closure

i) $a = \dfrac{F_{sc1}}{F_{max}}$ or $a = \dfrac{\delta_{sc1}}{\delta_{max}}$

F_{sc1} : Excavation force
F_{max} : Maximum excavation force
δ_{sc1} : Enforced displacement
δ_{max} : Maximum displacement
α : Reduction factor
(set to 0.15 in the present analysis)

① Excavation ②Enforced displacement

3.Consolidation settlement

ii) Final settlement
$\delta = \Delta\delta_1 + \Delta\delta_2$
Displacement fixed

Fig.A11.3 Numerical model for the shield tunnel construction (Akiba et al., 1999)

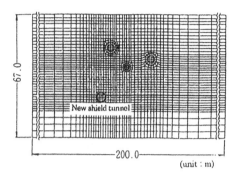

(unit : m)

Fig.A11.4 The FEM mesh (Akiba et al., 1999)

Table A11.1 Soil parameters used in prediction
analysis (Akiba et al., 1999)

Soil	Unit weight γ_t (tonf/m³)	E_{50} (tonf/m³)	Poisson's Ratio
(B)	1.90	700*	0.33
Yuraku-chou layer(A_{Su})	1.89	770	0.33
Yuraku-chou layer(A_{C2})	1.64	640	0.45(0.33**)
Yuraku-chou layer(A_{S2})	1.95	2100	0.33
Nanago layer(A_{S3})	1.95	2700	0.33
Nanago layer(A_{C1})	1.75	2400	0.45(0.33*)
Edogawa layer(D_S)	2.00	3500	0.33

*:E=70·N (Uto, 1967) , where N is the SPT blow count
**:( ) is the value for consolidation analysis based on
Mori and Akagi (1980)

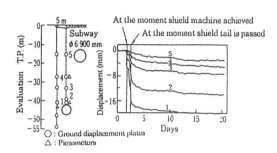

Fig.A11.5 Measured settlements and elapsed
time relationships
(Akiba et al., 1999)

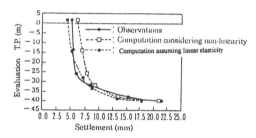

Fig.A11.6 Comparison between measurements
and computations
(Akiba et al., 1999)

APPENDIX 12. CASE NO.12 (LNG TANK AT CHITA)

A12.1 *Construction period*

1993-1997.

A12.2 *Location*

Aichi, Japan.

A12.3 *Type of project*

A cylindrical excavation supported by a diaphragm wall was made to construct a huge in-ground LNG tank with a capacity of 160,000 *kl*. The excavation is 70m in diameter and 53m in depth (Fig. A12.1).

A12.4 *Ground conditions*

The ground consists mainly of alternative layers of Neogene sand and cemented silt (Tokoname layers; denoted as T_s and T_m in Fig. A12.1, respectively). The stratification formed a slope structure and aligned along a NE-SW axis with a dip of about 5 degrees toward the NW.

The diaphragm wall with a thickness of 1.6 m was embedded in an impermeable layer at a depth of 118 m from the ground surface before excavation. During excavation, deep wells were employed to maintain the ground water level inside the diaphragm wall at a depth of final excavation.

A12.5 *Laboratory tests*

During excavation, block samples were obtained from layers of T_{m1}, T_{m2}, T_{s1} and T_{s2} (Fig. A12.1). Isotropically consolidated drained triaxial compression tests with local measurements subjected to single stage monotonic loading (denoted as SST), multi stage loading under different confining stress levels (MST), or loading with small amplitude unload/reload cycles (CST), were conducted on these samples. Dynamic measurements with ultrasonic log and bender elements were also performed on the samples (Table A12.1).

Dependency of initial Young's modulus E_0, defined at an axial strain ε_a of 0.001 %, on the effective mean principal stress σ'_m is shown in Fig. A12.2. The E_0 values of the cemented silt layer showed a stress level dependency at σ'_m smaller than 0.2 MPa, while they were almost constant at σ'_m larger than 0.2 MPa. Since this threshold stress level corresponds to the effective overburden stress, it was estimated that the reduction of confining stress due to sampling caused opening of potential cracks, resulting in decrease in the stiffness of samples under low confining stress. On the other hand, the E_0 values of the sand layer showed a stress level dependency under the tested conditions with σ'_m between 0.1 and 1.0 MPa.

Relationships between secant shear modulus G_{sec} normalized by the initial shear modulus G_0 and shear strain γ is shown in Fig. A12.3. In evaluating G_{sec}, G_0 and γ from the values of E_{sec}, E_0 and ε_a measured in the triaxial tests, the Poisson's ratio ν was assumed to be 0.35. Different tendencies were observed on the effects of strain levels on degradation of stiffness between the cemented silt and sand layers.

The above deformation characteristics were approximately formulated as shown in the figures, which were employed in the elastic analysis as mentioned later. Note that no significant anisotropy was observed in the deformation characteristics at ε_a < 0.4 % of vertically and horizontally trimmed specimens, while the peak and residual strengths of the vertically trimmed specimen of the cemented silt were larger than those of the horizontally trimmed specimens (Fig. A12.4).

The shear wave velocities V_s measured with ultrasonic log and bender elements on isotropically consolidated sand specimens are shown in Fig. A12.5. The V_s values measured with ultrasonic log were in general larger than those measured with bender elements, due possibly to combined effects of the heterogeneity of specimen and different frequency ranges of excitation employed for these tests (about 50 kHz for ultrasonic log and 5 to 10 kHz for bender elements), and also to partial disturbance of specimens caused by insertion of bender elements. As listed in Table A12.2, these V_s values were smaller than the results of in-situ seismic survey conducted before excavation, when compared at the same confining stress. However, considering the possibility that in-situ seismic survey may over-estimate V_s values of the sand layers which were sandwiched with stiffer cemented silt layers, it was estimated that the block samples retrieved from the sand layers were not subjected to any major disturbance.

A12.6 *Field tests*

Before excavation, a down-hole seismic survey was performed (Table A12.2). During excavation of the sand layers, a cross-hole seismic survey using two bore holes drilled from the excavated surface and bender element tests on trimmed samples for block sampling were performed (Table A12.1).

The V_s values measured in-situ during excavation are shown in Fig. A12.6. Those measured with bender elements were smaller than those measured with the cross-hole seismic survey, due possibly to reduction in the lateral stress during trimming of the block samples.

A12.7 *Comparison of tests results with full-scale behavior during excavation*

An axi-symmetric equivalent linear elastic analysis was conducted to simulate the full-scale behavior during excavation. Dependencies of the deformation characteristics of soils on confining stress and strain level were modeled based on the triaxial test results (Figs. A12.2 and A12.3), and the coefficient of earth pressure at rest K_0 was assumed to be 1.0.

The calculated rebound of soils below excavation is compared with the measured one in Fig. A12.7. Near the center of excavation (at R= 3 m in the figure), the values of calculated rebound agreed well with the those of measured one. On the other hand, near the diaphragm wall (at R= 28 m in the figure), they were significantly under-estimated by the analysis, due possibly to difficulties in modeling the interface between the diaphragm wall and surrounding soils.

Distribution of calculated maximum shear strain is shown in Fig. A12.8. It was on the order of 10^{-3} to 10^{-4}, which resulted in the degradation ratio of stiffness G_{sec}/G_0 (Fig. A12.3) about 0.55 through 0.75 in the affected area.

Radial displacements of the diaphragm wall were compared in Fig. A12.9. Good agreement between the measured and simulated results were obtained.

A12.8 *References for case No.12*

Mizuno, N., Nakamura, J., Saito, I. and Takano, T. 1998a. Deformation properties of ground and loadings diaphragm wall during a largescale excavation of sedimentary soft rock, Journal of Construction Management and Engineering, JSCE, No.595/VI-39, pp.1-15 (in Japanese).

Mizuno, N., Nakamura, J., Watanabe, K., Saito, I. and Nishio, S. 1998b. Deformation properties of Neogene sedimentary Tokoname layer, Journal of Geotechnical Engineering, JSCE, No.603/III-44, pp.179-190 (in Japanese).

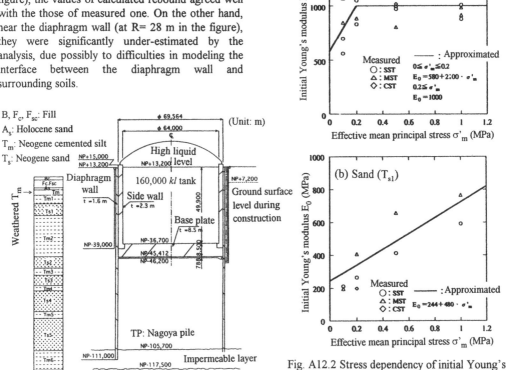

Fig. A12.2 Stress dependency of initial Young's modulus (Mizuno et al., 1998a)

Fig. A12.1 Cross section of LNG tank at Chita (Mizuno et al., 1998a, b)

Table A12.1 Schematic procedures to measure shear wave velocities of sand layers (Mizuno et al., 1998b)

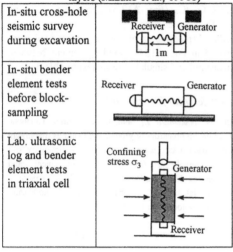

In-situ cross-hole seismic survey during excavation	Receiver Generator 1m
In-situ bender element tests before block-sampling	Receiver Generator
Lab. ultrasonic log and bender element tests in triaxial cell	Confining stress σ₃ Generator Receiver

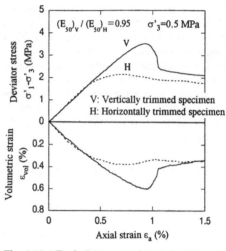

Fig. A12.4 Typical stress-strain relationships of cemented silt (Mizuno et al., 1998b)

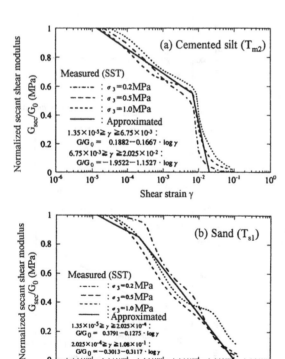

Fig. A12.3 Degradation of normalized secant shear modulus (Mizuno et al., 1998a)

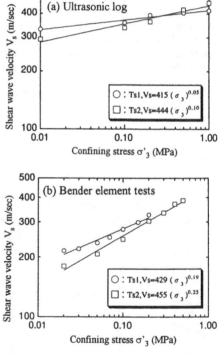

Fig. A12.5 Shear wave velocity measured on isotropically consolidated sand specimen (Mizuno et al., 1998b)

Table A12.2 Comparison of shear wave velocities measured with different methods (Mizuno et al., 1998a)

Method	V_s of T_{s1} at confining stress of 0.2 MPa	V_s of T_{s2} at confining stress of 0.5 MPa
In-situ seismic survey	430 m/sec	470 m/sec
Lab. bender element tests	330 m/sec	390 m/sec
Lab. ultrasonic log	390 m/sec	420 m/sec

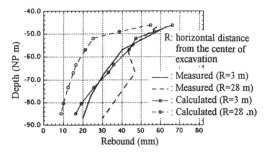

Fig. A12.7 Rebound of soils beneath excavation (Mizuno et al., 1998a)

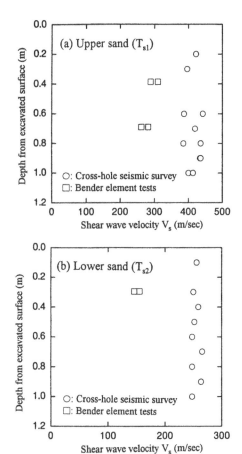

Fig. A12.6 Shear wave velocity measured in-situ during excavation (Mizuno et al., 1998b)

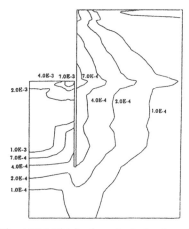

Fig. A12.8 Distribution of calculated maximum shear strain (Mizuno et al., 1998a)

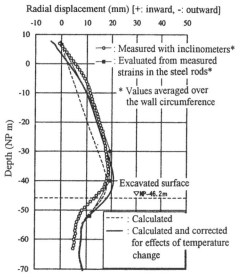

Fig. A12.9 Radial displacement of diaphragm wall (Mizuno et al., 1998a)

APPENDIX 13. CASE NO.13 (ROCK-FILL DAM)

A13.1 *Construction period*

1987-1992.

A13.2 *Location*

Tottori, Japan.

A13.3 *Type of project*

A rock-fill dam with a central impervious core zone was constructed. It has dimensions of 46.2m in height, 237.0m in length, 651,600m³ in volume, and 2,006m³ in reservoir volume. The cross-section of this dam is shown in Fig. A13.9.

A13.4 *Ground conditions*

The ground consists mainly of lapilli-tuff and lapilli-breccia, partly including nearly horizontal layers or lenses of tuff, pumice-tuff or lake bed sediment (mud stone). Fig. A13.1 shows the borehole log and the physical properties of the foundation materials.

A13.5 *Laboratory tests*

In an artificial model soil foundation, PS-logging, bore hole lateral loading tests, plate loading tests and triaxial compression tests (measuring axial strains locally with LDTs) were carried out. The model foundation was made in a rectangular solid concrete tank of 5m long, 5m wide and 3m deep by compacting a natural sand with use of a vibrating compactor. The model foundation consisted of 30 layers. Each sub-layer was made by compacting a loosely spread layer with a thickness of 15cm into a layer with a thickness of 10cm. Fig. A13.2 shows the grain size accumulation curve of the material. The results from field density tests by the sand replacement method showed a high homogeneity in the density and water content. The wet density was 1.83 g/cm³; the water content was 18.0 %; and D-value (=ρ_d/ρ_{dmax}) was 89.0 % on average.

A13.6 *Young's Modulus Formulation*

In order to evaluate the deformation properties, the layer depth was selected as the parameter controlling the soil stiffness rather than confining pressure, because of a convenience of its use in practice and a difficulty of accurately evaluating the confining pressures in the subsoil. As shown in Fig. A13.3, a highly linear relationship can be seen

between the elastic modulus obtained from downhole PS-logging and the depth. The relationship was best-fitted by the following equation;

$$E = E_0 + m \cdot d \tag{1}$$

Where E is the Young's modulus at a depth of d (m); E_0 and m are the elastic modulus at 0 m depth and a constant, respectively. The values of E_0 and m thus obtained are 1163 (kgf/cm²) and 498 (kgf/cm²/m), respectively.

It is known that when relevant set of data obtained from field tests and laboratory tests are plotted in a single figure showing the relationship between the stiffness and strain level, all of them fall within a reasonably narrow band (*e.g.*, Tatsuoka et al., 1991). In order to eliminate the influence of depth, the parameter "elastic modulus ratio" was introduced, which will herein be denoted as E'. It was defined here as the ratio of the elastic modulus data obtained from various field loading tests and the corresponding one by PS-logging (expressed as E_{PS}) at each depth. Since the total depth of the artificial foundation is only 3 m, it was very difficult to investigate the data separately for many different layers. Then it was tried to classify the data into two groups, shallower and deeper ones.

The relationships between E' and strain level in both shallower and deeper parts are shown together in Fig. A13.4. E_{PS} was obtained from the mean value of the data by downhole and crosshole methods. The results indicate negligible effects of depth. The bold line represents the fitted relation expressed as follows;

$$E' = 1 \qquad (\varepsilon \leqq 10^{-5}) \tag{2a}$$
$$E' = -k (\log \varepsilon + 5)^{0.20} + 1 \quad (\varepsilon > 10^{-5}) \tag{2b}$$

where ε is the strain level and k is a positive constant, the only unknown quantity. The k value can be obtained by substituting the E' and ε values obtained from the PS-logging and plate loading test (or bore hole lateral loading test) results into Eq. (2b). In this case, k was calculated to be 0.73. The value of ε at the boundary between Eqs. (2a) and (2b) is assumed to be constant (= 10^{-5}), based on the largest strain level in usual PS-logging. This boundary strain value may change to some extent for other soils.

A13.7 *Field loading tests*

Fig. A13.5 shows the borehole log and some

physical and strength properties of the subsoil materials. Fig. A13.6 shows the relationship between the elastic modulus and depth obtained from downhole PS-logging. It can be seen that Eq. (1) is in a good agreement with the results from the PS logging, except for two data points in the shallow layer. This exception would be due to much different material properties from other layers. Here the values of E_0 and m were calculated to be 2344 (kgf/cm^2) and 160 (kgf/cm^2/m), respectively.

The obtained relationship between E' and strain in the field is shown in Fig. A13.7. Here the confining pressures employed in the triaxial compression tests are equal to about one-half of the product of mean wet density and depth. The constant k was calculated to be 0.64 in this case. The bold solid line represents the calculated relationship of Eqs. (2a) and (2b), which represents the tendency of the test results well. An underestimation of stiffness at strains smaller than about 10^{-4} is not a serious problem in practical applications because it leads to safer design about strain problem.

Now the effectiveness of the above equations will be examined by analyzing the deformation of soil or soft rock foundations in the field. Further investigation is needed to find whether the stiffness of sedimentary soft rock becomes rather constant at deeper depths in general cases (*e.g.*, Ochi et al., 1993) and whether this trend can be applied to harder rock foundations.

A13.8 *Comparison between results of FEM settlement analysis and in-situ measurements*

In this analysis, the foundation was treated as a non-linear elastic body having the elastic modulus expressed by Eqs. (1) and (2). The dam body was treated as a non-linear elastic body or a non-linear elasto-plastic body. For comparison, another analysis was conducted assuming that the soil and rocks being a linear elastic body. Fig. A13.8 shows a flow chart of the analysis. The details are as follows;

1) Before the start of embanking, the value of E_{init} (the initial elastic modulus) in every part of the foundation is calculated as a function of depth using Eq. (1). The elastic modulus at a depth d_k is denoted by "E_{initdk}".
2) When the first layer of the dam body is embanked, the load mobilizes a strain ε_1 in the foundation. The value of ε_{1dk} (ε_1 at d_k) is calculated by using E_{initdk}.
3) E'_{1dk}, the elastic modulus ratio when the loading of the first layer is completed, is calculated based on ε_{1dk} using Eq. (2).

4) When the second layer of dam is embanked, the added and existing loads mobilizes a new strain ε_2 in the foundation. The value of ε_{2dk} (ε_2 at d_k) is calculated from $E_{1dk} (= E'_{1dk}{}^* E_{initdk})$.
5) E'_{2dk} is calculated based on ε_{2dk} using Eq. (2). E'_{2dk} is regarded as the elastic modulus ratio when the loading of the second layer is completed.
6) Similar procedures are repeated until the completion of the embanking.

In order to evaluate the effectiveness of this model, FEM embanking analyses based on this model and the "linear model" were carried out on a 46.2 m high rockfill dam. The settlements along the dam axis calculated by both methods were compared. Fig. A13.9 shows the FEM mesh. The dam and foundation was divided into seven different material zones, consisting of 4 for the dam body, 2 in the foundation, and 1 for the inspection gallery. The FEM model had 328 elements and 354 nodes. The horizontal displacements at the nodes along the left and right boundaries in the foundation were fixed, while the vertical displacements were allowed to occur. Both horizontal and vertical displacements were fixed at the bottom boundary. The foundation was set up in the first step before the embanking process, and the dam body was filled up in eight steps.

Three constitutive relations for the soil properties were used in the analyses. The first one was the linear elastic model, applied to the foundation in the analysis denoted as "linear model". Another one was the non-linear elastic model as explained before. The other one was an elasto-plastic model proposed by Drucker and Prager, which was applied to the dam body in both analyses. Table A13.1 shows the parameters used in the analyses. The elastic moduli of the foundation were obtained based on the results from plate loading tests and bore hole lateral loading tests. The values of the dam body were obtained from triaxial compression tests. Poisson's ratios were assumed empirically. Cohesions and angles of shear resistance were determined based on the triaxial compression test results. Wet densities of the foundations were specified to be zero to obtain only settlements due to stress increments by embankment loading. The constants m, E_0 and k are the ones used in Eqs. (1) and (2).

The other analytical conditions for the "linear model" and this non-linear model were the same, except that the latter took the elastic modulus of the foundation to be not constant. In the right of Fig. A13.9, the calculated settlements along the dam axis obtained by the both analyses and the observed one are shown. With the non-linear model, the settlement in the foundation increases exponentially as the depth becomes smaller, while it increases linearly with the linear model. At the surface of the foundation layer it reaches 2.1 cm with the non-linear model, about one-seventh that of the linear model (= 14.9 cm). The former model estimates fewer settlements than the latter one for the whole depth of foundation. It should be noted that the difference between calculated settlements from the two analyses is influenced by the adopted elastic moduli in the "elastic model", the depth of the foundation modeled in the analyses, and other factors. As a conclusion, these results suggest that this non-linear model can better simulate the settlement of soil or soft rock foundation.

A13.9 *References for case No.13*

Goto, S., Tatusoka, F., Shibuya, S., Kim, Y. S. and Sato, T. 1991. A simple gauge for local small measurements in the laboratory". Soils and Foundation, Vol.31, No.1, pp.169-180.

Ochi, K., Tsubouchi, T. and Tatsuoka, F. 1993. Deformation characteristics of sedimentary soft rock examined by excavation of deep shaft and field tests, Journal of Geotechnical Engineering, JSCE, No.463/III-22, pp.143-152.

Siddiquee, M. S. A., Tatsuoka, F., Inoue, A., Kohata, Y., Yoshida, O., Yamamoto, Y. and Tanaka, T. 1995. Settlement of a pier foundation for Akashi-Kaikyo Bridge and its numerical analysis, Proceedings of International Workshop on Rock Foundation of Large-Scaled Structures, pp.413-420.

Tagashira, H. and Yasunaka, M. 1998. Built-up analysis of a fill dam taking nonlinear elastic characteristics of its foundation into consideration, Proceedings of 9th Conference on JSDE, pp.19-21.

Tagashira, H., Kohgo, Y. and Asano, I. 2000. A simplified method of built-up analysis of fill-type dams considering nonlinear elastic characteristics of their foundation, Proceedings of 35th Annual Conference of JGS, pp.325-326 (in Japanese).

Tatsuoka, F. and Shibuya, S. 1991. Deformation characteristics of soils and rocks from field and laboratory tests, Proceedings of 9th Asian Regional Conference on SMFE, Vol.2, pp.101-177.

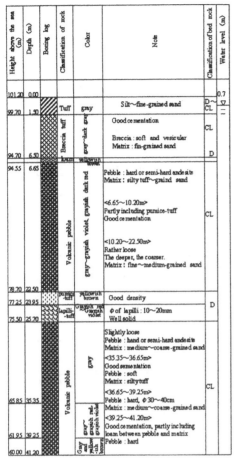

Fig. A13.1　Borehole log and physical properties of the dam foundation (Tagashira et al., 1998)

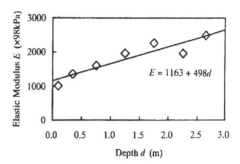

Fig. A13.3　Relationship between the elastic modulus and depth obtained from downhole PS-logging (Tagashira et al., 1998)

○ PS logging(Down hole, U)　● PS logging(Down hole, L)
◎ PS logging(Cross hole, U)　◇ PS logging(Cross hole, L)
□ Plate loading test(U)　　　× Plate loading test(L)
△ Bore hole test (66mmφ, U)　▲ Bore hole test (66mmφ, L)
▽ Bore hole test (60mmφ, U)　▼ Bore hole test (60mmφ, L)
○ LDT (σ_3=0.3, U)　□ LDT (σ_3=0.3, L)　▽ LDT (σ_3=0.5, U)
× LDT (σ_3=0.5, L)　◇ LDT (σ_3=0.7, U)　+ LDT (σ_3=0.7, L)
(*) U : the upper part,　L : the lower part
Bore hole test : Bore hole lateral loading test
LDT : LDT traixial compression test (Unit of σ_3 : ×98kPa)

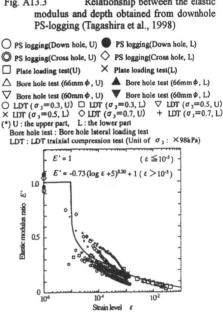

Fig. A13.4　Relationship between elastic modulus ratio and strain level at the artificial foundation (Tagashira et al., 1998)

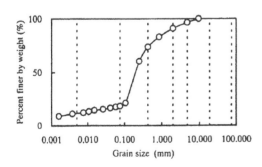

Fig. A13.2　Grain size accumulation curve of the artificial foundation material (Tagashira et al., 1998)

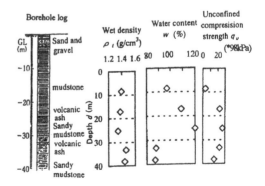

Fig. A13.5　Borehole log and the physical properties of the foundation materials (Tagashira et al., 1998)

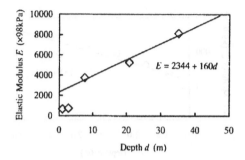

$$E = 2344 + 160d$$

Fig. A13.6 Relationship between the elastic modulus and depth obtained from downhole PS-logging (Tagashira et al., 1998)

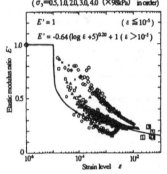

○△◇ PS logging (Depth : 3.5~12, 12~30, 30~40 (m) in order)
①~⑤ Bore hole lateral loading test (Depth : 8, 16, 24, 32, 40 (m) in order)
○◇△□× LDT traixial compression test
(σ_3=0.5, 1.0, 2.0, 3.0, 4.0 (×98kPa) in order)

$E' = 1$ ($\varepsilon \leqq 10^{-5}$)
$E' = -0.64 (\log \varepsilon + 5)^{0.20} + 1$ ($\varepsilon > 10^{-5}$)

Fig. A13.7 Relationship between elastic modulus ratio E' and strain level for the foundation (Tagashira et al., 1998)

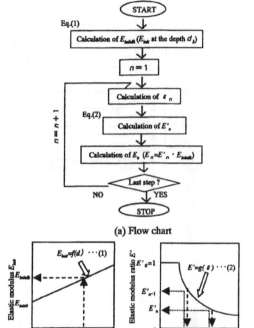

(a) Flow chart

$E_{init} = f(d)$ ···(1)

$E' = g(\varepsilon)$ ···(2)

(b) Relationship between elastic modulus and depth

(c) Relationship between elastic modulus ratio and strain level

Fig. A13.8 Procedure of the analysis (Tagashira et al., 1998) (Tagashira et al., 2000)

Table A13.1 Parameters used in the analyses of the dam (Tagashira et al., 2000)

	Elastic Modulus E (*98kPa)	Poisson's Ratio ν	Cohesion c (*98kPa)	Angle of Shear Resistance ϕ (°)	Wet density ρt (t/m³)	m	E_0 (*98kPa)	k
<Dam body>※ (Common in both models)								
Core	420	0.45	0.5	16.0	1.98	—	—	—
Filter	360	0.30	0.1	41.0	2.13	—	—	—
Transition	650	0.35	0.1	37.0	2.08	—	—	—
Rock zone I	650	0.30	0.1	41.0	2.05	—	—	—
Rock zone II	660	0.30	0.1	43.0	2.02	—	—	—
<Foundation> (Linear model)								
Upper layer	2,400	0.35	1.5	38.0	0.0	—	—	—
Lower layer	2,900	0.35	1.5	42.0	0.0	—	—	—
Lake bed sediments	1,900	0.35	1.5	40.0	0.0	—	—	—
Lower part of Upper layer	3,100	0.35	1.5	38.0	0.0	—	—	—
Inspection gallery	210,000	0.15	10.0	50.0	2.45	—	—	—
(Non-linear model)								
Upper layer	2,400	0.35	1.5	38.0	0.0	0.0	19,765	0.542
Lower layer	2,900	0.35	1.5	42.0	0.0	263.8	34,051	0.545
Lake bed sediments	1,900	0.35	1.5	40.0	0.0	0.0	28,033	0.539
Lower part of Upper layer	3,100	0.35	1.5	38.0	0.0	991.8	4,270	0.521
Inspection gallery	210,000	0.15	10.0	50.0	2.45	0.0	2,100,000	0.000

※ Drucker-Prager's model is applied.

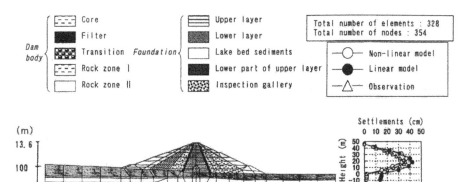

Fig. A13.9   Meshing for the FEM analyses and comparison of the calculated settlements and
observed one (Tagashira et al., 2000)

APPENDIX 14. CASE NO.14 (ROAD EMBANKMENT AT NIIGATA)

A14.1 *Construction period*

1980-1982 (original embankment) and 1996-1997 (replacement of abutment backfill with expanded polystyrene, EPS).

A14.2 *Location*

Niigata, Japan.

A14.3 *Type of project*

The construction of embankment and bridge for Hokuriku Expressway is described in Fig. A14.1. After applying and removing preloads (Table A14.1), the original embankment with a height of about 8 m was filled on soft ground about 60 m thick (Fig. A14.2). This embankment exhibited a total settlement of about 3.1 m for a period of 13 years after the expressway began operating in 1983. During this period, overlay of the road surface and repair of the bridge facilities had to be occasionally made to accommodate them to such a large residual settlement.

In order to reduce embankment load, the backfill of the bridge abutment was partly replaced with expanded polystyrene (hereafter referred as EPS) in 1996 and 1997 (Figs. A14.3 and A14.4; and Table A14.1).

A14.4 *Ground conditions*

The ground consists of alternative layers of Holocene alluvial clay and sand (denoted as Ac-1 to Ac-3 and As-1 to As-2 in Fig. A14.2, respectively) with SPT N-values of about 10, which are underlain by a dense sand layer (As-3) with an SPT N-value of about 40. Note that these SPT N-values were measured after filling the original embankment.

A14.5 *Laboratory stress-strain tests*

To determine the thickness of the EPS in the design stage, undisturbed samples were retrieved from clay layers beneath the original embankment. After being consolidated in an oedometer for 24 hours under an overburden stress equal to the pre-consolidation pressure, they were subjected to unloading to different extents and were left for 2 weeks under the reduced overburden stress. Based on these results, it was estimated that no additional settlement would take place if the overburden stress

were reduced by 30 %. Consequently, the thickness of the EPS was set as being equal to 4 m.

Unconfined compression tests and conventional oedometer tests were conducted on the undisturbed clay samples (Fig. A14.5). Based on these results, the values of the input parameters that were employed in the elasto-visco-plastic analysis as mentioned later were evaluated following the procedure proposed by Iizuka and Ohta (1987). Relevant correction for the effects of sample disturbance and other factors on the measured unconfined compression strength was also made (Ohta et. al., 1989).

A14.6 *Field tests*

During preloading and construction of the original embankment, settlement at the ground surface was monitored. After construction, this was restarted after a certain period of intermission (Fig. A14.6).

During and after replacement work with EPS, settlement at ground surface (Fig. A14.7), differential settlement of several clay layers (Fig. A14.8) and their excess pore water pressures (Fig. A14.9) were monitored (Fig. A14.2).

A14.7 *Comparison of numerical prediction with full-scale behavior during construction and replacement of embankment*

A 2-D elasto-visco-plastic analysis using an FE code named DACSAR (Iizuka and Ohta, 1987) was conducted to simulate the full-scale behavior during construction and replacement of the embankment. Input parameters for the clay layers, excluding the surface layer, were evaluated based on the results from unconfined compression tests and oedometer tests, as mentioned before. On the other hand, the sand layers, the embankment and the surface clay layer were assumed to be linear-elastic. Young's modulus E was evaluated from their SPT N-values using the following empirical formulation.

$$E = (2747 \times N) \text{ kPa} = (28 \times N) \text{ kgf/cm}^2$$

Considering the distribution of settlements that were measured prior to replacement work, a 2-D finite element mesh (Fig. A14.10) was prepared so that it could reasonably simulate the actual loading conditions due to the weight of the original embankment. This was subjected to a loading sequence consisting of preloading, filling of the original embankment and overlay work, followed by unloading and reloading histories due to replacement work with EPS and pavement work, respectively.

Computed settlements at the ground surface during and after the construction of the original embankment are compared with the measured ones in Fig. A14.6. For reference, computed results using a 1-D mesh are also shown. With respect to the settlement during preloading and the rebound after removing the preload, the results using the 2-D mesh showed a tendency towards overestimation, while those using the 1-D mesh were in good agreement with the measured data. On the other hand, with respect to the rate of settlement after construction of the original embankment, both results were consistent with the measured behavior. The overestimation in the 2-D analysis is considered to be due to the assumption of a linear relationship between the void ratio e and the logarithmic of overburden stress $\log(p')$. This may lead to an overestimation of the change in the e value of the clay layers located near the surface under an extremely small p' value before preloading and after removing the preload.

Computed vertical displacements at the ground surface during and after the replacement work with EPS using a 2-D mesh are compared with the measured ones in Fig. A14.7. The data measured near the EPS showed a slight rebound due to unloading by the replacement, while the computed results largely overestimated the amount of rebound. A similar discrepancy can be seen in Fig. A14.8, where relative settlement of several clay layers is compared between the computed and the measured results. It should be noted, however, that the significant reduction in the average settlement rate at ground surface by replacement with EPS could be reasonably simulated by the present analysis, as summarized in Table A14.2.

Excess pore water pressures in several clay layers are compared in Fig. A14.9. They were re-defined to be zero at the beginning of the first replacement work with EPS. In the early stages of the first replacement, generation of negative excess pore water pressure was recorded in the data measured near the EPS, excluding the Ac-1(U) layer. In the later stage, however, significant accumulation of excess pore water pressure on the positive side took place in the measured data. Although the computed results were able to simulate the generation of negative excess pore water pressure in the early stages, they failed to simulate the subsequent accumulation of positive excess pore water pressure.

It is to be noted that, in the present analysis, no adjustment was made to the input parameters that were evaluated based on laboratory and field test results. Even with such an adjustments, however, it would be difficult to fully simulate the peculiar behavior as mentioned above. Further investigations will be required on these issues.

A14.8 *References for case No.14*

Iizuka, A. and Ohta, H. 1987. A determination procedure of input parameters in elasto-viscoplastic fininte element analysis, Soils and Foundations, Vol.27, No.3, pp.71-87.

Kawaida, M., Morii, Y., Horikoshi, K., Iizuka, A. and Ohta, H. 2000. Analyses of embankment on soft ground during unloading and 'EPS' loading processes, Journal of Geotechnical Engineering, JSCE, No.645/III-50, pp.209-221 (in Japanese).

Ohta, H., Nishihara, A., Iizuka, A., Morita, Y., Fukagawa, R. and Arai, K. 1989. Unconfined compression strength of soft aged clays, Proc. of 12th ICSMFE, Vol.1, pp.71-74.

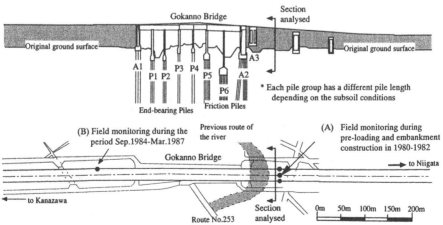

Fig. A14.1   Longitudinal section and plan of embankment and bridge for Hokuriku Expressway (Kawaida et al., 2000)

Table A14.1 Sequence of construction and maintenance works (Kawaida et al., 2000)

Period	Event	Height of embankment
Oct.1980-Dec.1981	Pre-loading	0→7.2→1.2 m
Dec.1981-Dec.1982	Embankment	1.2→8.8 m
Nov.1983-	Service start	
Mar.1985	Overlay	
Jul.1985	Overlay	
Sep.1985	Overlay	
Apr.1996-Mar.1997	Replacement with EPS (Lane to Niigata)	About 4 m
Apr.1997-Mar.1998	Replacement with EPS (Lane to Kanazawa)	About 4 m

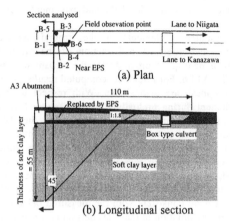

(a) Plan

(b) Longitudinal section

Fig. A14.3 Location of replacement with EPS (Kawaida et al., 2000)

Table A14.2 Average settlement rate at ground surface before and after replacement with EPS (Kawaida et al., 2000)

		Before EPS (1986-1987)	After EPS (1997.3-1997.6)
Settlement rate at ground surface (mm/year)	Measured	127	nearly 0
	Calculated	229	15

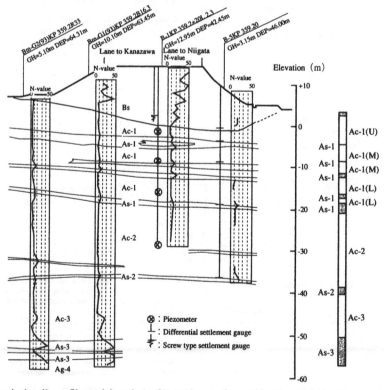

Fig. A14.2 Typical soil profile and location of transducers for monitoring during replacement work with EPS (Kawaida et al., 2000)

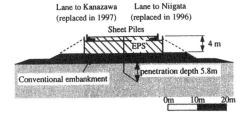

Fig. A14.4 Typical cross section of embankment replaced with EPS (Kawaida et al., 2000)

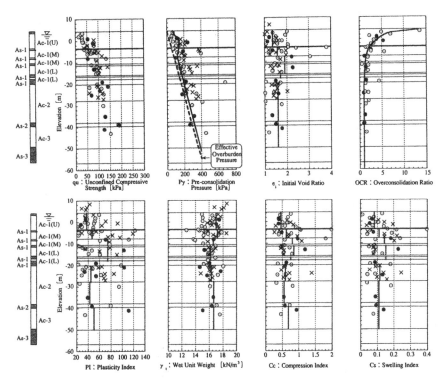

Fig. A14.5 Laboratory test results employed to evaluate input parameters for analysis (Kawaida et al., 2000)

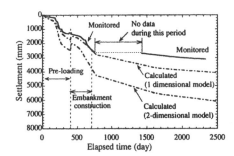

Fig. A14.6 Settlement at ground surface during and after construction of original embankment (Kawaida et al., 2000)

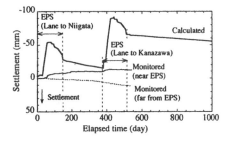

Fig. A14.7 Settlement at ground surface during and after replacement work with EPS (Kawaida et al., 2000)

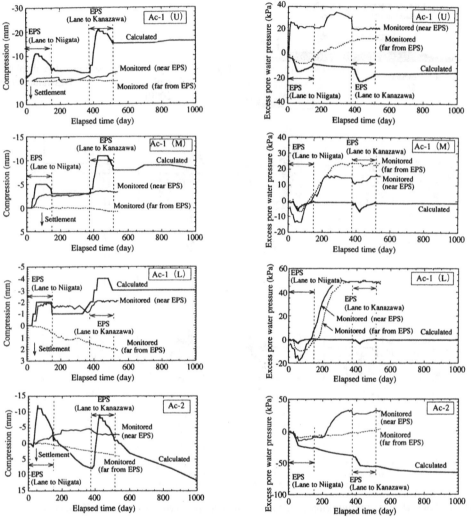

Fig. A14.8 Differential settlement of several clay layers during and after replacement work with EPS (Kawaida et al., 2000)

Fig. A14.9 Excess pore water pressures during and after replacement work with EPS (Kawaida et al., 2000)

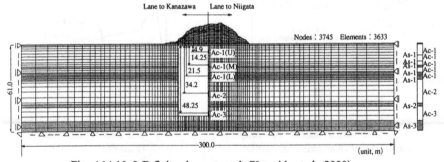

Fig. A14.10 2-D finite element mesh (Kawaida et al., 2000)

APPENDIX 15. CASE NO.15 (EXCAVATION FOR BUILDING CONSTRUCTION)

A15.1 *Construction period*

July 1990 – December 1991

A15.2 *Location*

Osaka City, Japan.

A15.3 *Type of project*

A large-scale ground excavation, with dimensions of 80 m×90 m in plane and 24 m in depth, was conducted for constructing a high rise building (Fig. A15.1). Due to existing buildings adjacent to the excavation, ground improvement by means of soil-cement columns and bracing by concrete slabs was adopted to minimize ground deformation. The excavation was proceeded in seven steps. In each step, excavation of the ground and construction of braces were repeated, and the whole excavation was finished in around 500 days. The ground water level inside the diaphragm wall was kept always 1 m below the cutting surface by relief wells (Figs. A15.1 and A15.2). In addition to these relief wells, deep wells were introduced to prevent heaving of the cutting surface. During the last (seventh) excavation step, the total water head at the points of these deep wells was kept 9 m below the ground surface.

Three-dimensional soil/water coupled FEM analysis, based on Biot's theory of three dimensional consolidation, was applied to this excavation project and the calculated ground deformation was compared with the measured values. This analysis was carried out after the excavation work was finished.

A15.4 *Ground conditions*

The ground conditions at the site are summarized in Figs. A15.2 and A15.3. The major Alluvial material was a high-plastic soft clay deposit (Umeda Layer) found between sandy layers. This Alluvial clay is normally consolidated with N-values of 2 to 6 and an unconfined compression strength q_u of around 0.1 MPa. The sand deposit at the ground surface includes loose layers with N-values less than 10.

The Diluvial deposits were comprised of an interbed of clayey layers and sandy layers. The Diluvial clays were over-consolidated with an OCR of 2 to 2.5, though their strength was relatively small with q_u of around 0.1 MPa. The Diluvial sand layers were very dense with N-values larger than 60.

A15.5 *Laboratory tests*

Oedometer tests, unconfined compression tests and triaxial tests were conducted on the material obtained from the site.

A15.6 *Field tests*

Standard penetration tests, PS logging tests and permeability tests were conducted at various ground elevations (Fig. A15.3).

A15.7 *Finite element modeling of the ground and the structure*

One quarter of the construction site that has boundaries of A-A' and B-B' cross-sections (Fig. A15.1) was analyzed using GRASP3D (a three-dimensional soil/water coupled FEM program) developed at Obayashi Corporation. The finite element mesh for the final (seventh) stage of the excavation (GL.−24m) is shown in Fig. A15.4. Displacement of the nodes on the boundaries were fixed at zero in a perpendicular direction.

The ground water level inside the diaphragm wall was fixed at 1m below the cutting surface in all stages of the excavation. In the final (seventh) stage of the excavation, the total water head at the positions of the deep well slits was fixed at GL.−9m. The ground water level at the outer boundaries was fixed at GL.−3.2m.

The elasto-plastic constitutive model proposed by Sekiguchi and Ohta (1977) was used to simulate the behavior of the clay layers except that at the bottom of the Diluvial deposit below GL.−56m. The model parameters for the clay layers were determined as shown in Table A15.1, mainly from the oedometer tests and with the aid of the empirical relationships between these parameters and the plasticity index of the soils (Iizuka and Ohta, 1987).

The sand layers and the clay layer at the bottom of the Diluvial deposit were simulated as an elastic material with an elastic modulus value from the PS logging tests, and a Poisson's ratio of 0.35. The coefficients of permeability k of the sand layers were $2×10^{-3}$ to $4×10^{-4}$ cm/sec from the in-situ permeability tests.

The concrete bracing slabs were simulated by 4-node elastic shell elements. 2-node elastic truss elements were adopted for the cast-in-place piles at the bottom of the excavation. The diaphragm wall was simulated by an elastic shell between elements with a small permeability ($k=1×10^{-7}$ cm/sec). The model parameters for these concrete members are given in Table A15.2.

A15.8 *Comparison between the analysis and the measurements*

The calculated and measured values for the lateral displacement of the diaphragm wall, the settlement of the ground surface behind the diaphragm wall and the bending moment of the diaphragm wall are plotted on Figs. A15.5, A15.6 and A15.7, respectively. The simulation using the elastic modulus of the sand layers from the PS logging tests (E_{PS}) predicted the real behavior of the ground and the diaphragm wall adequately. It should be noted that the stiffness of the soils from the triaxial tests and the unconfined compression tests was much smaller than that from the PS logging ones (Fig. A15.3). This fact indicates that the elastic modulus from the PS logging test, rather than from conventional laboratory tests, should be used for an analysis of this kind of excavation. In Fig. A15.3 the elastic modulus calculated from N-values using the empirical equation of Imai and Tonouchi (1982) is also plotted by a broken line. It can be seen that such a correlation between N-value and elastic modulus as above may be useful in the simulation of elastic modulus.

Figs. A15.8 and A15.9 show a 3-dimensional view of the deformation of the ground and the distribution of the bending moment of the diaphragm wall, respectively. It can be seen from Fig. A15.8 that the deformation of the ground is smaller around the corners of the excavation than elsewhere. Such kinds of 3-dimensional effects can also be seen in Fig. A15.9, indicating that the vertical bending moment is larger at the center of the diaphragm wall, while the horizontal bending moment is larger near the corner.

The relationship between the maximum ground surface settlement δ_v and the depth of excavation H calculated from this study was plotted along with measurements in a previous study (Clough et al., 1990) in Fig. A15.10. It may be seen that the prediction of ground settlement in the present study is consistent with the trend of the measurements in this previous study, which is represented by the equation $\delta_v / H = 0.15\%$.

A15.9 *References for case No.15*

Iizuka, A and H. Ohta. 1987. A determination procedure of input parameters in elasto-viscoplastic finite element analysis, *Soils and Foundations*, Vol. 27, No. 3, pp.71-87.

Imai, T and K. Tonouchi. 1982. The relationship between N-value and Vs and its applications. *Kisoko*, Vol. 10, pp. 70-76 (in Japanese).

Sekiguchi, H and H. Ohta. 1977. Induced anisotropy and time dependency in clay. *Ninth ICSMFE, Tokyo*, Proc. Specialty session 9, pp.229-239.

Sugie, S., T. Ueno, H. Ohta, A. Iizuka, N. Akino and J. Sakimoto. 1999. 3-dimensional soil/water coupled finite element analysis of ground behavior adjacent to braced cuts. *Tuchi-to-Kiso*, Vol. 47, No.7, pp.13-16 (in Japanese).

Sugie, S., T. Ueno, N. Akino and J. Sakimoto. 1999. Three-dimensional soil/water coupled FEM simulation of ground behavior adjacent to braced cuts. *Report of Obayashi Corporation Technical Research Institute*, No.59, pp.69-74(in Japanese).

Clough, G.W. and O'Rourke, T.D. 1990. Construction induced movements of insitu wall. *Design and Performance of Earth Retaining Structure*, edited by P.C. Lambe and L.A. Hansen, ASCE, pp.439-479.

Table. A15.1 Input parameters of clayey subsoils (Sugie et al., 1999)

	Depth (m)	C_C	e_0	ϕ' (deg)	ν'	K_0	OCR	k (cm/sec)
Alluvial clay	6~18	0.40 ~1.01	1.20 ~1.65	23	0.36 ~0.38	0.57 ~0.62	1.0	2×10^{-7} ~8×10^{-7}
	18~24	0.52	1.21 ~1.25	25	0.36	0.57	1.7	4×10^{-7} ~7×10^{-7}
Diluvial clay	30~34	0.45	1.00	26	0.36	0.57	2.0	2×10^{-7}
	36~43	0.45 ~0.48	0.94 ~0.97	29	0.34 ~0.36	0.52 ~0.57	2.5	8×10^{-8} ~2×10^{-7}

C_C:compression index, e_0: void ratio at pre-consolidation,
ϕ': effective friction angle, ν': Poisson's ratio,
K_0: coefficient of earth pressure at rest,
OCR: over consolidation ratio, k: coefficient of permeability

Table. A15.2 Input parameters of concrete structures (Sugie et al., 1999)

	E (MPa)	A (m²)	I (m⁴)	Type of element
Diaphragm wall	2.1×10^8	1.0	8.3×10^{-2}	shell
Bracing slab	2.1×10^8	0.4	5.3×10^{-3}	shell
Cast-in-place pile	2.1×10^8	2.726		truss

E: elastic modulus, A: cross-section area (per unit width for wall and slab, per one for pile),
I: moment of inertia (per unit width)

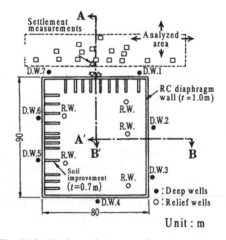

Fig. A15.1 Horizontal cross section view of the construction site (Sugie et al., 1999)

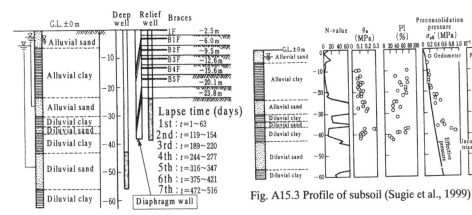

Fig. A15.2 Brace system and well installation for the excavation (Sugie et al., 1999)

Fig. A15.3 Profile of subsoil (Sugie et al., 1999)

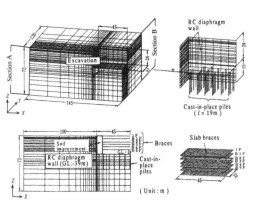

Fig. A15.4 Finite element modeling of the construction site (Sugie et al., 1999)

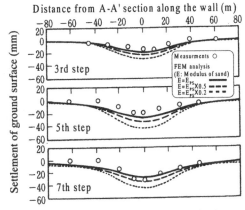

Fig. A15.6 Ground surface settlement behind the diaphragm wall (Sugie et al., 1999)

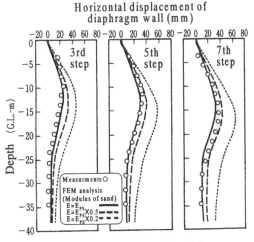

Fig. A15.5 Lateral deformation of the diaphragm wall (Sugie et al., 1999)

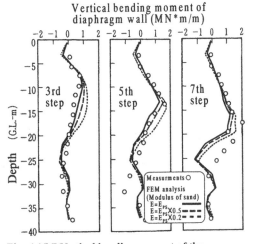

Fig. A15.7 Vertical bending moment of the diaphragm wall (Sugie et al., 1999)

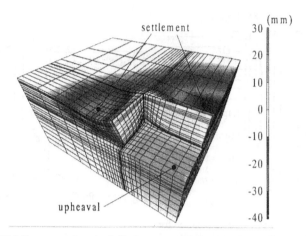

Fig. A15.8 Contours of vertical displacement of ground (courtesy of Sugie, S.)

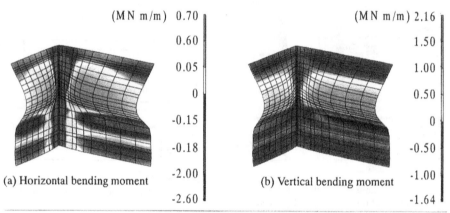

(a) Horizontal bending moment

(b) Vertical bending moment

Fig. A15.9 Contours of bending moment of diaphragm wall (courtesy of Sugie, S.)

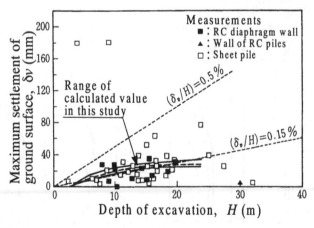

Fig. A15.10 Maximum ground surface settlement behind in-situ wall (Sugie et al., 1999) (modified from Clough et al., 1990)

APPENDIX 16. CASE NO.16 (EXCAVATION FOR SUBWAY STATION)

A16.1 *Construction period*

1998-1999.

A16.2 *Location*

Fukuoka, Japan

A16.3 *Type of project*

A countermeasure was taken for restraining the movement of an existing cable-setting tunnel due to the excavation work for new subway construction at Watanabe-Street, Fukuoka. The existing tunnel, which is one of the typical shield tunnels having an outer diameter of 3.15 m, was located at just 1.1 m below the surface of the excavation (Fig.A16.1). At the design stage, ground movement exceeding the allowable deformation of the existing tunnel due to the excavation work was concerned.

A16.4 *Ground condition*

A typical soil profile at the executed section is shown in Fig. A16.2. A 10 m-thick alluvial sandy layer with N values of 10 to 20 is underlain by diluvial clay and sandy layers with N values of 5 to 15 and 20 to 30, respectively, as shown in Fig. A16.2. The average water level is about 2 m below the ground surface.

A16.5 *Laboratory tests*

To predict ground deformation associated with the excavation stage for constructing a new subway station, unconsolidated-undrained triaxial compression tests (UU tests) and one-dimentional consolidation tests were performed on undisturbed samples retrieved from the diluvial clay layer, together with physical property tests on those from the diluvial sandy and clay layers. Typical experimental results of the UU tests are shown in Fig.A16.3. Based on these results, strength parameters for this layer were estimated, and the deformation moduli were compared with those determined by field tests such as borehole lateral loading tests.

A16.6 *Field tests*

Borehole lateral loading tests and standard penetration tests were conducted to mainly estimate the deformation and strength characteristics of each layer to be used in FE analysis as shown in Fig. A16.2. Settlement gauges, deformation gauges and inclinometers (No.1 to No.19 shown in Fig. A16.4) were also installed to the existing tunnel in order to measure the movement of the tunnel during the progress of the excavation.

A16.7 *Comparison of numerical prediction with field measurements due to excavation*

In order to restrain the deformation of the existing tunnel from the allowable value, the soil improvement by mixing with cement and the ground surrounding the tunnel was adopted as shown in Fig.A16.5. It was considered that, by using this method, the strength and deformation properties of the ground could be improved to sufficiently high values. The improvement depth was gradually changed to protect a sudden deformation of the ground adjacent to the excavation boundary. The prescribed relationship between the maximum bending moment of the tunnel and the soil improvement depth is shown in Fig.A16.6. These numerical results were obtained by using a spring model on the assumed elastic plate. It was found that when the improvement length exceeds 24 m, the magnitude of the bending moment is within the allowable value of 1052.5 kN·m specified in the design. In this case, the coefficient of the equivalent subgrade reaction in the soil improvement region can be smoothly changed from 147 MN/m^3 to 49 MN/m^3 by controlling the improved depth of the ground.

In order to determine the depth to be improved, 2-D FE analysis using a linear elastic model was conducted, in which the coefficients of the equivalent subgrade reaction were used in the soil improvement region. The cross-section of the FE analysis model is shown in Fig.A16.7. The distance between the existing tunnel and the improvement column was about 0.5 m and the improvement width was kept as 2 m. Considering that the ground movement due to excavation would be rebounding, the deformation moduli E_b of each layer were estimated based on the moduli obtained from unloading curves in the borehole lateral loading tests. These values were similar to the values of E_{50} from the triaxial UU tests. The soil constants used in the analysis are summarized in Table A16.1.

The predicted relationship between the coefficient of the equivalent subgrade reaction and the corresponding depth of the soil improvement column are shown in Fig.A16.8. It was found that an 18m-deep improvement would be needed to satisfy the required coefficient of subgrade reaction of 147 MN/m^2 in the design. Based on the computed results, the improvement area was determined as shown in Fig.A16.9. The comparison of the computed results with the field measurements of the vertical deformation of the existing tunnel during each stage of the excavation is shown in Fig.A16.10. The predicted deformation due to the rebound of the tunnel is agrees well with the measured one at the final excavation stage. Such numerical analysis as described in this report is considered to be a good tool to evaluate the effect of soil improvement on the deformation of an existing tunnel.

A16.8 *References for case No.16*

Ogata, T., Mandai, K., Kai, S. and Imanishi, H. 1999. Stability analysis of the existing shield tunnel due to vertical excavation, The foundation Engineering & Equipment, Vol.27, No.8, pp.37-39 (in Japanese).

Takai, T., Imanishi, H., Mandai, K., Tsuruoka, Y and Kai, S. 2000. Countermeasure for restraining the upward deformation of existing tunnel due to excavation, Annual meeting of JSCE of Kyushu branch, Vol.III-53, pp.476-477 (in Japanese).

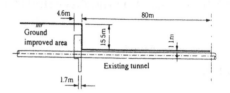

Fig.A16.1 Cross-section of the existing shield tunnel (Ogata et al., 1999)

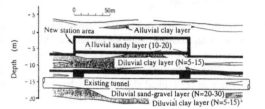

Fig.A16.2 Soil profile at the executed section (Ogata et al., 1999)

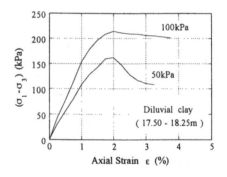

Fig.A16.3 Typical stress strain curves from UU tests (Ogata et al., 1999)

174

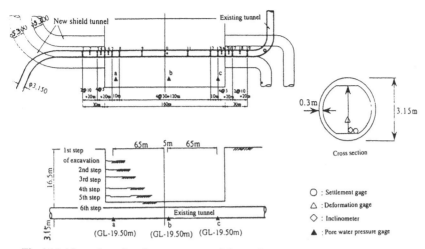

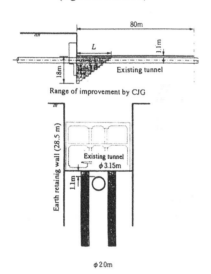

Fig.A16.4 Location of settlement gauges, deformation gauges and inclinometers
(Ogata et al., 1999)

Fig.A16.5 Soil improvement region in the ground
(Ogata et al., 1999)

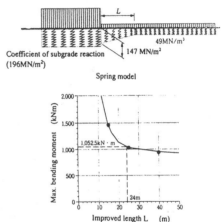

Fig.A16.6 Determination of soil improvement
length (Ogata et al., 1999)

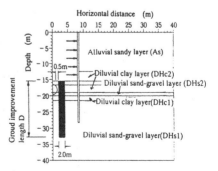

Fig.A16.7 Cross section of the FE analysis
(Ogata et al., 1999)

Table A16.1 Soil parameters used in FE
analysis (Ogata et al., 1999)

	N_{av}	c(kPa)	ϕ(degs.)	γ_t(kN/m^3)	E_0(MPa)	k(cm/s)	ν
As	15	20	30	18.5	10	0.02	0.3
DAs	24	20	20	19.0	8	0.003	0.3
DHc2	15	30	15	18.5	8	0.00007	0.49
DHs2	30	20	20	19.5	15	0.003	0.3
DHc1	20	30	15	18.0	20	0.0004	0.49
DHs1	35	20	20	19.0	25	0.002	0.3

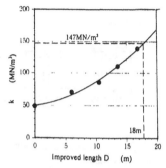

Fig.A16.8 Predicted relationship between the
coefficient of the equivalent subgrade
reaction and the corresponding depth of
the soil improvement column (Ogata et
al., 1999)

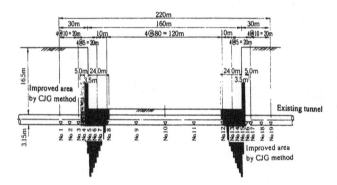

Fig.A16.9 Improvement area determined

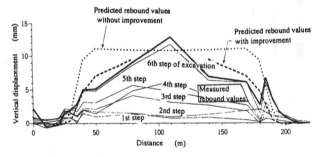

Fig.A16.10 Comparison of computed results
with the field measured ones for the
vertical deformation of the existing
tunnel (Ogata et al., 1999)

A17.1 Construction period

Phase I : Feb. 1990-Feb. 1992, Phase II : Nov. 1994-Feb. 1998

A17.2 Location

Tokyo, Japan

A17.3 Type of project

Two adjacent high-rise buildings with mat foundations (both 23-stories) were constructed with time delay, in Phase I and Phase II construction stages. The excavations were 26.4 m and 22 m from the ground surface in Phase I and II, respectively (Fig.A17.1).

A17.4 Ground conditions

The ground consists mainly of alternative layers of sand and clay. The upper soil layers to a depth of 16m from the ground surface are Holocene layers, while those below are Pleistocene (Fig.A17.2). During the excavations, in order to prevent the bottom of the excavation from excessive heaving, the ground water tables in the Pleistocene soil layers were lowered to depths shallower than 60m and 30 m from the ground surface, in Phase I and II, respectively.

A17.5 Laboratory stress-strain tests

Undisturbed sand and clay samples were obtained by the triple-tube sampling method at depths shown in Fig.A17.2. The strain-dependency of the shear modulus was determined by conducting cyclic triaxial tests (staged tests), and typical test results are shown in Fig.A17.3. For each sample, before starting the staged test, the pressure dependency of the initial shear modulus (G_0 at strain levels of about 2 to 5×10^{-5}) was measured in sequence at five to seven different confining pressures (Fig.A17.4). The axial displacement was measured by using a pair of local displacement transducers (Goto et al., 1991) in cyclic shear tests. The cyclic triaxial tests were conducted under the confining pressure close to the effective overburden stress at the sampling depth. Test results are shown in Table 1. The strain-dependency of shear modulus used for the analysis was modeled using the Ramberg-Osgood Model. A typical comparison of the model with the test results is shown in Fig.A17.3. The initial shear modulus was also calculated from the shear wave velocity measured by P-S wave logging tests performed in the field. Poisson's ratios were assumed to be 0.3 and 0.4 for sand and clay, respectively, as widely used in practice.

A17.6 Field observation

Heaving and settlement during excavation and building construction were measured by settlement gauges installed at different ground depths indicated in Fig.A17.5.

a) Observation results at the Phase I construction stage:
The maximum heave at Point A due to excavation was 6.4 mm. Afterwards, the ground was settled about 1.9 mm by constructing the building. At completion of building construction, the final heave was 4.5 mm (Fig.A17.6a).

b) Observation results at the Phase II construction stage:
The maximum heave due to excavation at Point F was 29.5 mm. Afterwards, the ground was settled about 15.9 mm by building construction, and at completion of building construction, the final heave was 13.6 mm (Fig.A17.6b).

It is clear that, both the maximum heave and the settlement after construction are larger in the Phase II construction stage than those shown in Phase I, even though the excavation depth in the Phase II construction stage was shallower than that in Phase I.

The difference in heave between the two construction stages is due mainly to the fewer stress changes revealed in the Phase I construction process, where an inverted construction method was used employing a greater lowering of the ground water table.

A.17.7 Comparison of test results with full-scale behavior during excavation

The procedure for estimating the heave and settlement is shown in Fig.A17.7 together with diagrams indicating the methods used to determine soil constants for estimation. The Steinbrenner's multi-layer elastic method was used as the numerical method.

Comparisons of the estimation with the full-scale behavior of the vertical displacement of Point A at completion of the excavation and building construction in the Phase I construction stage are shown in Figs.A17. 8 (1) and (2), respectively.

In Fig.A17.8 (1), the value of the initial Young's modulus E_0 was obtained based on the shear wave velocity measured in the field. At each depth, the estimated vertical displacement is smaller than the observation results.

Taking into consideration the difference in vertical displacement between the estimation and the observation as shown in Fig.A17.8 (1), the value of E_0 calculated from the shear wave velocity was modified to a reduced value of $0.75 \cdot E_0$. The estimation results based on $0.75 \cdot E_0$ were again compared with the observation as indicated in Fig.A17.8 (2). A better agreement of vertical displacement between the estimation and the observation was obtained. Fig.A17.8 (3) shows the Phase II comparison between the vertical displacement estimation based on $0.75 \cdot E_0$ and its actual observation. Both at the completion of the excavation and building construction, a fairly good agreement can be seen in the Phase II stage.

A17.8 References for case No.17

Aoki, M., Yoshimura, A., Kamiya, M., Inoguchi, K., Ishii, T. and Nakamura, K. 1997. Behavior of bearing stratum of skyscraper structures constructed separately (Part 1 and 2), AIJ. J. Annual Meeting, pp.1315-1318.

Goto, S., Tatsuoka, F., Shibuya, S., Kim, Y-S. and Sato, T. 1991. A simple gauge for local small strain measurements in the laboratory, Soils and Foundations, Vol.31, No.1, pp.169-180.

Aoki, M., Shibata, Y., Maruoka, M. and Tanimura, K. 1990. Heave and rebound of foundation soil during the underground excavation, AIJ. J. Annual Meeting, pp.1649-1652.

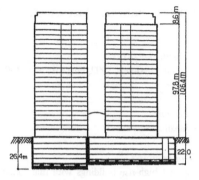

(a) Vertical cross section

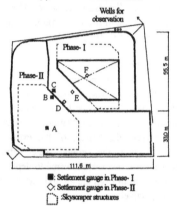

■: Settlement gauge in Phase- I
◇: Settlement gauge in Phase- II
☐ :Skyscraper structures

(b) Plan and locations of instruments

Fig.A17.1 Outline of structures (Aoki et al., 1997)

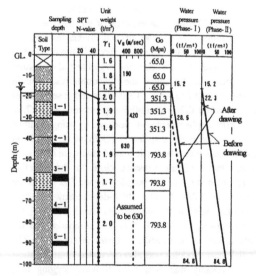

Fig.A17.2 Soil profile and properties used for estimation (Aoki et al., 1997)

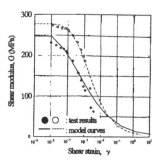 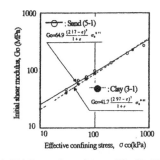

Fig.A17.3 Strain dependency of G (Aoki et al., 1997)   Fig.A17.4 Stress dependency of Go (Aoki et al., 1997)

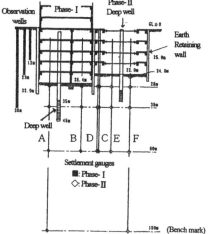

Fig.A17.5 Vertical cross section of underground structures
(Aoki et al., 1997)

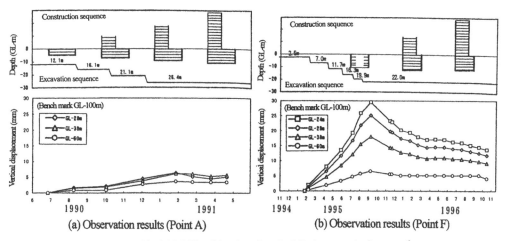

(a) Observation results (Point A)   (b) Observation results (Point F)

Fig.A17.6 Time histories of vertical displacement in the ground
(Aoki et al., 1997)

179

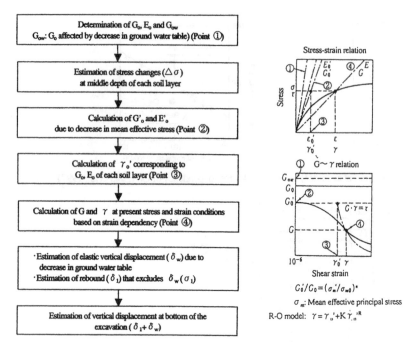

Fig.A17.7 Outline of estimation method
   (After Aoki et al, 1997)

Stress-strain relation

① G~γ relation

$G_0'/G_0 = (\sigma_m'/\sigma_{m0})^*$

σ_m: Mean effective principal stress

R-O model: $\gamma = \gamma_0' + K \dot{\gamma}_n'^R$

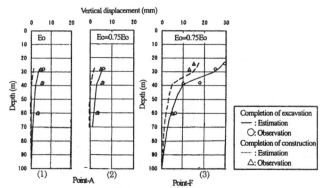

Vertical displacement (mm)

Completion of excavation
— : Estimation
○ : Observation
Completion of construction
······ : Estimation
△ : Observation

(1) Point-A   (2) Point-F   (3)

Fig.A17.8 Vertical displacement
   (Aoki et al., 1997)

Table 1   Sample properties and test results

Sample No.	Sampling depth (m)	Soil type	Ip	Fines content Fc (%)	Initial confining stress (kPa)	Initial void ratio	Experimental constants (*1)	
							A (MPa)	n
1-1	-28.0	Clay	22	97.0	294	1.08	50.1	0.61
2-1	-43.0	Sand	—	16.9	392	0.74	56.6	0.54
3-1	-59.0	Clay	28	97.5	490	0.96	41.7	0.61
4-1	-75.0	Sand	—	10.8	784	0.58	51.1	0.51
5-1	-90.0	Sand	—	4.6	784	0.64	64.9	0.53

*1: $G_o = A \dfrac{(2.17-e)^2}{1+e}\left(\dfrac{\sigma_c}{98}\right)^n$ σ_c : kPa for sand, $G_o = A \dfrac{(2.97-e)^2}{1+e}\left(\dfrac{\sigma_c}{98}\right)^n$ σ_c : kPa for clay

APPENDIX 18. CASE NO.18 (EXCAVATION FOR BUILDING CONSTRUCTION)

A18.1 *Construction period*

1989 – 1991

A18.2 *Location*

Tokyo, Japan.

A18.3 *Type of project*

A ground excavation with dimensions of 35 m×62 m in plane and 24 m in depth was conducted in order to construct a 90 m-high building (Fig. A18.1). Due to the existence of foundations for an expressway viaduct, a pedestrian conduit and a subway tunnel adjacent to the excavation, great care had to be taken in conducting the excavation work.

In particular, the expressway foundations were located at horizontal distances of only 2.65 m and 1.85 m respectively from the newly constructed side wall of the building and the temporary soil retaining wall (Fig. A18.2). The foundation piles were embedded to a depth of 15 m from the ground surface, and the excavation was deeper than these piles by 9 m.

For the temporary soil retaining wall, a soil-cement mixing wall with a maximum thickness of 80 cm having a core of H-shaped steel bars (H-588*300*12*20 mm, at a horizontal spacing of 60 cm) was constructed. During excavation, the inverted construction procedure was employed for the underground part of the building. The island cut procedure was also employed at the final excavation stage.

A18.4 *Ground conditions*

The ground conditions at the site are summarized in Fig. A18.3. A 9 m-thick volcanic ash soil (Kanto Loam) layer is underlain by a 7 m-thick sandy gravel (Musashino Gravel) layer with SPT N-values larger than 50. Below these layers, there exists a sand (Upper Tokyo) layer, which locally sandwiches a silt layer, underlain by a sandy gravel (Tokyo Gravel) layer having a limited thickness of 1 m. Below a depth of 25 m from the ground surface, there exists a clayey silt (Upper Edogawa) layer.

Ground waters in the Musashino Gravel and the Upper Tokyo layers were found to be under artesian conditions with a water head at a depth of 9 to 10 m from the ground surface.

A18.5 *Laboratory tests*

No laboratory tests that could be relevant to the estimation of the pre-failure deformation of soil layers were conducted.

A18.6 *Field tests*

Seismic surveys were conducted, and the shear wave velocities of soil layers between the Musashino Gravel and the Upper Edogawa layers were evaluated to be 350 to 450 m/sec.

A18.7 *Prediction of effects of excavation on existing foundations*

The procedures to predict the effects of excavation on the adjacent existing viaduct foundations are summarized in Fig. A18.4. Effects of the deformation of the temporary soil retaining wall were evaluated by applying a lateral earth pressure on the wall modeled as an elasto-plastic beam (step 1 in Fig. A18.4). This analysis was followed by a linear elastic FE analysis of the ground deformation (step 2) and a 2-D frame analysis of the foundation footing and piles (step 3).

At the same time, in step 4 of Fig. A18.4, the effects of unloading due to excavation on the ground deformation were evaluated based on the approximated solution by Steinbrenner assuming elasticity (Terzaghi, 1943). The computational procedure employed in step 4 is shown in Fig. A18.5. Finally, the effects of the ground excavation on the viaduct piers were evaluated based on the computed displacement of their foundations (step 5).

It should be noted that, in step 4, the effects of the reduction in confining stress on the initial shear modulus of the soils were considered as:

$$G_0'/G_0 = (\sigma_{m0}/\sigma_m)^{0.5}$$
$$= [\{1+(\sigma_{z0}/\sigma_z)\}^{0.5}/\{2*\sigma_{z0}/\sigma_z\}]^{0.5}$$

where G_0 and G_0' are the initial shear moduli before and after excavation, respectively; σ_m and σ_{m0} are the effective mean principal stresses before and after excavation, respectively; and σ_z and σ_{z0} are the effective vertical stresses before and after excavation, respectively. In deriving the above formulation, the coefficient of earth pressure at rest, K_0, was assumed to be:

$$K_0 = (1-\sin\phi)*OCR^{(\sin\phi)}$$

where OCR is the ratio of overconsolidation, set to 1.0 before excavation; and ϕ is the angle of internal friction, set to 30 degrees (Aoki et al., 1990).

In addition, the strain dependency of secant shear modulus was modeled by employing the Ramberg-Osgood model as:

$$\gamma = \tau/G_0' + K*(\tau/G_0')^R$$

where K and R are the material constants. Applicability of the above modeling was validated through comparisons with the results from cyclic triaxial tests on undisturbed samples retrieved from different sites (Fig. A18.6).

As a result, for the present excavation work, it was predicted that the viaduct foundations located closest to the temporary soil retaining wall would uplift by a total of 5.5 mm. This value was evaluated from the predicted settlement of 2.5 mm due to the deformation of the wall and the predicted uplift of 8 mm caused by unloading due to the excavation.

A18.8 *Comparison between prediction and measurement*

Predicted and measured values of the lateral displacement of the temporary soil retaining wall and the lateral earth pressures at the fourth excavation stage (see Fig. A18.2) are compared in Fig. A18.7. The measured wall displacement at its lower part was smaller than the predicted value, while the maximum wall displacement had a similar prediction and measurement. In the figure, distribution of the design lateral earth pressure modified based on the measured values is also shown. These measured values were employed to predict the displacement of the temporary soil retaining wall in the subsequent excavation stages.

The measured relationship between the depth of excavation and the maximum wall displacement agreed very well with the predicted one, as shown in Fig. A18.8.

In Fig. A18.9, the time histories of vertical displacements of the viaduct foundations that were measured at two locations are plotted along with the excavation depths. During excavation work, the foundations uplifted gradually, while their maximum uplift displacement values were slightly smaller than those predicted. It should be noted that, after the fifth excavation stage, the foundations started to settle, due possibly to the effects of dead load from the partly- constructed building body, which were not considered in the prediction.

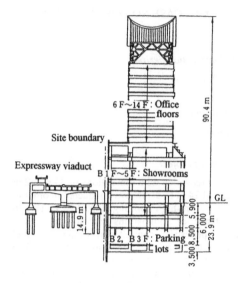

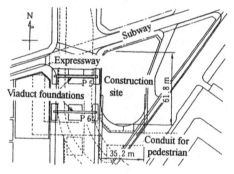

Fig. A18.1 Plan and cross-section of the building (Maruoka, 1995)

A18.9 *References for case No.18*

Aoki, M., Shibata, Y. and Muraoka, M. 1990. Estimation of ground deformations during construction period (part 1), *Annual meeting of AIJ*, pp.1649-1650 (in Japanese).

Muraoka, M. 1995. Excavation work for building construction near existing structures and its field observation, *Chishitsu-to-chosa*, Vol. 4, pp. 8-13 (in Japanese).

Terzaghi, K. 1943. Theoretical Soil Mechanics, Wiley, pp.423-425.

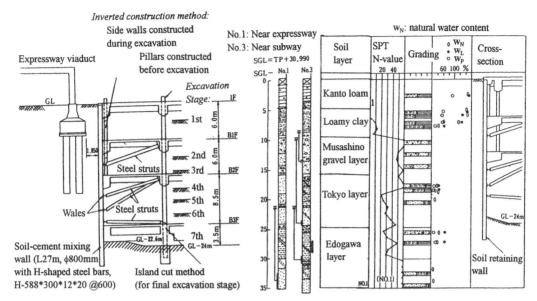

Fig. A18.2 Consequence of excavation stages and location of temporary soil retaining wall and viaduct foundation (Maruoka, 1995)

Fig. A18.3 Estimated soil profile (Maruoka, 1995)

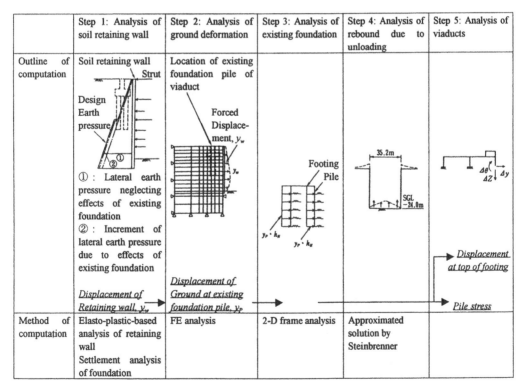

Fig. A18.4 Outlines of prediction of effects of excavation work on viaduct foundations (Maruoka, 1995)

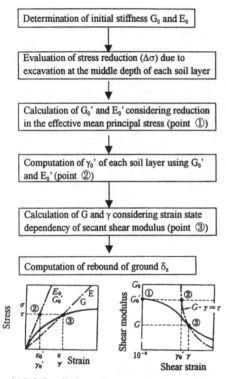

Determination of initial stiffness G_0 and E_0

↓

Evaluation of stress reduction ($\Delta\sigma$) due to excavation at the middle depth of each soil layer

↓

Calculation of G_0' and E_0' considering reduction in the effective mean principal stress (point ①)

↓

Computation of γ_0' of each soil layer using G_0' and E_0' (point ②)

↓

Calculation of G and γ considering strain state dependency of secant shear modulus (point ③)

↓

Computation of rebound of ground δ_z

Fig. A18.5 Prediction of rebound of ground caused by unloading due to excavation (Maruoka, 1995)

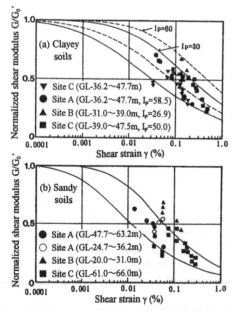

Fig. A18.6 Comparison of strain state dependencies of secant shear modulus (Aoki et al., 1990)

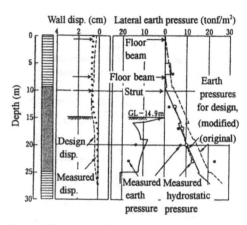

Fig. A18.7 Lateral displacement of temporary soil retaining wall and lateral earth pressures at fourth excavation stage (Maruoka, 1995)

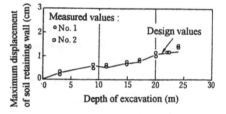

Fig. A18.8 Relationships between the depth of excavation and maximum lateral displacement of temporary soil retaining wall (Maruoka, 1995)

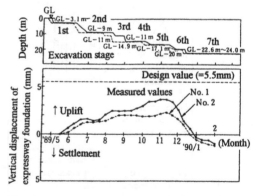

Fig. A18.9 Vertical displacement of viaduct foundations (Maruoka, 1995)

Individual papers

Advanced Laboratory Stress-Strain Testing of Geomaterials, Tatsuoka, Shibuya & Kuwano (eds),
© *2001 Taylor & Francis, ISBN 90 2651 843 9*

Viscous deformation in triaxial compression of a dense well-graded gravel and its model simulation

LeQuang AnhDan
University of Tokyo, Japan

Junichi Koseki
University of Tokyo, Japan

Fumio Tatsuoka
University of Tokyo, Japan

ABSTRACT: A series of triaxial tests were performed on large specimens of a very dense well-graded gravel to study into the viscous property. A considerably large amount of drained creep deformations were observed at creep stages during shearing. The creep behaviour was essentially the same between partially and fully saturated specimens. Creep strains became larger as the stress state approached the failure state, despite that such behaviour of very dense gravel has been often ignored in geotechnical engineering practice. The observed viscous behaviour could be reasonably simulated by one of the three-component rheological models, called the general TESRA model. Effects of stress path and the viscous property on the flow characteristics are discussed.

1 INTRODUCTION

It has been often considered that the creep deformation of very dense gravel is not important because of its high strength. For this reason on one hand and due to considerable experimental difficulties on another hand, the viscous deformation property of dense gravel has been studied only to a very limited extent. It will be shown in this paper that the viscous property of very dense gravel could become surprisingly important as the stress state approaches the failure state. This property is therefore very important in many geotechnical engineering problems, including long-term residual deformation and displacements of reinforced soil structures, rock-fill dams and pile and raft foundations of important structures constructed on and in gravel deposits.

In order to have a better insight into this issue, a series of triaxial tests on a very dense well-graded gravel were performed to investigate into:

1) the viscous property of dense gravel during the process of stress unloading as well as stress loading;
2) effects of the degree of saturation on the viscous property;
3) whether the measured viscous property can be simulated by a relevant rheological model; and
4) whether a unique potential for irreversible strain increments exists; or more specifically;

a) whether the direction of irreversible strain increment vector depends on the stress path;
b) whether the direction of irreversible strain increment vector changes with time during a creep test;
c) whether two different potentials for irreversible strain increments exist for, respectively, compression yielding and shear yielding.

2 TESTING PROCEDURES

The specimens were rectangular prismatic with dimensions of 58 cm high and 23 cm x 23 cm in cross-section (**Fig. 1**). The axial load was measured with a load cell placed inside the large triaxial cell in order to eliminate the effects of piston friction (Tatsuoka, 1988). Axial strains ε_v that were free from the effects of bedding error at the top and bottom ends of specimen, which were not lubricated, were measured directly on the specimen lateral surface by using a

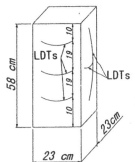

Fig.1: Large prismatic specimen

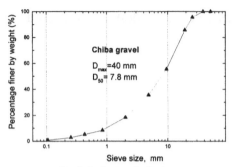

Fig. 2: Grain size distribution curves

pair of vertical local deformation transducers (LDTs) (Goto et al. 1991; Hoque et al. 1997). Lateral strains ε_h that were free from the effects of membrane penetration at lateral surfaces of specimen were measured also by using four pairs of horizontal LDTs.

A well-graded crushed sandstone (called Chiba gravel) was used, which could be categorized as a sandy gravel (**Fig. 2**). Very dense specimens were prepared by manual compaction at a water content of 5.5 % in 14 layers with a compacted thickness of 4 cm per layer. A 15 mm-thick steel platen, which was made to fit to the inside cross-sectional area of the specimen mold, was placed on the surface of respective precedent compaction layer. The compaction was achieved by vertically hitting a 11 kg wooden tamper having a 20 cm diameter at the bottom end onto the steel platen. Before placing the material for the next layer, the surface of the compacted precedent layer was scrapped to a small depth to ensure a good interlocking between vertically adjacent layers. By this compaction method, such a high density as 2.2 - 2.3 g/cm^3 or a void ratio of as low as about 0.2 was achieved (**Table 1**).

Results from three tests DR3, DR4 and DR5 using partially saturated specimens and another test DR9 using a fully saturated specimen will herein be reported (see Table 1). Loading of tests DR3, DR4 and DR5 was made by stress control. The axial

strain rate on average during loading and unloading in these tests were about 0.009%/min in test DR3, 0.016%/min in test DR4 and 0.019%/min in test DR5. The measured stress paths of these tests are shown in **Fig. 3**. The stress path in test DR9 was essentially the same as the one in test DR3. In test DR9, however, the axial strain rate was kept constant at a smaller value, about 0.005%/min, to ensure fully drained conditions throughout the test. In all the tests, creep tests were performed, each lasting for about 6 hours, under drained conditions at different stress levels.

Table 1: Initial testing conditions

Test	Loading condition	Dry density, void ratio	Degree of saturation	Stress path
DR3	Stress control	2.229 g/cm^3 0.198	74.2%	I.C. from 49 kPa to 490 kPa, and then T.C. at σ_h=490 kPa
DR4	Stress control	2.224 g/cm^3 0.198	73.4%	I.C. from 49 kPa to 980 kPa, and then A.C. at $\Delta\sigma_v/\Delta\sigma_h$= -1
DR5	Stress control	2.226 g/cm^3 0.199	73.8%	I.C. from 49 kPa to 260 kPa, and then A.C. at$\Delta\sigma_v/\Delta\sigma_h$= 4.4
DR9	Strain control	2.231 g/cm^3 0.197	100% B-value =0.97	I.C. from 49 kPa to 490 kPa, and then T.C. at σ_h=490 kPa

I.C.: isotropic consolidation, A.C.: anisotropic consolidation, T.C.: triaxial compression.

3 RESULTS AND DISCUSSIONS

3.1 General trend:

The overall relationships between the stress ratio R= σ_v/σ_h and the vertical strain ε_v and shear strain γ from tests DR3, DR4 and DR5 are shown in **Figs. 4a & b**, while the time histories of γ are compared

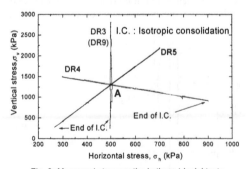

Fig. 3: Measured stress paths in three triaxial tests

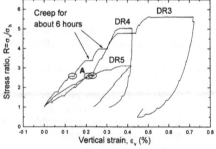

Fig.4a: Overall relationships between stress ratio and vertical strain

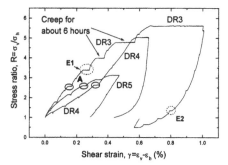

Fig.4b: Overall relationships between stress ratio and shear strain

in **Fig. 5**. The following trends of behaviour can be seen from these figures:

1) A considerable amount of creep strain took place regardless of the stress paths examined. Creep strain became larger as the stress state approached the failure state and the material became more dilatant.

2) When the loading was restarted at the original loading rate following each creep stage, the stress-strain curve exhibited a very stiff response, followed by marked yielding. Then, the stress-strain curve tended to rejoin the respective primary one that would have been obtained if loading had been continued without an intermission

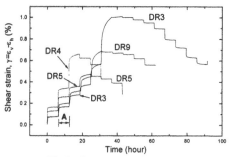

Fig. 5a: Time histories of shear strain
(partially and fully saturated specimens)

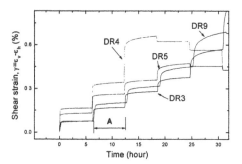

Fig. 5b: Intitial part of Fig. 5a

of creep loading. **Fig. 6a** shows detailed behaviour typical of the above. The size of high stiffness zone in the stress space developed by creep deformation increased with the increase in the creep strain. The behaviour described above can be fully explained by the relevant rheological model, as shown later in this paper.

3) In all of four tests, at each creep stage during stress-unloading, the shear strain decreased with time (i.e., creep recovery), as typically shown in **Fig. 6b**. The amount of creep recovery became larger as the stress ratio became smaller. Immediately after unloading was restarted following each creep stage, the stress-strain curve exhibited a stiff response (Fig. 6b) in a similar manner to the one observed during loading (Fig. 6a).

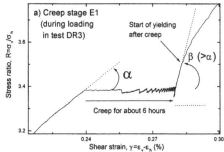

Fig. 6a: Local stress-strain behaviors before, during and after a creep test (loading part)

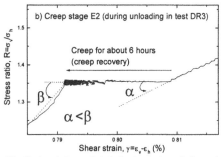

Fig. 6b: Local stress-strain behaviors before, during and after a creep test (unloading part)

3.2 Comparison of viscous property between partially and fully saturated specimens

Essentially the same viscous property was observed with sand whether the specimens is air-dried or water-saturated (Di Benedetto et al. 2001; Tatsuoka et al. 2001b). It is likely therefore that the interaction between the pore water and the fabric of gravel particles is not the major cause for the observed viscous property of gravel described above. One may say, however, that the effect of suction is significant on such viscous property of partially saturated gravel as described above. To confirm whether the observed viscous behaviour is due to the intrinsic viscous property of the gravel tested, one test was performed on a fully saturated specimen (test DR9) under otherwise the same test conditions as test DR3. The overall relationships between the stress ratio R= σ_v/σ_h and the shear strain γ of a pair of partially and fully saturated specimens, from tests DR3 and DR9, are compared in **Fig. 7**. With nearly two times differences in the strain rate during monotonic loading between the two tests, the overall stress-strain behaviour, including that during creep stages, is very similar between the two tests. It is very likely that effects of the different strain rate were insignificant, if any. No pore water was expelled from the partially

relatively high confining pressure, the viscous deformation property of the gravel is essentially the same between the partially and fully saturated specimens. These results indicate that the effects of suction on the significant viscous behaviour that was observed with the partially saturated specimens were not important, if any. Further study will be necessary, however, to confirm whether it is also the case at lower confining pressure.

3.3 Numerical simulation

A number of different constitutive models have been proposed to simulate a wide variety of stress-strain-time behaviour of geomaterials. Tatsuoka et al. (2001a; 2001b) proposed a new type of model, called the TESRA model, which could simulate surprising well the stress-strain-time behaviour of

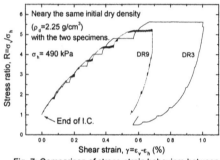

Fig. 7: Comparison of stress-strain behaviors between partial and fully saturated specimens

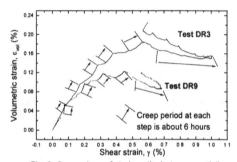

Fig. 9: Comparison of strain paths between partially and fully saturated specimens

saturated specimen throughout the test. **Fig. 8** compares the total shear strain increment that took place at each creep stage from these two tests (and the others), while **Fig. 9** compares the strain paths from the two tests. It may be seen that the partially satu-

poorly graded sand (Di Benedetto et al. 2001; Tatsuoka et al. 2001b). Anh Dan (1999) showed, however, that the TESRA model is not relevant to this type of gravel used in the present study and a more general model, called the general TESRA model, be relevant to this type of gravel. Due to the page limitation, only the essence of these models will herein be described.

The stress is decomposed into viscous and non-viscous components, σ^f and σ^v, as:

$$\sigma = \sigma^f(\varepsilon^{ir}) + \sigma^v(\varepsilon^{ir}, \dot{\varepsilon}^{ir}) \tag{1a}$$

$$\sigma^v = \int_{\varepsilon_1^{ir}}^{\varepsilon^{ir}} [d\sigma^v]_{(\tau)} = \int_{\varepsilon_1^{ir}}^{\varepsilon^{ir}} [d\{\sigma^f \cdot g_v(\dot{\varepsilon}^{ir})\}] g_{decay}(\varepsilon^{ir} - \tau) \tag{1b}$$

where $d\sigma^v$ is the viscous stress increment taking place before, at $\varepsilon^{ir} = \tau$; and g_v is the non-linear viscous function, given as:

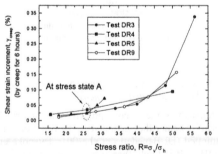

Fig. 8: Shear strain increment taking place during each creep test

rated specimen was more contractive. It is likely, however, that under the present test condition at a

$$g_v(\dot{\varepsilon}^{ir}) = \alpha \left[1 - \exp\left\{1 - \left(\frac{|\dot{\varepsilon}^{ir}|}{\dot{\varepsilon}_r^{ir}} + 1\right)^m\right\} \right] \tag{2}$$

190

where α, m and $\dot{\varepsilon}_r^{ir}$ are the positive material parameters. Based on experimental data, the following form of the decay function was proposed (Tatsuoka et al., 2001b):

$$g_{decay}(\varepsilon^{ir} - \tau) = r^{(\varepsilon^{ir} - \tau)} \qquad (3)$$

where $\varepsilon^{ir} - \tau$ is the strain difference between the respective previous state and the current state; and the parameter r is a constant, which is equal to 1.0 with the new isotach model; constant and smaller than unity with the TESRA model; and not constant with the general TESRA model, given as:

$$r = \frac{r_i + r_f}{2} + \frac{r_i - r_f}{2} \cdot \cos\left[\pi \cdot \left(\frac{\varepsilon^{ir}}{c}\right)^n\right] \qquad (\varepsilon^{ir} \le c)$$

$$r = r_f \qquad (\varepsilon^{ir} > c) \quad (4)$$

The non-viscous stress component σ^f is the lower bound attained when $\dot{\varepsilon}^{ir} = 0$ with the new isotach model. This relationship is called the reference function which is not the lower bound with the TESRA model and general TSRA model. In the present study, the reference relation was modelled by the following empirical equation:

$$R = R_0 + c_R.A_1.\left\{1 - \exp\left(-\frac{\varepsilon}{t_1.c_\varepsilon}\right)\right\} + c_R.A_2.\left\{1 - \exp\left(-\frac{\varepsilon}{t_2 c_\varepsilon}\right)\right\}$$

(5)

where R_0 is the initial value of $R = \sigma_v/\sigma_h$, which is 1.0 in the present study. The parameters c_R, c_ε, A_1, A_2, t_1 and t_2 were determined by fitting procedures.

In the present study, the general TESRA model for shear strain deformation, in the form of one-dimension model, was applied to the results from tests DR3 and DR9 (drained triaxial compression

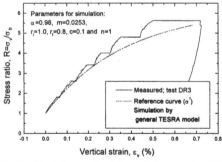

Fig. 10a: Simulation results of stress-strain curves (test DR3)

tests at $\sigma_h = 490$ kPa). **Figs. 10 & 11** compare the overall measured and simulated relationships between the stress ratio R and the vertical strain ε_v and the time histories of ε_v for, respectively, tests DR3 & DR9.

The same values were used for the parameters for the viscous and decay functions for simulating two

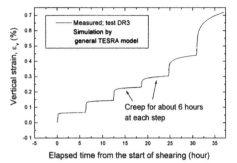

Fig. 10b: Simulation results of strain history (test DR3)

tests. Simulation of tests DR3 and DR9 was done by strain control simulation during monotonic loading, while the simulation of creep was done by stress control. The elapsed times at the start and end of each creep test in the simulation were set to be equal to the respective value in the experiment.

It may be seen from Figs. 10 and 11 that the general TESRA model simulates very well the test results, in particular both creep deformation and post-

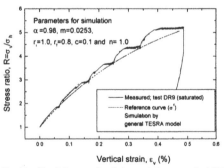

Fig. 11a: Simulation results of stress-strain curves (test DR9)

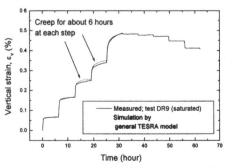

Fig. 11b: Simulation results of strain history (test DR9)

creep behaviour. Simulation of the behaviour during stress-unloading will be reported in the near future by the authors.

3.4 Effect of stress path on the flow property

The stress paths of tests DR3, DR4 and DR5 pass through a common stress point (A at $q = \sigma_v - \sigma_h = 800$ kPa & $p' = (\sigma_v + 2\sigma_h)/3 = 770$ kPa; or $\sigma_v = 1300$ kPa & $\sigma_h = 500$ kPa) (see Fig. 3). A creep test was performed at point A in the three tests. It can be seen from Fig. 8 that the creep shear strain at point A is similar among the three tests. Tatsuoka et al. (2001b) showed that at creep stages during otherwise monotonic plane strain compression loading at $\sigma_h = 396$ kPa on saturated Toyoura sand, the creep shear strain increment became larger with the increase in the initial shear strain rate. The initial shear strain rate was 0.0147 %/min in test DR3, 0.0196 %/min in test DR4 and 0.0267 %/min in test DR5. These rather similar initial shear strain rates would explain the rather similar creep shear strain rates among three tests. On the other hand, the creep volumetric strain rate is very different among the three tests.

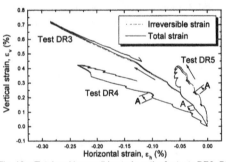

Fig. 12a: Total and irreversible strain paths for tests DR3, DR4 and DR5 (overall)

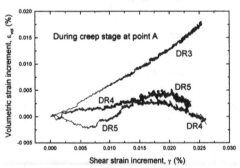

Fig. 12b: Strain paths at point A

The strain paths of these three tests are shown in

Fig. 12a, where the solid lines represent the total strain paths while the dotted lines represent the irreversible strain paths. Irreversible strain increments at each stress state were obtained by subtracting quasi-elastic strain increments from the measured total strain increments. Quasi-elastic strain increments at each stress state were obtained based on the elastic deformation property evaluated by conducting very small-amplitude cyclic loading tests in the course of each test. The following trends of behaviour may be seen from Fig. 12a:

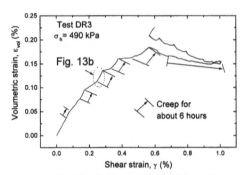

Fig. 13a: Strain path in test DR3 (overall)

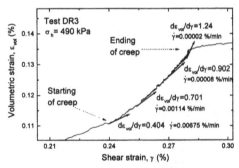

Fig. 13b: Strain path in test DR3 (enlargement)

1) At the same stress point A, the direction of strain increment vector immediately before the start of creep stage largely depends on the stress path direction.
2) Along each stress path, when the respective creep stage starts, the slope of strain path starts rotating at different rates among the three tests.
3) Even with the time elapsing at the creep stage, the slope does not tend to become the same among the three tests. This point could be examined by directly comparing the strain paths during the creep stage at point A (Fig. 12b).
4) When loading was restarted at the original loading rate along the original stress path after the creep stage at point A (and another), the strain

path rotated again to recover the original direction that would have been obtained if loading had been continued without an intermission of creep.

These results indicate that the potential concept is not relevant to model the flow property of gravel. This point is examined more in details below with stress paths in terms of volumetric and shear strains, $\varepsilon_{vol} = \varepsilon_v + 2\varepsilon_h$ and $\gamma = \varepsilon_v - \varepsilon_h$ (see **Figs. 13, 14 & 15**). Note that as the stress was kept constant at each creep stage, the total and irreversible strain increments are the same with each other; $\dot{\varepsilon}_{vol} / \dot{\gamma}_{vol} = \dot{\varepsilon}_{vol}^{ir} / \dot{\gamma}_{vol}^{ir}$.

<u>Test DR3 (Fig. 13a)</u>: As the creep started in the course of both stress-loading and stress-unloading,

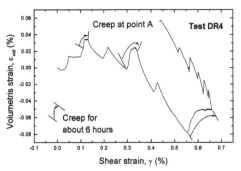

Fig. 14: Strain path in test DR4

the gravel started becoming more contractant (i.e., the ratio of irreversible strain rate $\dot{\varepsilon}_{vol}^{ir} / \dot{\gamma}_{vol}^{ir}$ started increasing) (see also Fig. 13b).

<u>Test DR4 (Fig. 14)</u>: As the creep stage started, the gravel became suddenly much more contractive, or became largely contractive from dilative behaviour during monotonic loading immediately before. So, the rotation of the irreversible strain rate vector $\dot{\varepsilon}_{vol}^{ir} / \dot{\gamma}_{vol}^{ir}$ at the start of creep stage was very sharp, while the amount of rotation increases as the stress ratio at the creep stage became higher. However, as the time elapsed at the creep stage, the ratio of irreversible strain rate $\dot{\varepsilon}_{vol}^{ir} / \dot{\gamma}_{vol}^{ir}$ did not increase monotonically, but the vector started rotating in the opposite direction with the gravel becoming less contractant, or becoming dilatant again. This behaviour is typically seen at point A. For this reason, the rotation when loading was restarted at point A was not significant. This behaviour took place in a more obvious way at the last creep stage during stress-loading.

<u>Test DR 5 (Fig. 15)</u>: As the creep started, opposite to the cases of tests DR3 and DR4, the gravel be-

came less contractive with a decrease in the ratio of irreversible strain rate $\dot{\varepsilon}_{vol}^{ir} / \dot{\gamma}_{vol}^{ir}$ (see **Fig. 15b**). The ratio $\dot{\varepsilon}_{vol}^{ir} / \dot{\gamma}_{vol}^{ir}$ during the creep stage was low, similar to test DR4. As loading was restarted, the ratio $\dot{\varepsilon}_{vol}^{ir} / \dot{\gamma}_{vol}^{ir}$ quickly returned to the original value.

The direction of irreversible strain increment vector is not a unique function of stress state (i.e., a potential function does not exist at a given stress point). In addition, the manner of rotation of this vector when, during and after a creep stage is very complicated and not well understood. So, we cannot propose at present a rule or model that can fully

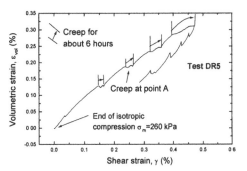

Fig. 15a: Strain path in test DR5 (overall)

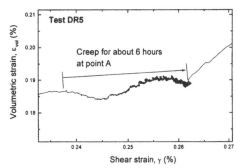

Fig. 15b: Strain path in test DR5 (enlargement)

simulate all aspects of the observed flow characteristics.

4 CONCLUSIONS

The following conclusions can be derived from the test results presented above:

1) At creep stages during otherwise monotonic increase and decrease of deviator stress in triaxial compression tests on a dense well-graded gravel at $\sigma_h' = 490$ kPa, significant creep deformation took place. The creep shear strain increased as the stress state approached the failure.

2) Under otherwise the same loading conditions, the viscous deformation characteristics as well as the stress-strain behaviour were essentially the same between partially saturated and fully saturated specimens, showing insignificant effects of suction on the observed behaviour.

3) In contrast with positive creep shear strain increments taking place at creep stages during stress-loading, negative shear strain increments (called the creep recovery) took place at creep stages during stress-unloading.

4) The viscous shear deformation characteristic observed during triaxial loading along a fixed stress path with $\sigma'_h = 490$ kPa could be simulated very well by one of the three-component rheological models, called the general TESRA model.

5) The viscous aspect of the flow property was investigated under more general stress conditions by performing triaxial compression tests along three different stress paths, and creep tests were performed during global stress-loading and stress-unloading. When the creep stage started, the irreversible strain increment vector started rotating at different rates from the one observed at the end of monotonic loading. At the respective creep stage, the vector continuously changed with time at different rates according to the loading stress path immediately the start of creep. These results show that no unique flow rule exists at a given stress state.

REFERENCES

AnhDan LeQuang (1999): "The applicability of a rheological model for dense gravel", Internal report, University of Tokyo.

Di Benedetto,H., Tatsuoka,F. and Ishihara,M. (2001): "Time-dependent deformation characteristics of sand and their constitutive modelling", Soils and Foundations (submitted)

Goto,S. Tatsuoka,F., Shibuya,S., Kim,Y.-S. and Sato,T. (1991): "A simple gauge for local strain measurements in laboratory", Soils and Foundations, Vol. 31, No. 1, pp. 169-180.

Hoque,E., Sato,T. and Tatsuoka,F. (1997): "Performance evaluation of LDTs for the use in triaxial tests", Geotechnical Testing Journal, ASTM, Vol.20, No.2, pp.149-167

Tatsuoka,F. (1988): "Some recent developments in triaxial testing systems for cohesionless soil", Advanced Triaxial Testing of Soil and Rock, ASTM, STP 977, pp. 7-67.

Tatsuoka,F., Santucci de Magistris,F., Hayano,K., Momoya,Y. and Koseki,J. (2001a): "Some new aspects of time effects on the stress-strain behaviour of stiff geomaterials", The Geotechnics of Hard Soils – Soft Rocks (Evangelista and Picarelli eds.), Balkema, Vol.2, (in print).

Tatsuoka,F., Ishihara,M., Di Benedetto,H. and Kuwano,R., (2001b): Time-dependent deformation characteristics of geomaterials and their simulation, Soils and Foundations (submitted).

Advanced Laboratory Stress-Strain Testing of Geomaterials, Tatsuoka, Shibuya & Kuwano (eds),
© *2001 Taylor & Francis, ISBN 90 2651 843 9*

Modelling of stress and strain relationship of dense gravel under large cyclic loading

K. Balakrishnaiyer & J. Koseki
University of Tokyo, Tokyo, Japan

ABSTRACT: A modelling approach to characterize the stress~strain relationship of gravel under monotonic as well as large cyclic loading is presented with the use of a series of drained triaxial test data on reconstituted dense gravel, using $\sin(\phi)_{mob}$ as stress parameter and plastic shear strain γ^p as strain parameter. Monotonic loadings were modelled according to the generalized hyperbolic equations (GHE). Masing's 2nd rule, proportional rule and the drag rule were implemented for modelling of cyclic loading. The stress and strain parameters were further modified as $\sin(\phi)_{mob} / \sin(\phi)_{peak}$ and $\sum (d\gamma^p / \gamma_r)$, respectively, aiming at achieving a unique stress~strain relationship independent of stress paths. $(\phi)_{peak}$ is the peak mobilized angle of internal friction and γ_r $[=\tau_{peak}/G_O]$ is a reference shear strain. τ_{peak} and G_O are peak shear strength and quasi-elastic shear modulus, respectively. The simulated stress-strain behaviours were well in accordance with those measured under various triaxial stress paths.

1 INTRODUCTION

In general soils under foundations of super structures are often subjected to various kinds of loadings under normal load conditions. They can be monotonic loading/unloading, creep loading, small vibrations and also relatively large vibrations as those developed due to earthquake forces. Cyclic loading, among others gets important, especially in earthquake prone regions, where the soil can undergo large cyclic loading, and the stability is needed to be ensured for the serviceability of the super structure and foundation, even after the structure survived without much damage. The deformation characteristics of soils undergo continuous change during repetitive loading, and it is needed to formulate the deformation character-ristics with the consideration of change in stress and strain states and histories.

It is obvious that in the real field the loading conditions are not generally uniform or harmonic. Formulation of models and simulation of real soil behaviors, however, cannot be initiated from a complex stress and strain condition, and hence rather simpler stress fields can pave a way towards the prediction in real field with feasible solutions. Unlike the case of modelling the stress~strain characteristics during monotonic loading, cyclic loading needs much more consideration on the selection of the stress~strain parameters. For example, the relationships between the deviator stress q $[= \sigma_v - \sigma_h]$ and the axial strain ε_v in triaxial compression and extension tests at constant confining pressure σ_h or vertical stress σ_v become largely unsymmetrical as the stress state approaches the peak state in both stress conditions. This may be explained by the influence of the distance between the current and the failure states on the deformation characteristics. However, the relationship between $\sin(\phi)_{mob}$ $[= (\sigma_v-\sigma_h)/(\sigma_v+\sigma_h)]$ and the shear strain γ $[=\varepsilon_v-\varepsilon_h]$ becomes essentially symmetrical and hence this procedure is equivalent to obtaining relationships between q and γ from triaxial compression and extension tests at constant mean stress σ_m $[=(\sigma_v + \sigma_h)/2]$ (Tatsuoka et al., 1999). It would be preferable to employ stress~strain relationships as symmetrical as possible, even if further modification may be required, as shown later. In the present study, therefore, the basic stress and strain parameters considered were, $\sin(\phi)_{mob}$ $[= (\sigma_v - \sigma_h) / (\sigma_v + \sigma_h)]$ and plastic shear strain γ^p $[= \varepsilon_v^p - \varepsilon_h^p]$, respectively.

Quasi-elastic strains were separated from the total strains and the plastic strain is considered for the simulation, as the accumulation of plastic strain is the key-parameter in the modelling approach presented.

2 FRAMEWORK OF THE MODEL

2.1 Basic skeleton curve

Soil deformation, in general, is non-linear, and this non-linearity needs to be considered in modelling.

The deformation characteristics during unloading and reloading are linked and correlated with the basic skeleton curves obtained during monotonic loading. For modeling a given highly non-linear stress~strain relation of soil or rock, the use of hyperbolic equation is very popular. Hyperbolic formulation is widely used, because of it's simple form using only two parameters, the initial stiffness and peak strength, which are having clear physical meanings. Among others, Tatsuoka and Shibuya (1991) demonstrated possible models for the monotonic loading and their applicability to soil responses with the evidences of laboratory testing results. Starting from a simple hyperbolic formulation, they proposed a generalized hyperbolic equation (GHE) as follows;

$$y = x / [1/C_1(x) + x/C_2(x)] \qquad (1)$$

Where, x & y are strain and stress parameters, respectively. C_1 & C_2 are modification parameters as functions of strain parameter, which were considered as two cosine functions as given in Tatsuoka & Shibuya (1991).

2.2 Modelling of hysteretic curves

2.2.1 Masing's second rule and proportional rule
In the case of modeling hysteretic curves, the well-known Masing's 2nd rule (Ohsaki, 1980) may be used as a basic idea. However, simulation of reliable and precise experiment results suggests that it needs modifications to simulate the real soil behavior more precisely. Regarding plastic properties, Tatsuoka et al. (1997) have reported that, in modelling stress-strain characteristics during cyclic loading, there would be a better response by modifying (generalizing) Masing's 2nd rule to proportional rule due to the effect of possible non-symmetry in basic skeleton curves obtained for compression and extension sides under triaxial or plane strain conditions. The principles in Masing's 2nd rule and proportional rule are basically the same, but additionally, possible non-symmetry of skeleton curves is accounted for in proportional rule.

2.2.2 External and internal rules (without drag)
The proportional rule can be separated into external and internal rules depending on the strain level of the current turning point and previous strain histories (Masuda et al., 1999; Balakrishnaiyer & Koseki, 2000; Tatsuoka et al., 2001). They will be herein explained briefly. External and internal rules are basically similar, but vary in the way of deriving the proportional parameter n. For deciding the type of rule to be applied for the next loading, two limiting strain levels, called as the maximum and minimum strains are referred each time of reversing stress

direction. If the turning occurs at this limiting strain then external rule is applied. Otherwise, i.e. when the turning point occurs within the limit, the internal rule is applied. Note that these maximum and minimum strains, denoted as X_{min} & X_{max}, are not the real maximum or minimum strains experienced throughout the strain history. Rather, a particular strain history rule is applied to define those limiting strain levels for this purpose. In Figure 1, for example, two distinct basic skeletons f(X) and g(X) are considered for triaxial compression (TC) and triaxial extension (TE), respectively. Note that, in case of perfect symmetry, g(X) ≡ -f(X) and n = 2, as assumed in Masing's 2nd rule. Using the GHE, the basic skeleton curves can be modelled. With these known basic skeletons, subsequent unloading and reloading characteristics are predicted.

The point A in Figure 1 is assumed as a reversing point after following a virgin loading path from the origin O. Then $X_A > X_{max}$, since X_{min} and X_{max} had been defined as 0.0 at the beginning. Thus, the unloading from A would be external rule and X_A is assigned to new X_{max}. The current (X_{max}, Y_{max}), i.e. point A in this case, is joined with the origin of loading skeleton curve and extended to unloading skeleton curve to get the new (X_{min}, Y_{min}).

The external unloading curve from A can be expressed as Y =g_1(X) in the following form;

$$\frac{g_1(X) - Y_A}{n_1} = g\left(\frac{X - X_A}{n_1}\right) \qquad (2a)$$

where, the proportional parameter n_1 is given by,

$$n_1 = -\left(\frac{Y_A - Y_B}{Y_B}\right) \text{ or } -\left(\frac{X_A - X_B}{X_B}\right) \qquad (2b)$$

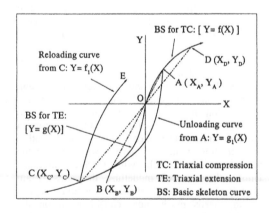

Figure 1. Graphical explanation for Masing's 2nd rule and proportional rule (external rule)

X and Y are the strain and stress parameters while subscripts denote the corresponding points. Point B is obtained by joining the reversing point A to the origin of the loading skeleton curve and extending that straight line to meet the basic skeleton curve in the other side at B. According to this, the modeled unloading curve bounds for point B and meets the basic skeleton curve tangentially at B and then follows the basic skeleton curve until reaching next turning point. In the case of Masing's second rule the approach is similar to the above, but the proportional parameter would be always equal to 2.0 as symmetry in basic skeleton curves is assumed (i.e. $X_B = -X_A$ and $Y_B = -Y_A$).

When reaching next reversing point along the unloading path, say point C, then X_C is compared with the current X_{min}, and the minimum of X_C and X_{min} is assigned to new X_{min} and the corresponding Y coordinate is assigned to new Y_{min}. After redefining, if X_C is equal to X_{min}, then external rule is applied for reloading from C (Fig. 1). If $X_{min} < X_C < X_{max}$, then internal rule is applied for reloading, as explained later. If reloading from C was by external loading, it would be bound for a point D on the loading skeleton. Point D is obtained by joining current reversing point C with the origin of loading skeleton and extending it to meet the loading skeleton again at D. Another set of equations will be obtained from eqs. (2a) & (2b), by substituting A, B, g(X), $g_1(X)$ and n_1 with C, D, f(X), $f_1(X)$ and n_2, respectively. A new (X_{max}, Y_{max}) is to be defined at the next reversing point E, by joining (X_{min}, Y_{min}) with the origin of skeleton curve in unloading direction and extending it to the skeleton curve in loading direction. In this case it would be same as the target point D. Comparing the X-coordinate of that intersecting point (=D) with X_E, the new (X_{max}, Y_{max}) is defined and the rule for next unloading from E is decided accordingly. This makes the instantaneous points of (X_{max}, Y_{max}) and (X_{min}, Y_{min}) be located always on a straight line, when the proportional rule is applied without drag.

The modeling for the case of internal loading from point C, is expressed in Figure 2. The target point is assumed to be the previous reversing point, i.e. in this case it was point A on loading skeleton curve. The reloading curve up to the target point A would be given as $f_1(X)$,

$$\frac{f_1(X) - Y_C}{n_3} = f\left(\frac{X - X_C}{n_3}\right) \qquad (3a)$$

where, the proportional parameter n_3 is given as,

$$n_3 = \left(\frac{Y_A - Y_C}{Y_D}\right) = \left(\frac{X_A - X_C}{X_D}\right) \qquad (3b)$$

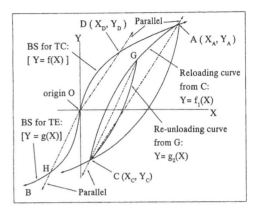

Figure 2. Graphical explanation for Masing's 2nd rule and proportional rule (internal rule)

To obtain the point D, a straight line parallel to CA is drawn through the origin of the loading skeleton so that it will meet the loading skeleton at D. The internal reloading curve $f_1(X)$ would meet the loading skeleton at A and then switches to the loading skeleton curve when loading continues further. It should be noted that unlike the case of external rule, in the case of internal rule the rejoining with another curve at the target point would not be tangential. To make it a smooth rejoining similar to real condition, including some additional features to the internal curve may be possible. However, this needs to be studied in detail and was not considered in the present model.

When the next turning occurs at point G before reaching the target point A, internal rule would be applied for unloading from G, because at G, $X_{min} < X_G < X_{max}$ (=X_A). The target point would be C, which is the previous reversing point for G. A line parallel to GC is drawn through the origin of unloading skeleton g(X) so that it will meet the curve at point H. A new proportional parameter n_4 is obtained from the coordinates of G, C & H and the re-unloading curve from G, i.e. $g_2(X)$, is modeled according to eqs. (3a) & (3b) by substituting C, G, H and g(X) for A, C, D and f(X), respectively. Once the target point is reached by internal unloading then the curve switches to the new curve, that is the external unloading curve from A. It should be noted that the target point for any reloading by internal rule before reaching C would be G and after reaching C would be A (not G). In any case, after the previous reversing point has passed during shearing, all the memory of cyclic loading experienced during a period starting from and ending at that reversing point is erased.

197

2.2.3 Drag rule

Drag rule is introduced to simulate drained cyclic behavior in which the stress amplitude increases during cyclic loading with a constant strain amplitude, or the strain amplitude decreases during cyclic loading with a constant stress amplitude, both at a decreasing rate with the increase in the number of loading cycles. The entire skeleton curve for one loading direction is assumed to be dragged along the strain axis according to the occurrence of accumulation of plastic strain in the opposite direction.

Tatsuoka et al. (1997) discussed the importance of further modification to the proportional rule in the same manner. They reported that there would be significant influences by cyclic strain hardening, damage by straining etc., and suggested that additional rules are required to model these features and to account for behaviors under more general stress conditions.

In this regards, an approach by applying a horizontal shift to the basic skeleton curves i.e. dragging the basic skeleton curve along the X-axis (strain parameter axis) was implemented for dense Toyoura sand under plane strain condition (Masuda et al., 1999). A similar approach was explored for dense gravel by Balakrishnaiyer & Koseki (2000).

It was assumed that the amount of drag β applied to one basic skeleton curve in one loading direction is a function of the plastic shear strain accumulated in the opposite loading direction (Masuda et al. 1999; Balakrishnaiyer & Koseki, 2000). The amount of drag was calculated according to the formula given in Figure 3. F & β_{max} are kept constants.

2.2.4 External and internal rules (with drag)

The selection of internal or external rule with drag rule is basically the same as the concept explained in 2.2.2 for the case without drag. The additional features with drag would be the shifting of origins of basic skeletons along the X-axis. Therefore, at each reversing point, the corresponding amount of drag is evaluated, and the shifted skeletons and their shifted origins are considered to define new X_{min} or X_{max}. It should be noted that with drag rule, the origins of skeleton curves are continuously moving apart until reaching the maximum drag limit (β_{max}), and never comes closer to each other. Unlike the case without drag, with drag the target point for external loading or unloading is not related to either X_{min} or X_{max}. The target point for the external unloading from point A in Figure 4, for example, would be point B while X_{min} at A is X_P.

The target points for internal loading would be the relocated position of the previous turning point. The term relocation means the new position attained by the influence of the application of drag to the skeletons. The target point F for the internal reloading from C', for example, is obtained by extending the line C'A to the instantaneous loading skeleton until it meets at F. Typical examples of X_{max} and X_{min} at different points are given in Figure 4. A more detailed explanation on drag rule is given in Tatsuoka et al (2001).

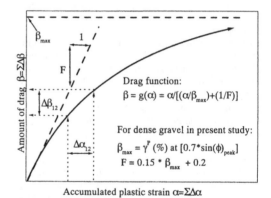

Drag function:
$$\beta = g(\alpha) = \alpha/[(\alpha/\beta_{max})+(1/F)]$$

For dense gravel in present study:
$$\beta_{max} = \gamma^P \ (\%) \ at \ [0.7*\sin(\phi)_{peak}]$$
$$F = 0.15 * \beta_{max} + 0.2$$

Figure 3. The assumed relationship between parameters α and β in drag rule

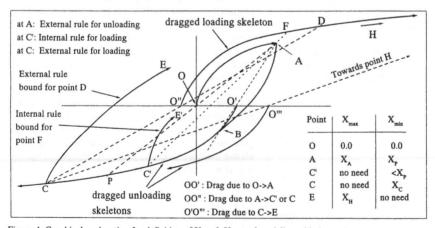

Figure 4. Graphical explanation for definition of X_{max} & X_{min} and modeling with drag rule.

3 SIMULATED RESULTS-1

Typical set of test data of dense gravel subjected to large cyclic loading during shearing at constant σ_h of 883 kPa, and its simulation is presented in Figs. 5a,b,c. Refer to Balakrishnaiyer & Koseki (2000) for testing procedures. When the simulated results with Masing's 2nd rule and proportional rule (both without and with drag) are compared, it is obviously visible that proportional rule significantly improved the simulation and application of drag rule further improved the results. Similar trends could be observed with other selected stress paths such as σ_v or σ_h or σ_m constant paths.

4 UNIQUE STRESS~STRAIN RELATIONSHIPS

Though it was possible for getting a reasonable prediction from the above-mentioned model with the $\sin(\phi)_{mob}$ - γ^p phase, it was not a unique relationship among various specimens with different stress paths and initial conditions (Fig. 6a). When considering a better practical application of the model, necessity for finding a unique stress~strain relationship between a set of stress and strain parameters independent of stress paths and initial conditions is coming up (Tatsuoka et al., 1999).

The primary objective of this new formulation is, based on the results from monotonic and cyclic loading triaxial tests on gravel performed along some specific stress paths (e.g. σ_h constant or σ_v constant or σ_m constant or $R = \sigma_v / \sigma_h$ constant), to develop a hysteretic model which can predict stress and strain relationships of gravel subjected to cyclic loading applied;

   a) along general stress paths (continuous rotation of principal stress direction is out of the scope at present)

   b) with different initial void ratios

   c) with different and varying degrees of damage to the soil particle structure

In an attempt to eliminate at least some of the possible effects, a new set of stress and strain parameters was investigated as given below;

$$Y = \sin(\phi)_{mob}/ \sin(\phi)_{peak}$$
$$X = \Sigma \left(d\gamma^p / \gamma_r\right) \qquad (4)$$

where, $\sin(\phi)_{peak}$ is a peak value of $\sin(\phi)_{mob}$; $\gamma_r = \tau_{peak}/G_0$; τ_{peak} is the peak shear stress; G_0 is the quasi-elastic shear modulus.

The normalizing parameters $\sin(\phi)_{peak}$, τ_{peak} and G_0 are not constants but functions of current stress state and corresponding failure state coefficients under triaxial compression (TC, $\sigma_v > \sigma_h$) or triaxial extension (TE, $\sigma_v < \sigma_h$) condition. The normalizing parameters $\sin(\phi)_{peak}$ and τ_{peak} are taken as variables depending on the current mean stress σ_m assuming that the failure state coefficients would be evaluated

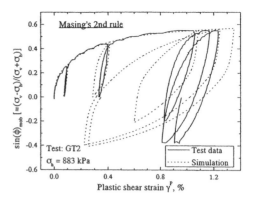

Figure 5a. Typical comparison of test results with simulation (symmetric skeletons; i.e. Masing's 2nd rule).

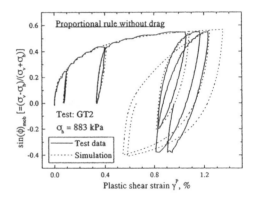

Figure 5b. Typical comparison of test results with simulation (non-symmetric skeletons; i.e. proportional rule).

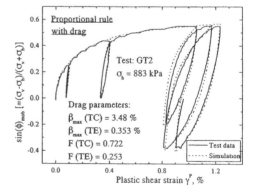

Figure 5c. Typical comparison of test results with simulation (non-symmetric skeletons with drag).

along a stress path where the current mean stress σ_m $[=(\sigma_v+\sigma_h)/2]$ is kept constant. σ_m constant path could be equal to or approximately equal to the closest distance from the current stress state to the failure envelope.

In Figure 6b, a graphical explanation is given how the modifying parameters could be evaluated. The figure includes different stress paths applied under triaxial stress condition. For example, consider any instantaneous stress state P. If we assume a set of linear lines for the failure envelops in TC & TE, then by following constant σ_m path the instantaneous failure state (Point Q) can be reached as given in Figure 6b. Note that two different sets of failure envelopes as shown in Fig. 6c, are used to account for the influence of the direction of loading and the initial dry density. If the stress path is towards TC failure, then the corresponding failure envelop for TC will be referred and if the stress path is towards TE failure, then the failure envelop for TE will be referred. From the coordinates of the projected point (e.g. Q) the instantaneous τ_{peak} and $\sin(\phi)_{mob}$ values relevant to stress state P can be obtained as follows;

$$\tau_{peak} = (\sigma_{vQ} - \sigma_{hQ}) / 2 \qquad (5a)$$

$$\sin(\phi)_{peak} = (\sigma_{vQ} - \sigma_{hQ}) / (\sigma_{vQ} + \sigma_{hQ}) \qquad (5b)$$

Similarly, for any given instantaneous stress state, with the given failure envelops and the stress path direction, the modifying parameters τ_{peak} and $\sin(\phi)_{peak}$ can be obtained.

The quasi-elastic shear modulus G_0 is representing the maximum shear modulus for the current stress state. In the case of triaxial tests or plane strain tests the major principal stresses are vertical stress σ_v and horizontal stress σ_h. Depending on compression or extension under triaxial or plane strain condition, the σ_v and σ_h values or σ_h and σ_v values will be assigned to major and minor principal stresses, respectively. Under this circumference, from Mohr's circle at any instant the plane where the maximum shear stress occurs would be 45° inclined to the vertical or horizontal direction. Then;

$$G_0 = G_{45} = \frac{d\tau_{45}}{d\gamma^e} = \frac{1}{2} \cdot \frac{d\sigma_v - d\sigma_h}{d\varepsilon_v^e - d\varepsilon_h^e} \qquad (6a)$$

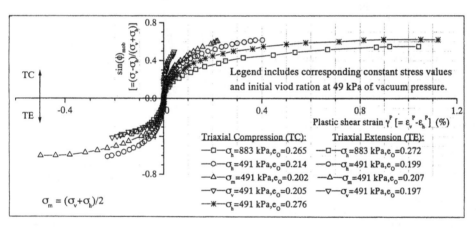

Figure 6a. $\sin(\phi)_{mob}$ and γ^P relationships under different stress paths in triaxial condition.

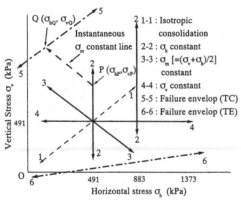

Figure 6b. Sketch of different stress paths and the failure envelopes used under TC and TE conditions (not in scale).

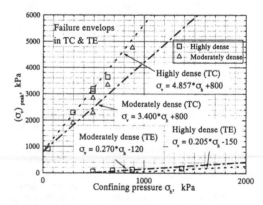

Figure 6c. Linearly fitted failure envelopes in TC and TE for highly ($e_0 = 0.20 \sim 0.23$) and moderately ($e_0 = 0.26 \sim 0.28$) densed reconstituted gravel.

where, superscript e refers to quasi-elastic component. A condition of $d\sigma_h = -d\sigma_v$ is assumed (equivalent to constant σ_m) to have a consistent way in deriving relevant modifying parameters for the unique curve considering the σ_m constant path. Then,

$$G_O = \frac{1}{2} \cdot \frac{2d\sigma_v}{(d\varepsilon_v^e - d\varepsilon_h^e)} = \frac{d\sigma_v}{d\varepsilon_v^e - d\varepsilon_h^e} \quad (6b)$$

By substituting respective quasi-elastic formulations for $d\varepsilon_v^e$ and $d\varepsilon_h^e$ corresponding to triaxial condition, we can get the formula for the instantaneous quasi-elastic shear modulus G_O as given in eq. (6c).

$$G_O = \frac{1}{\left(\dfrac{1+\nu_{vh}}{E_v}\right) + \left(\dfrac{1+2\nu_{hv} - \nu_{hh}}{E_h}\right)} \quad (6c)$$

where E and ν refer the quasi-elastic Young's moduli and Poisson's ratio, respectively.

In the above-mentioned modification with a new set of stress and strain parameters, some of the mostly influential factors on the deformation characteristics of granular materials may possibly be accounted for. For example, using $\sin(\phi)_{mob}$ as the stress parameter, effect of various stress paths especially confining stress effect under triaxial stress condition may be accounted for, whereas normalizing the stress parameter by $\sin(\phi)_{peak}$ may account for the effect of relative difference between the current stress state and the peak stress state as well as the difference in the shear strengths in TC and TE. This may be relevant for granular materials where the angle of internal friction contribute to the shear strength properties more predominantly than it's cohesion.

The inherent and stress-induced anisotropies in the deformation characteristics of the compacted material can be possibly accounted for with the normalization of strain parameter by the reference strain γ_r. The instantaneous maximum quasi-elastic shear modulus G_O [$=G_{45}$], which is obtained from relevant quasi-elastic formulations, includes parameters for inherent anisotropy and stress state-induced anisotropy.

Note that the effect of damage or destructuration can also be accounted for by introducing a stress parameter-dependent damage function to the quasi-elastic shear modulus. In the present study, however, as the general extent of damage to the elastic properties was not significant, damage was assumed negligible.

Application of the above set of stress~strain parameters very much improved the uniqueness of the basic skeleton curves obtained for selected different stress paths as given in Figure 6d.

5 SIMULATED RESULTS-2

A unique relationship was approximated as two distinct skeleton curves from the results in Figure 6d, and proportional rule and drag rule were applied to simulate test results with complicated stress paths as shown in Fig. 7a. In this case, R-constant stress path (i.e. keeping $R = \sigma_v/\sigma_h$ constant) was also included. It is worthy to note that, R-constant stress path cannot be properly modeled with the initial set of stress~strain relationship because $\sin(\phi)_{mob}$ will be unchanged when stress ratio R is constant. Figures 7b, c show the simulation of such test results by using the new set of stress and strain parameters and the back-calculated relationships between $\sin(\phi)_{mob}$ and γ^P, respectively. It can be seen that the simulated and back-calculated results are well in accordance with the test results, suggesting that the above modeling approach is relevant.

6 CONCLUSIONS

It was clear with the empirical evidences that moving to proportional rule and also applying a shift (drag) to the referred basic skeleton (GHE) considering the previous strain history resulted into a

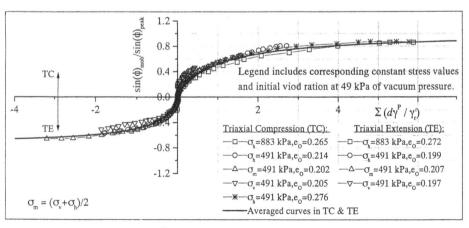

Figure 6d. $\sin(\phi)_{mob} / \sin(\phi)_{peak}$ and $\sum (d\gamma^P / \gamma_r)$ relationships under different stress paths and the assumed average curve.

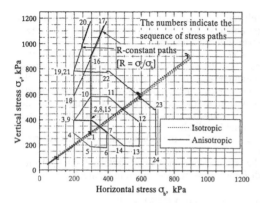

Figure 7a. Stress path applied for the special test

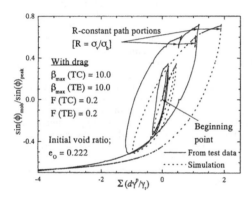

Figure 7b. Comparison of experimental data and corresponding simulation data under unique relationship.

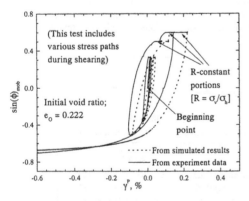

Figure 7c. Comparison of $\sin(\phi)_{mob} \sim \gamma^p$ relationship back-calculated from simulation with that from test data.

better prediction of the deformation behavior under cyclic loading than direct application of Masing's 2nd rule for the simulations.

A significant improvement towards a unique stress-strain relationship was observed when the stress~strain parameters were modified and an average relationship could be formulated as unique relationship.

A modelling approach similar to the one employed with stress-strain parameters $\sin(\phi)_{mob}$ and γ^p, could be carried out for the new set of modified stress and strain relationship. The response was reasonably good when the simulated results were compared to the test data. This new set of relationships could be applicable to R-constant paths also.

REFERENCES

Balakrishnaiyer, K. & Koseki, J. 2000. Modeling of stress-strain relationships of a reconstituted gravel subjected to large cyclic loading. Proc. of the third Japan-UK workshop. *Implications of Recent Earthquakes on Seismic Risk* Vol. 2: 105-114.

Masuda, T., Tatsuoka, F. & Yamada, S. 1999. Sand behavior in cyclic lateral plane strain loading and modelling the hysteretic stress-strain relations. Proc. of IS on *Pre-failure deformation characteristics of geomaterials*. Balkema, Vol. 1: 675-680.

Ohsaki, Y. 1980. Some notes on Masing's law and non-linear response of soil deposits. Journal of the Fac. of Engineering, Univ. of Tokyo, Vol. XXXV, No. 4: 513-536.

Tatsuoka, F. & Shibuya, S. 1991. Modelling of Non-Linear Stress-Strain Relations of Soils and Rocks. *Seisan-Kenkyu*, Journal of IIS-University of Tokyo, Part I & II, Vol. 43, Nos. 9 & 10: 409-412 & 13-15.

Tatsuoka, F., Jardine, R.J., Lo Presti, D., Di Benedetto, H. & Kodaka, T. 1997. Characterizing the Pre-Failure Deformation Properties of Geomaterials. Theme Lecture, 14th IC on SMFE, Balkema, Vol. 4: 2129-2164.

Tatsuoka, F., Modoni, G., Jiang, G.L., Anh Dan, L.Q., Flora, A., Matsushita, M. & Koseki, J. 1999. Stress-behavior at small strains of unbound granular materials and its laboratory tests. Proc. of the Int. workshop on *Unbound Granular Material*. Balkema, 17-61.

Tatsuoka, F., Masuda, T. & Siddiquee, M.S.A. 2001. Modelling the stress-strain behaviour of sand in cyclic plane strain loading. Paper submitted for possible publication in *Soils & Foundations*.

Advanced Laboratory Stress-Strain Testing of Geomaterials, Tatsuoka, Shibuya & Kuwano (eds),
© 2001 Taylor & Francis, ISBN 90 2651 843 9

Tied-Back Wall Behavior and Selection of Geomechanical Parameters for Numerical Analysis

M.M. Berilgen, I.K. Ozaydin & V.Cuellar
Yildiz Technical University, Istanbul, Turkey

O. Inan
Geotest Consulting Co., Izmir, Turkey

ABSTRACT: : In this paper the selection of geomechanical parameters for the analysis of tied-back wall behavior under the influence of static loads is discussed. For this purpose the observed behavior of a model anchored wall in the laboratory and a field anchored wall are compared with the findings of the numerical analyses. For numerical analysis the finite element method is used, the soil medium is modelled as a nonlinear elasto-plastic material and interface elements are used for modelling the soil-anchor interfaces. In numerical modelling the construction stages are taken into account. For the laboratory model the geomechanical parameters are obtained from triaxial tets results, whereas for the field anchored wall they are estimated from CPT test results.

1 INTRODUCTION

In recent years, due to their important advantages over other methods, tie-back walls have become a common construction method for the support of steep and deep excavations. This is because, firstly, with the aid of a prestressing force applied through soil anchors, the stability of excavation sides is ensured and deformation is kept under control; secondly, tie-back walls provide for better working area in the excavation and; thirdly, such walls usually constitute the most economical support system for deep and wide excavations under all kinds of ground conditions. But, while having a lot of advantages over other excavation support system methods, their mechanics are quite complicated because they are composed of three different materials (soil, concrete and steel). Many researchers have dealt with the investigation of tie-back wall mechanics (Ostermayer & Scheele, 1977, Davis & Plumell, 1982, Desai, et al., 1986). This particular study aims at investigating tie-back wall behavior via numerical analysis and selection of the material parameters for such analysis.

2 LABORATORY MODEL TESTS

Model tests were carried out in a 70 cm x 95 cm x 75 - cm model sand tank manufactured with glass side walls (Figure 1). In the base of the tank, a groove was formed for the placement of the an-

chored diaphragm wall that allowed the rotation of the wall without any displacement at its base.

A quartz sand with minimum and maximum grain sizes of 0.05 and 0.50 mm, a uniformity coefficient of C_u=3, a gradation coefficent of C_c=1.33, a specific density of G_s=2.65 and minimum and maximum void ratios of 0.50 and 0.80 respectively, was used in the model tests. The sand was deposited in the tank with the aid of a raining system which allowed for the formation of a uniform sand bed with controlled density (Steenfelt,1973). A neoprene air bag placed on the surface of the sand deposit behind the wall was used to apply varying levels of surcharge loading.

The movable diaphragm wall was made of 7 mm thick glass which was placed in the groove at the

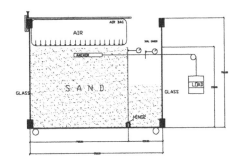

Figure 1: Model test system

base so that the wall was laterally fixed but could still freely rotate around its base. A hole in the middle of the wall at a 10 cm depth from the surface of the the the backfill allowed for the placement of a 2 mm diameter steel wire attached to the anchor placed in the sand. A hanger system attached to the steel wire was used to apply different levels of pull-out force on the anchor with dead weights. The anchor was made of an aluminum pipe 25 mm in diameter and 200 mm in length. Three sets of electrical strain-gauges were glued on the outer surface of the anchor pipe, at 25, 100, and 175 mm distances from the pulling head of the anchor to measure the axial and tangential deformations. The surface of the aluminum pipe was covered with epoxy glued sand grains to obtain a rough surface.

3 3 NUMERICAL ANALYSIS

The analysis of the behavior of the anchored diaphragm walls is basically a soil-structure interaction problem involving a system with various geometrical and behavioral complexities. The use of numerical analysis techniques seem to be more appropriate for these type of soil-structure interaction problems. In this study, the behavior observed in the laboratory model tests (both anchor pull-out tests and anchored wall tests) is analysed using a computer program called PLAXIS which utilizes the finite element method.

3.1 *Analysis of Laboratory Model Tests*

The laboratory anchored wall model was numerically modelled as a 2D soil medium using I80 plain strain finite elements with six nodes, the wall with beam elements and the anchor with 1D bar elements (Figure 2). Between anchor/soil and wall/soil zero thickness interface elements were used and the excavation stages were taken into account. For the soil elements elasto-plastic material behavior and the Mohr-Coulomb failure criteron are assumed. For sand deposit having 30% relative density, the soil parameters were determined from the triaxial tests as :

$$E_{50} = 18750 \quad (kN/m^2)$$
$$c = 0.0 \quad (kN/m^2)$$
$$\phi = 34°$$

where E_{50} is the secant modulus corresponding to 50% of the ultimate strength, c is cohesion, and ϕ is the angle of shearing strength with Poisson's ratio assumed to be $v=0.30$. As assumed in the PLAXIS manual, the use of E_{50} rather than the initial modulus (E_i) for loose medium dense sands was considered to be more appropriate. In order to take into account the variation in the secant modulus of sand with

depth, a relationship similar to the one proposed by Janbu (1963) for the initial modulus was utilized.

$$E_{50} = E_{50}^{ref}\left(\frac{\sigma_3'}{p_a}\right)^m \tag{1}$$

where E_{50}^{ref} = reference secant modulus (determined to be equal to 18750 kN/m^2)
  σ_3' = minor principal stress
  p_a = atmospheric pressure (101.2 kN/m^2)
  m = material constant
The shearing strength angle of the zero thickness interface elements used at wall/soil and anchor/soil interfaces were taken as

$$\tan\phi_i = R_{int} \tan\phi_{soil} \tag{2}$$

where ϕ_{soil} is the shearing strength angle of the soil and R_{int} is a reduction factor (taken as 0.30 for soil/wall interfaces and 1.0 for soil/anchor interfaces in these analyses).

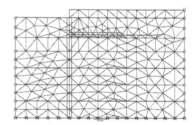

Figure 2: Tie-back wall model test FE mesh

Firstly, the behavior of the wall with no anchorage (Figure 3) was analysed. The computed wall displacements were compared with the measured values for different stages of excavation. Next, the behavior of the anchored wall was analysed and the computed wall displacements compared with the measured values for various stages of excavation (Figure 4). The results of these analyses were also compared with the results of an elastic analysis performed with the computer program LUSAS reported previously in Berilgen and Ozaydin,(1997). As can be observed from Figure 4, the findings of the elesto-plastic analysis are comparable with the measured values. The deformed shapes of the non-anchored and anchored walls are shown in Figures 5 and 6, respectively.

3.2 *Analysis of a Field Case*

A field case history, a 9 m deep excavation of 640 m in linear length covering an area of 20,900 m^2 in Izmir (Turkey) will be considered. A slurry wall with a single row of soil anchors was used to support

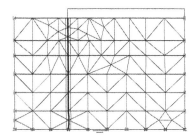

Figure 3. Tie-back wall model test FE mesh

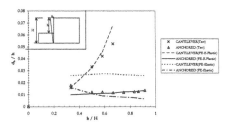

Figure 4. Comparison of displacements

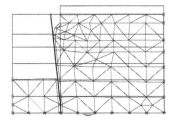

Figure 5. Displaced model of cantilever wall

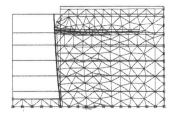

Figure 6. Displaced model of tie-back wall

Depth 0		γ_{wet} (kN/m³)	N_{30}	q_c (MPa)	c_u (kPa)	$\emptyset$ (°)	E_u (Mpa)
		17			0	35	60
GWL							
	FILL W/BLOCKS	17			0	40	100
-6.5							
-9.0	CLAY (CL)	18	7-	0.	5		10
-11	SAND (SM)	17	30	5		36	20
-15	CLAY (CL)	19	15-	0.	100		30
-23	GRAVEL (GW)	18	50	8		40	95
-29	CLAY (CL)	20	Refusal	3.	200		100
-35	GRAVEL (GM)	18	Refusal	8		45	120

Figure 7. Soil profile at the field case site

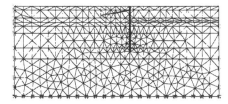

Figure 8. Tie-back wall FE mesh

the sides of the excavation. The soil profile at the
site and soil parameters for the various layers are
shown in Figure 7.

A reinforced concrete wall 0.80 thick and 25 m
deep was constructed using the slurry wall technique
to resist earth pressures and avoid seepage of water
into the excavation due to the high ground water
level in the area. In order to reduce lateral displace-

ment under the surrounding streets and buildings, 375 kN capacity prestressed soil anchors were used at 1.25 horizontal spacings. Inclinometers were placed in the wall to monitor the displacements as the excavation proceeded. The lower end of the inclinometers were placed at 30 m depth from the top of the wall and assumed to be fixed. The soil parameters required for the finite element analysis were estimated from the results of the point cone resistance (q_c) value given in Figure 6. The results of the CPT point resistance were compared with the SPT blow count numbers and some values were corrected accordingly. The constrained modulus values for the sand and gravel layers were obtained from the relationship proposed by Eshaamizaad and Robertson (1996):

$$M_0 = k_M p_a \left(\frac{\sigma'_{vo}}{p_a} \right)^n \tag{3}$$

where σ'_{vo} is the effective vertical overburden pressure, p_a is the atmospheric pressure, and n is the stress exponent which can be taken as equal to 0.200 for normally consolidated sand and 0.128 for overconsolidated sand. The coefficient k_M is a dimensionless modulus number, the value of which can be obtained from the graphs given by Eshaamizaad and Robertson (1996) depending on the overconsolidation ratio and the q_c/p_a ratio. The friction angle of granular soils were estimated from the CPT and SPT results.

For cohesive soils the undrained soil parameters are estimated from the CPT point cone resistance and the SPT blow count numbers. The undrained shear strength (s_u) is was computed firstly from the relationship given below

$$s_u = \frac{(q_c - \sigma_{v0})}{N_k} \tag{3}$$

where σ_{vo} is the total vertical overburden pressure and N_k is an emprical cone factor which is reported to vary between 11 and 19 for marine clays (Lunne ve Kleven, 1981). An average value of 15 is assumed for N_k in this study. The undrained Young's Modulus (E_u) for cohesive soils is estimated from

$$E_u = n \cdot s_u \tag{4}$$

where n is the stiffness ratio the value of which depends on the shear stress level and the overconsolidation ratio. The values of n were chosen from the curves given by Duncan and Buchignani (1976) for 25% of failure stress.

The finite element model (Figure 8) consisted of 860 plane strain soil elements with six nodes. Construction stages (first level excavation, anchoring and final level excavation) were taken into account in the numerical analysis. The wall was moelled with beam elements, node to node anchor elements (spring elements) were used for anchors and axial elements for anchor roots. Zero thickness interface elements were used at the wall/soil interfaces and the shearing strength of interface elements was taken to be equal to 70% of the neighbouring soil elements. At the final stage of excavation, the ground water level was taken at the excavation base (which was initially at 2 m depth from the ground surface). The variation of the lateral displacements with depth as computed for different stages of construction are shown together with the measured displacements from the inclinometer readings in Figure 9, and the deformed mesh of the finite element mesh at the final construction stage is shown in Figure 10. As can be observed from Figure 9, the computed and measured wall displacement are in quite good agreement.

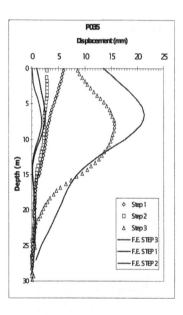

Figure 9. Comparison of wall dispacements

4 CONCLUSIONS

In order to obtain results comparable with observed behavior in the numerical analysis of anchored walls, taking into account the elasto-plastic nature of soil behavior and the selection of appropriate

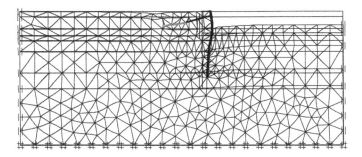

Figure 10.   Tie-back wall FE mesh

mechanical parameters are of prime importance. In this study the selection of soil parameters to be used in the numerical analysis of anchored walls was investigated and the observed behavior of a laboratory model test and a field wall was utilized to check the suitableness of the selected parameters. The anchored walls were numerically modelled using the finite element technique, the soil assumed to behave elasto-plastically, and the construction stages were taken into account in the analyses. Interface elements were employed at the wall/soil and anchor/soil interfaces to model the relative displacements. By comparing of the numerical findings with the measured behavior, the following conclusions were reached:

1.) The initial soil conditions and construction stages were observed to play an important role in the analysis. Also, the results of the analysis were observed to be affected significantly with the elasto-plastic nature of the soil behavior and the relative displacements at the wall/soil and anchor/soil interfaces.

2.) Use of Mohr-Coulomb failure criterion, the associated flow rule and isotropic strain hardening for the elasto-plastic material behavior of soil seems to give results to comparable the observed behavior.

3.) In making use of triaxial test results to obtain soil parameters, the effect of the confining pressure needs to be taken into account as proposed by Janbu (1963) or by a similar relationship. In this investigation it has been shown that such a relationship can be used successfully to obtain soil parameters for the laboratory model test for rather low stress levels.

4.) For field cases CPT results can be utilized to obtain soil parameters. But verifying the CPT results with other field tests (i.e. SPT) and/or laboratory test results leads to more realistic values. Some of the correlations used in practice and utilized in this investigation were

shown to give satisfactory results. It is quite clear that the results of numerical analyses were quite sensitive to the reliability of the field CPT results and the correlations used to obtain the material properties.

5.) Amongst the correlations proposed to obtain soil parameters from CPT point cone resistance, the ones taking into account the loading history of the soil deposit and the state of stress in the field are shown to give more realistic analytical results.

REFERENCES

Berilgen, M.M., and Ozaydin, I.K. (1997). Investigation of Soil-Structure Interaction Tied-Back Walls. In G.S. Littlejohn (ed), Ground Anchors and Ground Structures ; Proc. Intern. Confer., London, 20-21 March 1997. Inst. of Civil Eng.London:Thomas Telford

Davis, A.G. and Plumell, C., (1982) "Full-scale test on ground anchors in fine sand", Proc., ASCE, J. of Geotech. Engr. Div., Vol.108, No.GT3, pp.335-353.

Desai, C.S., Muqtadir, A. and Scheele, F. (1986) "Interaction analysis of anchor-soil system", Proc., ASCE, J. of Geotech. Engr. Div., Vol.112, No.5, pp.537-553.

Duncan, J.M. and Buchignani (1975). An engineering manual of settlement studies. Department of Civil Engineering, University of California, Berkeley

Eslaamizaad, S. and Robertson, R.K. (1996). Cone penetration test to evaluate bearing capacity of foundation in sands. Proc.49th Canadian   Geotechnical Confer., St John's, Newfoundland September.

Janbu, N. (1963). Soil compressibility as determined by odemeter and triaxial tests. Proc. European Confer. On Soil Mech. and Found. Engineering, Wiesbaden: 19-25.

Lunne, T. and Kleven, A. (1981) Role of CPT in North Sea foundation engineering, Session at the ASCE National Convention:Cone Penetration Testing and Materials, St Lois:76-107. American Society of Engineeris (ASCE).

Ostermayer, H. and Scheele, F. (1977) "Research on ground anchors in non-cohesive soils", Specialty Session No.4, 9th Inter. Conf. on SMFE, Tokyo, pp.92-97.

PLAXIS User Manual, Version 7, (1998), Rotterdam:Balkema

Steenfelt, J.S. (1973).   Sand laying techniques. Technical Report, Danish Geotechnical Institute.

Advanced Laboratory Stress-Strain Testing of Geomaterials, Tatsuoka, Shibuya & Kuwano (eds),
© *2001 Taylor & Francis, ISBN 90 2651 843 9*

Small strain stiffness under different isotropic and anisotropic stress conditions of two granular granite materials

Gomes Correia,A.
Technical University of Lisbon, Portugal
AnhDan,L.Q., Koseki,J. and Tatsuoka,F.
University of Tokyo, Japan

ABSTRACT: A series of cyclic triaxial tests were performed using two types of well-graded gravels of crushed aggregate from granite (U_c= 53 with D_{max}= 31.5 mm; and U_c= 28 with D_{max}= 12.5 mm) from Portugal at different stress states along isotropic and anisotropic stress paths (K=$\Delta\sigma_v/\Delta\sigma_h$= 1 to 16) and in triaxial compression at a constant σ_h. Axial and lateral strains were measured locally on-sample. Effects of stress state and loading history, as well as the type of material, on the elastic stiffness defined for a strain amplitude of the order of 0.001 % were evaluated. The vertical elastic Young's modulus E_v was essentially a unique function of the vertical (axial) stress, while it was rather independent of the horizontal (lateral) stress. This results is consistent with those from previous similar tests on sands and gravels. The effects of 21,000 cycles of a relatively large amplitude of deviator stress, causing relatively large strain amplitudes, on the elastic Young's modulus were found negligible, while a triaxial extension loading seems to decrease the vertical elastic Young's modulus. Despite a similar void ratio, the E_v values of the two similar gravels having different grading characteristics differed by a factor of more than two under otherwise the same test conditions.

1. INTRODUCTION

The stiffness at small strains of unbound granular materials is one of the important design parameters in many geotechnical engineering issues. In many cases, the strain level involved is very low. For example, it is usually less than about 0.05 % in the road sub-base subjected to traffic load. In such cases, it is necessary to evaluate the stiffness at these small strains by well controlling the essential factors (Tatsuoka et al., 1999a, c), including:

1) strain level (as the unbound granular materials usually exhibit noticeable strain-non-linearity even at these small strains);
2) stress state in terms of mean pressure and deviator stress (or vertical and horizontal stresses and shear stresses); and
3) preloading history (monotonic and cyclic).

With respect to the first factor 1), the elastic deformation characteristics can be evaluated only at very small strains, say less than 0.001 %, while the accurately evaluated elastic stiffness values could be the essential reference for the stiffness values at larger strain levels.

With respect to the second factor 2), a relevant model of hypo-elastic type, such as the one described in Tatsuoka et al. (1999b& c), can properly describe the stress state-dependency of the elastic stiffness of unbound granular material. The use of this type of model is essential when the field stress state is subjected to large changes in the pressure level, the stress ratio, the direction of the principal stresses and so on. It is known that the conventional isotropic model assuming that the elastic Young's modulus is isotropic with respect to the directions of the principal stresses in relation to the structure of a given material and is a unique function of the confining pressure σ_3 is too unrealistic (Tatsuoka et al. 1999a, b & c). With respect to the third factor 3), the effects of a large number of cycles of cyclic load, such as the one by traffic load, on the small strain stiffness of unbound material is only poorly understood. Tatsuoka et al. (1999a) reported some test results with a poorly-graded fine sand.

As another basic factor, the void ratio has been considered to be the relevant density index for the elastic stiffness of unbound granular materials (e.g. Hardin and Richart 1963). Such a proposal as above is however based on the test results mainly on poorly graded granular materials, but it is poorly understood whether it is also the case with densely compacted well-graded gravels, for example.

In view of the above and for the objective of improving the current design of road pavement in particular, a series of cyclic loading triaxial tests were performed on two types of realistic road base materials. This paper describes first how the vertical elastic Young's modulus E_v depends on the instantaneous stress state, based on results obtained from cyclic tests at various isotropic and anisotropic stress states along different stress paths (i.e., constant stress increment ratios or constant confining pressure in triaxial compression). The influence of a preloading history consisting of 21,000 cycles of a relatively large amplitude deviator stress at a constant confining pressure was also studied. Furthermore, the effects of grain size distribution characteristics on the E_v value were evaluated.

2. TRIAXIAL TESTING

Test materials

Two types of granular materials having the grain size distributions shown in Figure 1 were used. Some physical properties of these two materials are listed in Table 1. The first material is an aggregate of granite having a maximum grain size of 31.5 mm, which is actually used to construct road base layers. The second one is a scaled material of the original one described above, having a maximum size of 12.5 mm and the same percentage of grains less than 0.075 mm as the original one. The scaled material was fabricated from the original one to match the size of a smaller specimen with a diameter of 7 cm, which was used in other tests in this research program. One of the objectives of the present study is to investigate whether the stress-strain behaviour at small strains of an original coarse-grained mixture can be properly evaluated by triaxial tests using relatively small specimens sizes, accessible in engineering practice, using a scaled model of a given original material. This research is still in progress and the result is not presented here.

Triaxial apparatus

A relatively large triaxial apparatus was used with a square prismatic specimen (58 cm high and 23 cm times 23 cm in cross-section), while both axial and lateral strains were measured with a set (in total ten) of local deformation transducers (LDTs; Goto et al. 1991) (Figure 2). Lateral strains, which were free from the effects of bedding error and/or membrane penetration at the lateral surface of specimen, were measured locally on the flat lateral surfaces of specimen in the same way as the local measurement of axial strains. Four pairs of LDTs were used for the lateral strain measurement to alleviate effects of inevitable inhomogeneous deformation due to the flexible lateral boundary of specimen. The loading system, which consisted of a hydraulic system for the axial load and a pneumatic system for the lateral pressure, was able to apply various stress paths. The testing method described above has been used successfully for sands (Hoque et al. 1996) and for gravels (Jiang et al. 1997).

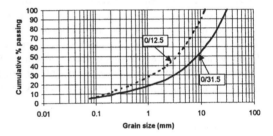

Figure 1: Grain size distribution curves of investigated unbound granular materials

Test program

The two materials were compacted to a very dense state, representative of typical road base granular layers (Table 1). The specimen was compacted in fourteen sub-layers manually by means of a 12 kg-weigh wooden hammer. The average compacted thickness of each layer was around 4.1 cm. A high uniformity of specimen density was achieved by controlling the weight and thickness of each sub-layer. With the second and scaled material (0/12.5), much larger efforts were necessary to achieve nearly the same void ratio as the first and original material

Table 1. Physical properties of investigated unbound granular materials

Material	D_{50} (mm)	U_c	Modified Proctor		G_s	Compaction conditions		
			w(%)	ρ_d (g/cm^3)		w_0(%)	ρ_{d0} (g/cm^3)	e_0
0/31.5	8.5	53	5.9	2.310	2.71	3.9	2.193	0.236
0/12.5	3.5	28	6.2	2.125	2.71	4.1	2.216	0.223

(0/31.5). In this way a density higher than the one by the modified Proctor (Table 1) was achieved with the scaled material (0/12.5). The top surface of the compacted specimen was made smooth, flat and perpendicular to the loading axial by using a small amount of finer material, which was necessary to ensure as much as possible uniform stress conditions in the specimen (not to totally remove effects of bedding error in the axial strain measurement).

After assembling all the triaxial system, the specimen, which was moist as it was compacted, was isotropically consolidated to a cell pressure of 40 kPa. Then, the following stress paths, shown in Figure 3, were applied to each specimen under drained conditions:

1. After ageing the specimen for about 30 min at various stress levels along the isotropic stress path for pressure up to 120 kPa, five unload/reload cycles of very small vertical stress increment and then other five cycles of small horizontal stress increment (i.e., very small unload/reload cycles) were applied. The amplitude of deviator stress was controlled so that the strain amplitude in single amplitude be less than about 0.0015 % to evaluate the elastic Young's moduli (Figure 4).

2. Also after ageing the specimen for about 30 min at various shear stress levels along the triaxial compression stress path at $\sigma'_h = 40$ kPa (Figure 3), very small unload/reload cycles were applied in the same way as in step 1. Figure 5 shows typical global stress-strain relationship with five very small unload/reload cycles at several stress state obtained from a test on the original material (0/31.5). Unload/reload cycles were applied during global unloading to minimise the effects of viscous deformation on the measured behaviour.

3. Isotropic compression measuring the elastic deformation property as in step 1 was repeated. In the first test on the original material (0/31.5), however, an accidental triaxial extension loading to an axial strain increment of -0.3 % took place before applying this step; so this step was skipped.

4. 21,000 cycles of a constant deviator stress amplitude (160 kPa) along the triaxial compression stress path at a confining pressure of 40 kPa (as shown in Fig. 3) was applied at a frequency of 0.5 Hz, as a large cyclic preloading history simulating typical stress paths in the road base.

5. Steps 1, 2 and 3 were repeated.

6. A set of compression tests were performed along isotropic and anisotropic stress paths at different ratios $R = \Delta\sigma_v/\Delta\sigma_h$ starting from p'=29.4 kPa and q=0 kPa, presented in Figure 6. These stress paths simulate typical stress paths in the road base subjected to traffic load. At each stress state along the respective stress path, denoted as B1,

B2 and so on, the specimen was aged about 30 min, followed by application of very small unload/reload cycles as in step 1.

7. Finally, the specimen was isotropically compressed to a cell pressure of 120 kPa, at which a triaxial compression stress path was applied at a constant cell pressure. At several stress states during the triaxial compression stress path, very small unload/reload cycles were applied after ageing the specimen for about 30 min.

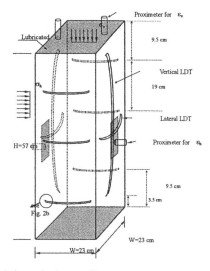

Figure 2: Schematic figure of instrumented rectangular prismatic specimen (Hoque et al. 1996)

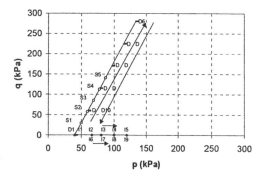

Figure 3: Isotropic and triaxial compression stress paths; $q=\sigma_v-\sigma_h$ and $p'=(\sigma_v+2\sigma_h)/3$

The very small unload/reload cycles of vertical stress were applied at a frequency of 0.05 Hz, while the very small cycles of horizontal stress were realised at a smaller frequency (0.005 Hz) in order to assure accurate and stable control of stress path. Figure 4

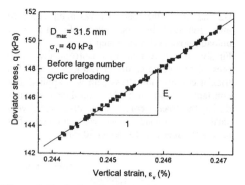

Figure 4: Typical vertical stress and vertical strain relation obtained during a very small vertical stress path

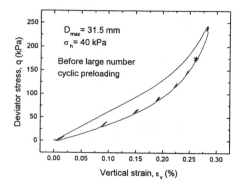

Figure 5: Overall stress strain curve with very small load/unload cycles.

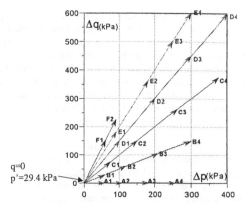

Figure 6: Isotropic and anisotropic compression stress paths at different ratios starting from q=0 and p'=29.4 kPa

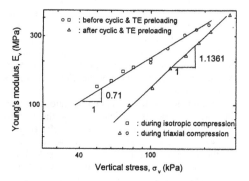

Figure 7: Vertical Young's modulus values of the original material (0/31.5) during first isotropic compression and those during triaxial compression before and after cyclic and TE preloading.

shows a typical result from a vertical cyclic loading test at an anisotropic stress state. It may be seen that the hysteresis loops are nearly closed. The Young's modulus $E_v= (d\sigma'_v/d\varepsilon_v)_{d\sigma'_h=0}$ was obtained from linear regression of such stress-strain curves. The stress-strain property analysed in this paper is only the vertical elastic Young's modulus E_v, which are the key reference parameter for the material behaviour at lager strains.

3. VERTICAL ELASTIC YOUNG'S MODULUS OF 0/31.5 GRAVEL

Figure 7 shows the vertical elastic Young's modulus E_v values, plotted against the confining pressure $\sigma'_v=\sigma'_h$, obtained by applying very small unload/reload cycles during the first isotropic compression stage. The E_v values evaluated along the triaxial compression stress path (σ'_h= 40 kPa) are plotted against σ'_v. It may be seen that E_v value is a unique function of the instantaneous value of σ'_v. This result is consistent with those from similar pre-

vious tests (e.g., Jiang et al. 1997; Hoque and Tatsuoka 1998; Balakrishnaiyer et al. 1998). Other similar results were obtained from this study as shown later. Based on this experimental finding, in Fig. 7, the E_v values (obtained before applying cyclic and TE preloading) is fitted by the following power law, not including the lateral stress σ'_h as a variable:

$$E_v= C\cdot\sigma'^m_v \qquad (1)$$

The power m was equal to about 0.7.

The E_v values that were evaluated along the triaxial cmpression stress path (σ'_h= 40 kPa) after applying 21,000 cycles of a large deviator stress amplitude of 160 kPa at σ'_h= 40 kPa, followed by accidental triaxial extension (TE) loading to an axial strain increment of -0.3 %, are also plotted in figure 7. It may be seen that the E_v value dropped noticeably by a preloading history consisting of cyclic loading and

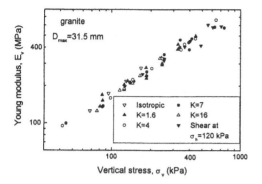

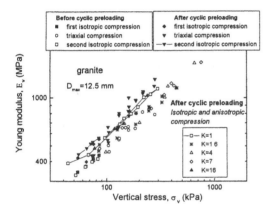

Figure 8: Vertical Young's modulus of the original material (0/31.5) during first isotropic and anisotropic compression after cyclic and TE preloading.

Figure 9: Vertical Young's modulus of the scaled material (0/12.5) before and after cyclic preloading.

TE straining. This drop in the E_v value is probably due to damage to the microstructure by the preloading history described above. Furthermore there is also an important influence of the preloading history on the power m, which has increased noticeably after cyclic pre-loading (Fig 7). It is shown below, on the other hand, that such a reduction in the E_v value and a change in the power m as above was not observed in the second similar test on the scaled material (as shown below), in which the accidental triaxial extension loading did not take place. It is likely therefore that the major cause for the damage cited above was the triaxial extension loading.

Figure 8 shows the relationships between the elastic vertical Young's modulus E_v and the effective vertical stress σ'_v evaluated along the isotropic and anisotropic stress paths with different stress incre-

ment ratios $R = \Delta\sigma_v/\Delta\sigma_h$ and along the triaxial compression path after applying the cyclic preloading history, followed by accidental TE laoding. It may be seen that all the E_v-σ'_v relationships for different stress ratios follow nearly the same relation, showing again that E_v is a unique function of σ'_v, independent of σ'_h.

4. INFLUENCE OF GRADATION ON THE ELASTIC YOUNG'S MODULUS

Figure 9 summarises the results for the scaled material (0/12.5). In this case, the preloading history consisted of only 21,000 cycles of a deviator stress amplitude of 160 kPa at a confining pressure of 40 kPa. It may be seen that the effects of the cyclic preloading history on the E_v value are negligible. It may also be seen that the E_v value is a unique function of σ'_v, independent of σ'_h, which is consistent with the behavior of the original material (0/31.5).

Kohata et al. (1997), Hoque and Tatsuoka (1998), and Tatsuoka et al. (1999a) showed that with a poorly graded fine quartz-rich sand (Hostun sand), the E_v value dropped noticeably and the power m increased, as seen in Fig. 7, only by applying a large number of cycles of a relatively large deviator stress amplitude. Further study will be necessary to find reasons for such different behaviors between poorly and well graded unbound materials.

By comparing Figures 7 & 8 and Figure 9, it can be seen that at nearly the same void ratio, 0.22 – 0.23, under otherwise the same test conditions, the E_v value of the scaled material (0/12.5) is larger by a factor of more than two and three than the original material (0/31.5), respectively, before and after cyclic and TE preloading. This unexpectedly large difference indicates that it is necessary to reconsider the empirical relationship between the elastic deformation characteristics and the density index of unbound granular materials.

Hardin and Richart (1963) showed that in their resonant-column (RC) tests on relatively poorly graded sands having void ratios for a range from around 0.6 to around 1.0, the relationship between the shear wave velocity (V_s) and the void ratio is essentially unique, independently of the particle size and some variations in the uniformity. They also showed that the empirical relationship is noticeably different for rounded and angular particle shapes. Also based on results from a comprehensive series of RC tests, Iwasaki and Tatsuoka (1977) showed that the empirical equations proposed by Hardin and Richart (1963) be not applicable to well-graded un-

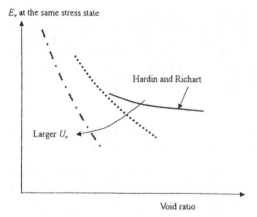

Figure 10: Schematic diagram showing the effects of U_c on the E_v and void ratio relationship.

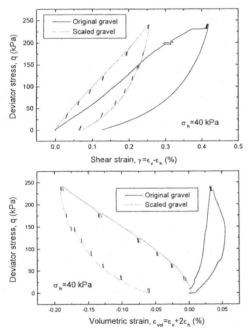

Figure 11 Comparison of the relationship between the deviator stress and the shear and volumetric strains of the original and scaled materials.

bound materials; i.e., for the same void ratio, the elastic shear modulus is smaller with sands having larger coefficients of uniformity (U_c) than those having smaller values of U_c (i.e., more uniformly graded ones). However, they used only medium dense sands in their RC tests. At a later stage, Kokusho and Yoshida (1997) measured V_s values in dense gravel placed in a large container under controlled vertical stresses. The density of each gravel ranged from a

very loose to a very dense state. They reported that the relationships between the V_s value and the void ratio of gravels having largely different U_c values are not unique under otherwise the same conditions, being largely affected by the gradation characteristics. That is, as schematically shown in Figure 10, the rate of increase in E_v with a decrease in the void ratio became larger with the increase in U_c and at the respective minimum void ratio, the E_v value became larger than the value predicted by the Hardin-Richart empirical equations to a larger extent for a gravel having a larger U_c. According to the two features described above, with two materials that are rather well-graved but having different U_c values, at the same void ratio close to the minimum of void ratio of the one having a smaller U_c, the E_v value of the material having a smaller U_c value could be significantly larger than that of the other having a larger U_c value. It is seen therefore that the experimental results presented in Figs. 7, 8 and 9 are consistent with those reported in the literature described above.

Figure 11 compares the relationship between the deviator stress and the shear and volumetric strains from the two triaxial compression tests (step 2). It may be seen that for the nearly the same void ratio, the scaled material (0/12.5) exhibits a much larger pre-peak stiffness with a much more significant trend of positive dilatancy, while the original material (0/31.5) exhibits a noticeably contractive behavior at the initial stage. These different behaviors are well consistent with a large difference in the E_v values shown above.

5. CONCLUSIONS

The following conclusions can be derived from the test results presented above:
1) For the two types of well-graded gravel having different grading characteristics, the Young's modulus (E_v) defined for a vertical (axial) strain of the order of 0.001 % is a rather unique function of the instantaneous vertical (axial) stress, essentially independent of the lateral stress.
2) The effects of applying a large number of cycles (21,000) of a relatively large deviator stress amplitude on the small strain stiffness were negligible with the scaled material. With the original material, it is likely that a triaxial extension loading to a relatively large axial strain increment (-0.3%) decreased the E_v value.
3) For the same void ratio, the dense specimens of the two materials exhibited substantially different values of E_v under otherwise the same test conditions; the E_v value of the scaled material having a smaller coefficient of uniformity (U_c) was larger

by a factor of larger than two than that of the original material having a larger U_c value. This result indicates that it is not possible to estimate accurately the elastic stiffness of a given material having a relatively large U_c based on an empirical equation relating the elastic modulus to the void ratio. It is also true that the elastic Young's modulus E_v of a given original coarse-grained material could be significantly over-estimated when it is assumed equal to the E_v value of a scaled material having a smaller U_c value but having the same void ratio as the original material.

REFERENCES

Balakrishnaiyer,K., Anh Dan,L.Q., Tatsuoka,F., Koseki,J. & Modoni,G. (1998): Deformation characteristics at small strain levels of dense gravel, The Geotechnics of Hard Soils – Soft Rocks, Proc. of Second Int. Conf. on Hard Soils and Soft Rocks (Evamgelista & Picarelli eds.), Balkema, Vol.1, pp.423-431.

Goto.S, Tatsuoka,F., Shibuya,S., Kim,Y.-S., & Sato,T.(1991), "A simple gauge for local small strain measurements in the laboratory", Soils and Foundations, Vol.31, No.1, pp.169-180

Hardin,B.O. & Richart,F.E.Jr. (1963): Elastic wave velocities in granular soils, J. ASCE 89-SM1, 33-65.

Hoque,E., Tatsuoka,F. & Sato,T. (1996): Measuring anisotropic elastic properties of sand using a large triaxial specimen, ASTM Geotechnical Testing Journal, ASTM, 19-4, 411-420.

Hoque,E., & Tatsuoka,F. (1998), "Anisotropy in the elastic deformation of materials", Soils and Foundations, Soils and Foundations, Vol.38, No.1, pp.163-179.

Iwasaki,T., & Tatsuoka,F. (1977), "Effects of grain size and grading on dynamic shear moduli of sands", Soils and Foundations, Vol.17, No.3, pp.19-35.

Jiang,G.L., Tatsuoka,F., Flora,A. & Koseki,J. (1997), "Inherent and stress state-induced anisotropy in very small strain stiffness of a sandy gravel", Géotechnique, Vol.47, No.3, Symposium In Print, pp.509-521.

Kohata,Y., Tatsuoka,F., Wang,L., Jiang,G.L., Hoque,E. & Kodaka,T. (1997), "Modelling the non-linear deformation properties of stiff geomaterials", Géotechnique, Vol.47, No.3, Symposium In Print, pp.563-580.

Kokusho,T. & Yoshida,Y. (1997); SPT N-value and S-wave velocity for gravelly soils with different grain size distribution, Soils and Foundations, 37-4, 105-113.

Tatsuoka,F., Modoni,G., Jiang,G.L., Anh Dan,L.Q., Flora,A., Matsushita,M., & Koseki,J. (1999a): Stress-strain behaviour at small strains of unbound granular materials and its laboratory tests, Keynote Lecture, Proc. of Workshop on Modelling and Advanced testing for Unbound Granular Materials, January 21 and 22, 1999, Lisboa (Correia eds.), Balkema, pp.17-61.

Tatsuoka,F., Correia,A.G., Ishihara,M. & Uchimura,T. (1999b): Non-linear resilient behaviour of unbound granular materials predicted by the cross-anisotropic hypo-quasi-elasticity model, Proc. of Workshop on Modelling and Advanced testing for Unbound Granular Materials, January 21 and 22, 1999, Lisboa (Correia eds.), Balkema, pp.197-204.

Tatsuoka,F., Jardine,R.J., Lo Presti,D., Di Benedetto,H. & Kodaka,T. (1999c), "Characterising the pre-failure deformation properties of geomaterials", Theme Lecture for the Plenary Session No.1, Proc. of XIV IC on SMFE, Hamburg, September 1997, Volume 4, pp.2129-2164.

Advanced Laboratory Stress-Strain Testing of Geomaterials, Tatsuoka, Shibuya & Kuwano (eds),
© *2001 Taylor & Francis, ISBN 90 2651 843 9*

Viscous and non viscous behaviour of sand obtained from hollow cylinder tests

H.Di Benedetto, H.Geoffroy & C. Sauzéat
DGCB (URA CNRS 1652), ENTPE, Vaulx en Velin, France

ABSTRACT: A torsional hollow cylinder device called "T4CstaDy" was developed to study the behaviour of sand in a wide range of strains (from approx. 10^{-6} to 10^{-2}). Experimental data have shown the existence of delayed or viscous effects for Hostun dry sand. A general framework based on a three component element is explained to model the viscous and non-viscous behaviour. As regards the non-viscous part, eight terms for compliance tensor may be determined and the existence of a limit tensor has been observed for very small amplitude cycles. A new hypoelastic model for this limit tensor is proposed to take into account some experimental observations (symmetry of this tensor, anisotropy, etc.). Simulations of moduli are presented for different tests, with and without rotation of axes, performed at a confining pressure of 300 kPa. For the viscous part, a new component named "viscous evanescent" is described. It allows the modelling of the viscous properties of Hostun sand, for example, when sharp strain rate changes or creep periods occur. A simulation of a typical test is presented.

1 INTRODUCTION

The behaviour of sands is strongly non-linear and irreversible as soon as the strain amplitude is "large" (more than approx. 10^{-4} m/m).

Recent investigations with the help of newer and more accurate apparatus have shown that time dependent behaviour (or delayed or viscous effects) exist for sands. This new research (Di Prisco & al. 1996, Hoque 1996, Lade & al. 1997, Di Benedetto 1997, Di Benedetto & al. 1999a, Jardine & al. 1999, Tatsuoka & al. 1997, 1998 and 1999, Santucci de Magistris & al. 1998, Ibraim 1998, Cazacliu & al. 1998a, Yasin 1998, among others) shows how complex and unknown the time dependent behaviour of sand still is. It has also been concluded that, although the viscous effects for unbound granular materials are of a small amplitude for most stress paths, their influence could be of the first order in some cases. For example, they can rigidify the material or create a perturbation allowing instability. In addition, creep deformation has to be taken into account in some practical cases (Tatsuoka & al. 1999).

In this paper we present a new type of law for sand, based on a tree component model (Di Benedetto & al. 1997a), which takes into account observed viscous and non-viscous behaviour.

The experimental investigations were made with a hollow cylinder prototype device "T4C StaDy", which was developed at ENTPE. "T4C StaDy" allows the investigation of small to large strain domains. It is introduced in the next paragraph.

Comparisons between the experimental results and the numerical simulations with this law are proposed. Non-viscous and viscous effects are also considered.

2 HOLLOW CYLINDER APPARATUS : T4C STADY

The hollow cylinder apparatus "T4C StaDy" allows the investigation of sand behaviour from very small to large strain domains. Independent application of torsion, compression / extension and confinement loading is possible for monotonous or cyclic quasi-static paths. Thus, the "T4C StaDy" is 3 dimensional stress path loading apparatus. It was developed thanks to a co-operation with "Electricité de France" (EDF).

The systems of measurement and control have a high degree of accuracy, while still allowing one to follow the loading up to large strains. Investigation of rheological behaviour is possible on a wide strain domain (from approx. 10^{-6} to 10^{-2}) by considering any combination of compression / extension and torsion, starting from any point on the global stress-strain curve. At each of these points the evolution of, i) stress or strain with time, and ii) rigidities and hysteretic damping can be given.

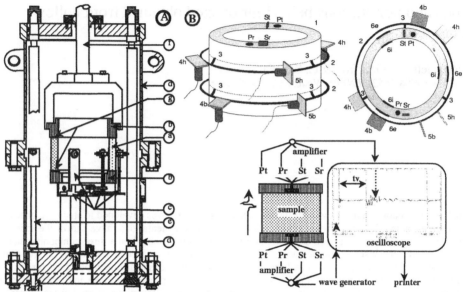

Figure 1. T4C apparatus. A. Cross section: "a". sample, "b". platens, "c". supports of radial sensors, "d". confinement cell, "e". support of ortho-radial sensors, "f". axis connected to the press piston, "g". piezoelectric sensors. B. Position of piezoelectric sensors (St, Sr, Pt and Pr) and of the 12 non-contact sensors (radial, 6i and 6e-vertical, 4h and 4b- and shear strain, 5h and 5b). C. Assembly diagram and instrumentation for the dynamic testing system.

Moreover, a system of piezoelectric sensors allows a dynamic investigation by wave propagation, at any stress and strain state accessible with the apparatus. A comparison is then possible between the behaviour obtained for quasi-static and dynamic loading in the very small strain domain. Comparison between cyclic and dynamic moduli is not given in this paper but good agreement was obtained by Cazacliu 1996, Cazacliu & al. 1998b, Di Benedetto & al. 1999b, 2000, Geoffroy & al 2000.

A hollow cylindrical sample (Fig. 1A-"a") is made with dry Hostun RF sand, prepared by air pluviation (Cazacliu 1996, Sauzéat 1997). The sample is a 12 cm in height and has an outer diameter of 20 cm and an inner diameter of 16 cm. It is placed between two rigid platens (Fig. 1A-"b") made of aluminium alloy. The base pedestal has a hole in the center in order to place the radial displacement sensors supports (Fig. 1A-"c"). The top cap, connected to the press actuator, is mobile in rotation and translation. Its hollow shape facilitates the disassembling of the inner mould after turning out. Two Neoprene membranes (0.5 cm thickness) constitute the lateral sides of the hollow cylinder sample.

A confining cell made of aluminium alloy (Fig. 1A-"d") ensures the confinement of the sample (maximum 100kN/m²). Non-contact displacement transducers are fixed on mobile supports. This allows their replacement during the test (see next

paragraph). The supports are adjusted from the outside of the confining cell.

The strain values are obtained by a local system of measurement. This type of measurement is required for a correct small strain determination (Tatsuoka & al. 1994, Lo Presti & al. 1999). Twelve non-contact sensors, using the eddy current principle, aim at aluminium targets fixed at different points on the sample. The sensors have a small range (1 or 2mm) in order to ensure sufficient resolution. At different steps during the test the sensors are repositioned in order to remain inside the measurement range. The global evolution of the strains is obtained by shifting the values for the different steps during the data analysis.

The location of the twelve non-contact sensors and the principle of measurements are shown in Figure 1B.

The axial strain (ε_z) and the shear strain (γ) are obtained by the measurement of axial displacement and the rotation of the two rings (Fig. 1B-"2"), which are fixed, at levels "b" and "h", on the sample by 3 flexible strips (Fig. 1B-"3"). The levels "b" and "h" are respectively at approximately 25 and 90 mm from the base pedestal. The targets of the four vertical sensors (Fig. 1B-"4h" and "4b") and of the two orthoradial sensors (Fig. 1B-"5h" and "5b") are small rectangular aluminium plaques fixed on the rings.

Six measurements of the radial displacements, at inner positions (Fig. 1B-"6i") and outer positions

(Fig. 1B-"6e") are provided. The three locations are at mid-height of the sample, on three rays situated at 120°. The targets of the radial sensors are sheets of aluminium paper, placed inside the membranes directly in contact with the sand. Thus, the non-contact sensors aim at these targets through the membrane. These sensors give the radial (ε_r) and the orthoradial strains (ε_θ). The strains (ε_z), (γ), (ε_r) and (ε_θ) are obtained taking the strain field to be homogeneous in the central part of the sample.

The resolution of local strain measurements is estimated at $\pm\ 7\ 10^{-7}$ for the axial strain, $\pm\ 8\ 10^{-7}$ for the shear strain, $\pm\ 5\ 10^{-7}$ for the orthoradial strain and $\pm\ 5\ 10^{-6}$ for the radial strain.

3 GENERAL FORMALISM

To take the experimental observations into account the strain is considered to be the sum of a non-viscous (or instantaneous) part and a viscous (or deferred) part:

$$\delta\underline{\varepsilon} = \delta\underline{\varepsilon}^{nv} + \delta\underline{\varepsilon}^{v} \qquad (3.1)$$

If both terms are divided by the time increment (δt)

$$\underline{D} = \underline{D}^{nv} + \underline{D}^{v} \qquad (3.2)$$

where:
$\delta\underline{\varepsilon}$ ($\underline{D}$) is the strain increment (strain rate),
$\delta\underline{\varepsilon}^{nv}$ ($\underline{D}^{nv}$) is the non viscous strain increment (strain rate),
$\delta\underline{\varepsilon}^{v}$ ($\underline{D}^{n}$) is the viscous strain increment (strain rate)
The symbol δ denotes an objective increment.

3.1 Non viscous (or instantaneous) part

Numerous laws have been proposed to describe non-viscous behaviour, such as elasticity, plasticity, elastoplasticity, hypoplasticity and interpolation type, among others. In the following, an analogical body -EP- is introduced to traduce any of these different formalisms used for modelling the time independent behaviour.

It can be shown (Darve 1978) that the general form of the non viscous strain increment is :

$$\delta\underline{\varepsilon}^{nv} = \underline{M}\ (h, dir\delta\underline{\sigma})\delta\underline{\sigma} \qquad (3.3)$$

Or after dividing by δt :

$$\underline{D}^{nv} = \underline{M}(h, dir\overset{\bullet}{\underline{\sigma}})\overset{\bullet}{\underline{\sigma}} \qquad (3.4)$$

where:
$\underline{M}$ is the "non viscous" tensor

$dir\overset{\bullet}{\underline{\sigma}} = dir\delta\underline{\sigma} = \dfrac{\delta\underline{\sigma}}{\|\delta\underline{\sigma}\|} = \dfrac{\overset{\bullet}{\underline{\sigma}}}{\|\overset{\bullet}{\underline{\sigma}}\|}$ is the "direction" of $\delta\underline{\sigma}$ or $\overset{\bullet}{\underline{\sigma}}$

$\|\delta\underline{\sigma}\| = (\Sigma\ \delta\sigma_{ij}^2)^{.5}$ denotes the norm of $\delta\underline{\sigma}$.
" $\bullet$ " is an objective derivative.

h represents the whole history parameters also called memory, hardening or state parameters.

The introduction of $dir(\overset{\bullet}{\sigma})$ expresses irreversibility and the parameters h, which may be scalars, vectors or tensors, describe the stress history dependence.

A general law of the interpolation type is presented in Di Benedetto 1987. In this paper only asymptotic behaviour and simplified hypoelastic (including loading and unloading switch function possibility) expression of the "EP" law is considered.

3.2 Viscous (or deferred) part

A general framework, presented in Di Benedetto 1987, has been proposed to describe the viscous part of the law. The analogical body explaining the law is given in Figure 2. Two particular bodies, EP2 and V, are used to model the viscous behaviour. EP2 is a non-viscous body of the EP type, which is described in the previous paragraph. The V type body is a purely viscous body. The behaviour of the V body is given by the equation:

$$\underline{D}^{v} = N(h2)\ \sigma^{v} \qquad (3.5)$$

$$\underline{D}^{v} = M_f(dir(\overset{\bullet}{\underline{\sigma}}{}^f), h2)\ \overset{\bullet}{\underline{\sigma}}{}^f \qquad (3.6)$$

$$\sigma = \sigma^f + \sigma^v \qquad (3.7)$$

Where σ^v is the "viscous stress" acting on "V" and σ^f is the "creep stress" acting on "EP2". N is the "viscous" tensor and M_f is the "creep" tensor. M_f has the same properties as the non-viscous tensor M. $dir(\overset{\bullet}{\underline{\sigma}}{}^f)$ is the "direction" of the creep stress increment. h2 are the history parameters for the viscous part, which can be different from the parameters h of the non viscous part.

Let us note that σ^f and σ^v, introduced above, are preferential history parameters of the viscous deformation description. They can be included in h2. Their introduction is not arbitrary because a physical interpretation for the linear monodimen-sional cases can be found (Di Benedetto 85 and 87).

When viscous effects can be neglected, the viscous stress is always null and only EP1 and EP2 act. They can be combined and become one unique EP model.

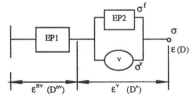

Figure 2. Analogical representation of the law : "EP" non viscous behaviour, "V" viscous behaviour.

It can be shown that this 3-component formalism is a tridimensional generalisation of classical theories such as viscoelasticity and classical viscoplasticity.

The simulations presented in this paper consider small strain, where asytmptotic linear behaviour of the hypoelastic type can be used. It has been shown from the experimental data that EP1 (Figure 2) is an hypoelastic body for small strain cycles. The developed ENTPE hypoelastic law is presented in paragraph 5.3.

4 HOW TO MEASURE INSTANTANEOUS AND DEFERRED PROPERTIES

A test consists in repeating three stages at different and successive levels of stress (Figure 3):

Stage 1 A quasi-static loading (evolution loading) is applied to the sample, which reachs a pre-deformed state. Subsequently we will call this state the "investigation point". The evolution loading corresponds to one of the stress paths, which is possible to describe with the hollow cylinder apparatus. For the considered test campaign 3 types of evolution loading were applied: classical triaxial loading (type C), pure torsion from an isotropic state (type T) and torsion after axial loading (type K) (cf. Table 1)

Stage 2 At each investigation point, the behaviour of the sample in very small and medium strain domains is characterised by application of quasi static cyclic loading with a strain amplitude ranging from approx. 10^{-6} to 10^{-4}. According to the direction of loading, the cyclic paths are distinguished as i) axial, ii) torsional, and iii) bi-axial. During the axial cyclic loading, axial stress varies at constant shear and confinement stresses. The torsional cyclic loading consists of a shearing of the sample at constant axial and confinement stresses. Lastly, during the bi-axial cyclic path, axial and shear stresses vary at constant confinement stress, while describing a segment in the stress space.

Stage 3 Velocity measurements for waves emitted by the bender and compression piezoelectric elements (dynamic loading) were also performed at each investigation point. No dynamic test results are presented in this paper.

At each investigation point, the stress is kept constant (between forty minutes and two hours). During this creep period, strains still increase. This reveals the existence of rate dependent or viscous behaviour for the tested sand (Di Benedetto 1997). The analysis of this creep period provides information on the viscous part of the law ($\delta\varepsilon^v$ or D^v, Equations 3.1 or 3.2), presented in part 6.

Generally, the viscous strain rate becomes "very small" after few ten minutes. After that, the "small" amplitude cycles are nearly stabilised, depending upon the stress level, and provide directly the non-viscous properties (Di Benedetto & al. 1997b, Cazacliu & al. 1998a, Di Benedetto & al. 1999b). The measured strain increment is then the non-viscous one: $\delta\varepsilon^{nv}$ or D^{nv} (Equation 3.1 or 3.2). The analysis and modelling of the non-viscous part is presented in the following paragraphs.

5 NON VISCOUS BEHAVIOUR MODELLING

5.1 *Rheological analysis for T4C StaDy tests*

In its initial state, the sand is assumed to be transversely isotropic (Figure 4). Moreover, the stress and strain fields are supposed to be homogeneous and consequently the tangential components rz and rθ are null during the entire experiment. Thus :

$$\underline{\underline{\sigma}} = \begin{pmatrix} \sigma_{rr} & 0 & 0 \\ 0 & \sigma_{\theta\theta} & \sigma_{\theta z} \\ 0 & \sigma_{z\theta} & \sigma_{zz} \end{pmatrix} = \begin{pmatrix} \sigma_r & 0 & 0 \\ 0 & \sigma_\theta & \tau \\ 0 & \tau & \sigma_z \end{pmatrix} \quad (5.1)$$

$$\underline{\underline{\varepsilon}} = \begin{pmatrix} \varepsilon_{rr} & 0 & 0 \\ 0 & \varepsilon_{\theta\theta} & \gamma_{\theta z}/2 \\ 0 & \gamma_{z\theta}/2 & \varepsilon_{zz} \end{pmatrix} = \begin{pmatrix} \varepsilon_r & 0 & 0 \\ 0 & \varepsilon_\theta & \gamma/2 \\ 0 & \gamma/2 & \varepsilon_z \end{pmatrix} \quad (5.2)$$

When considering experimental measurements,

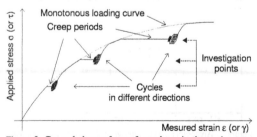

Figure 3. General shape of a performed test in the strain-stress axes.

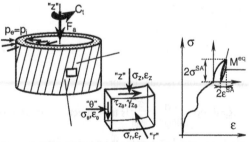

Figure 4. Stress and strain state in the hollow cylinder sample and definition of the equivalent matrix terms : M^{eq}

the loading increment has a finite length and a secant matrix is obtained. This is dependent on the increment direction and the history of loading, but also on the strain (or stress) increment amplitude.

For a cyclic loading corresponding to a segment in the loading space, half the amplitudes of strain (ε^{sa}) and stress (σ^{sa}) are traditionally considered (Figure 4). When the cycle is stabilised the equivalent matrix M^{eq} is introduced:

$$\begin{pmatrix} \varepsilon_r^{sa} \\ \varepsilon_\theta^{sa} \\ \varepsilon_z^{sa} \\ \gamma^{sa}/\sqrt{2} \end{pmatrix} = \begin{pmatrix} M_{rr}^{eq} & M_{r\theta}^{eq} & M_{rz}^{eq} & M_{r\gamma}^{eq} \\ M_{\theta r}^{eq} & M_{\theta\theta}^{eq} & M_{\theta z}^{eq} & M_{\theta\gamma}^{eq} \\ M_{zr}^{eq} & M_{z\theta}^{eq} & M_{zz}^{eq} & M_{z\gamma}^{eq} \\ M_{\gamma r}^{eq} & M_{\gamma\theta}^{eq} & M_{\gamma z}^{eq} & M_{\gamma\gamma}^{eq} \end{pmatrix} \begin{pmatrix} \sigma_r^{sa} \\ \sigma_\theta^{sa} \\ \sigma_z^{sa} \\ \tau^{sa}\sqrt{2} \end{pmatrix} \quad (5.3)$$

The elements of the equivalent matrix M^{eq} correspond to secant terms, which depend on the amplitude of the considered cycle. The application of particular cyclic loading paths with "T4C StaDy" apparatus allows us to determine the last two columns of the equivalent matrix M^{eq}.

For example, an axial cyclic loading :

$$\delta\sigma^{ax} = {}^t (0 \quad 0 \quad \sigma_z^{sa} \quad 0) \quad (5.4)$$

allows us to determine four components :

$$M_{rz}^{eq} = \frac{\varepsilon_r^{sa}}{\sigma_z^{sa}}, M_{\theta z}^{eq} = \frac{\varepsilon_\theta^{sa}}{\sigma_z^{sa}}, M_{zz}^{eq} = \frac{\varepsilon_z^{sa}}{\sigma_z^{sa}}, M_{\gamma z}^{eq} = \frac{\gamma^{sa}}{\sigma_z^{sa}\sqrt{2}}$$

In the same way, from a torsional cyclic loading:

$$\delta\sigma^T = {}^t (0 \quad 0 \quad 0 \quad \tau^{sa}\sqrt{2}) \quad (5.5)$$

one obtains four further components :

$$M_{r\gamma}^{eq} = \frac{\varepsilon_r^{sa}}{\tau^{sa}\sqrt{2}}, M_{\theta\gamma}^{eq} = \frac{\varepsilon_\theta^{sa}}{\tau^{sa}\sqrt{2}}, M_{z\gamma}^{eq} = \frac{\varepsilon_z^{sa}}{\tau^{sa}\sqrt{2}}, M_{\gamma\gamma}^{eq} = \frac{\gamma^{sa}}{2\tau^{sa}}$$

Let us emphasize that the diagonal terms M_{zz}^{eq} and $M_{\gamma\gamma}^{eq}$ are related respectively to the equivalent Young's modulus in direction z and to the equivalent shear modulus between θ and z, by the relations:

$$M_{zz}^{eq} = \frac{1}{E_z^{eq}} \text{ and } M_{\gamma\gamma}^{eq} = \frac{1}{2G_{\theta z}^{eq}}$$

5.2 Experimental results

The analysis of experimental campaigns, consisting of a series of small cycles performed with the hollow cylinder "T4C StaDy" at different strain amplitudes and stress increment directions, shows the existence of a limit behaviour, which is "quasi" elastic (for EP1, Figure 2). It could be concluded that, when loading after unloading or unloading after loading, the non viscous tensor M is equal to the equivalent tensor M^{eq}. When the strain amplitude is smaller

than approx. 10^{-6}, M_{eq} tends towards a limit tensor M^e, which can be assimilated to an elastic one. More details are given on the experimental evidence of the existence of this "quasi" elastic limit tensor M^e for sand in Di Benedetto & al. 1997, 1999b and Cazacliu & al. 1998a. A new hypoelastic model was developed in the ENTPE laboratory to express the anisotropic tensor M^e.

5.3 ENTPE's new hypoelastic model

More details on this model are given in Di Benedetto 2000, based on Hardin & al.'s 1989 development, which gave the following expression for the non-viscous tensor M ($=M^e$):

$$M_{HAR} = \frac{OCR^k}{F(e)} S_\nu S_f \Sigma_p \quad (5.6)$$

$$\text{With : } \Sigma_p = \frac{1}{P_a^{1-n}} \begin{pmatrix} \dfrac{1}{\sigma_1^n} & 0 & 0 & 0 \\ 0 & \dfrac{1}{\sigma_2^n} & 0 & 0 \\ 0 & 0 & \dfrac{1}{\sigma_3^n} & 0 \\ 0 & 0 & 0 & \dfrac{1}{\sigma_2^{n/2}\sigma_3^{n/2}} \end{pmatrix} \quad (5.7)$$

when expressed in the principal stress axes. And:

$$S_\nu = \begin{pmatrix} 1 & -\nu & -\nu & 0 \\ -\nu & 1 & -\nu & 0 \\ -\nu & -\nu & 1 & 0 \\ 0 & 0 & 0 & 2(1+\nu) \end{pmatrix} \quad (5.8)$$

$$S_f = \begin{pmatrix} \dfrac{1}{S_r} & 0 & 0 & 0 \\ 0 & \dfrac{1}{S_\theta} & 0 & 0 \\ 0 & 0 & \dfrac{1}{S_z} & 0 \\ 0 & 0 & 0 & \dfrac{1}{S_{\theta z}} \end{pmatrix} \quad (5.9)$$

One of the drawbacks of this model is the non-symmetry of the non-viscous tensor, while experiments from "T4C StaDy" (Di Benedetto & al 2000) tend to confirm the symmetry. In addition, some evolution of terms, which are not on the diagonal of the tensor M, are not very well simulated, in the case of the rotation of the principal axes and even without rotation of these axes. To improve the simulation, we propose a new model, which has the following expression:

$$M_{ENTPE} = (M_{HAR} + {}^t M_{HAR})/2 \quad (5.10)$$

Where $^t M$ denotes the transpose of M. The proposed hypoelastic law is then symmetrical. For the presented simulation, the value of "ν"(Equation 5.8) is 0.2, and the 4 terms of the tensor S_f (Equation 5.9), which introduce the initial anisotropy, are all equal.

In the particular cases of loading without the rotation of the principal stress axes, the Young and shear moduli are given by:

$$E_i = \frac{F(e)}{OCR^k} S_i P_a^{1-n} \sigma_i^n \qquad (5.11)$$

$$G_{ij} = \frac{F(e)}{OCR^k} \frac{S_{ij}}{2(1+v)} P_a^{1-n} \sigma_i^{n/2} \sigma_j^{n/2} \qquad (5.12)$$

In the general case with rotation of the axes, which can be applied by the hollow cylinder device, the expressions of these moduli are more complicated (Duttine 2000, Di Benedetto et al 2000). Some examples of simulation in both cases (with and without rotation of axes) are presented in the next paragraph.

5.4 Simulation with the ENTPE hypoelastic model

An experimental campaign was carried out on medium dry Hostun sand. Hostun sand is a silica sand, whose characteristics are given in Flavigny & al. 1990. The sample was made by pluviation with either loose or dense sand, determined with an initial void ratio e_0. It was then confined to a pressure P_0 of 300 kPa. From this initial isotropic state a loading path was applied (cf §4). The chosen loading paths were classical triaxial loading for C type tests and pure torsion for T type tests. For K type tests, an axial loading was applied to bring the sample to an anisotropic state (determined by the coefficient $K_0 = \sigma_r / \sigma_z = 0.5$). A pure torsion was then applied to the sample from this anisotropic state.

Table 1 gives some characteristics for 7 tests performed at a confining pressure of 300 MPa.

In the course of these tests, several investigation points were considered (Figure 3). In Table 1, the stress state of the sample can be found for each of the investigation points (values of P_0, σ_z and τ). The internal friction angle ϕ was also calculated for each of these points.

As explained previously, the application of cyclic loading allows us to determine some terms of the M^e tensor (i.e. the limit of the M^{eq} tensor for very small cyclic strain amplitudes). The values of both elastic moduli E_z and $G_{\theta z}$ are also presented in Table 1.

Simulations with the hypoelastic ENTPE model of these moduli are presented in Figures 5 to 10. In order to compare the results for each type of test, the ratios E_z/E_{z0} and $G_{\theta z}/G_{\theta z0}$ were plotted against a parameter representative of the stress value at the investigation points: σ_z/σ_{z0} for C type tests, and τ/P_0 for T type and K type tests. The "0" index was used for the values at the reference state. For C type and T type tests, this reference is the initial isotropic state. For K type tests, the anisotropic state with a null shear stress (τ=0 kPa) was chosen. Considering that the evolution of the void ratio e is very small during tests, the ratio E_z/E_{z0} and $G_{\theta z}/G_{\theta z0}$ are then dependant only on the stress state (Equation 5.11 and 5.12 for C type tests).

Some results from other tests, that are not detailed here, are also plotted on the same graph (Figures 5 to 10).

One can conclude that the simulations traduce correctly the experimental results even for loading with rotation of axes.

Table 1. Characteristics of 7 tests at a confining pressure Po = 300 kPa, and obtained values at the different investigation points (cf Figure 2): initial void ratio e_0, confining pressure Po($=\sigma_r=\sigma_\theta$), axial stress σ_z, shear stress $\tau_{\theta z}$, Coulomb friction angle ϕ, Young modulus E_z (the value in bold, for each test, is E_{z0}), shear modulus $G_{\theta z}$ (the value in bold, for each test, is $G_{\theta z0}$)

Test	Points	e_0	Po (kPa)	σ_z (kPa)	τ (kPa)	ϕ (°)	E_z (MPa)	$G_{\theta z}$ (MPa)
C300.74	0; 1; 2; 3	0.74	300	300; 450; 600; 750	0	0; 12; 19; 25	**377**; 442; 502; 559	**139**; 153; 164; 177
C300.86	0; 1; 2	0.86	283	300; 450; 600	0	0; 12; 20	**327**; 384; 445	**109**;115;121
C300.95	0; 1; 2; 3; 4; 5	0.95	300	300; 498; 698; 798; 849; 898	0	0; 14; 24; 27; 29; 30	**260**; 347; 435; 484; 513; 535	**72**; 90; 98; 105; 109; 112
T300.72	0; 1; 2; 3; 4; 5; 6	0.72	300	300	0; 50; 100; 50; 0; -75; -150	0; 10; 19; 30; 0; 14; 30	**415**; 413; 396; 348; 389; 383; 292	**136**; 144; 131; 113; 131; 124; 92
T300.90	0; 1; 2	0.90	300	300	0; 50; 100	0; 10; 19	**254**; 268; 225	**94**; 100; 89
K300.74	0; 1; 2; 3; 4; 5; 6	0.74	300	300; 450; 600; 600; 600; 600; 600	0; 0; 0; 75; 150; 0; -75	0; 12; 19; 22; 28; 19; 22	404; 481; **553**; 550; 532; 549; 549	116; 142; **156**; 150; 139; 150; 145
K300.99	0; 1; 2; 3; 4; 5; 6; 7; 8; 9	0.99	300	300; 600; 600; 600; 600; 600; 600; 600; 600; 600	0; 0; 75; 150; 0; -75; -150; 0; 75; 150	0; 19; 22; 28; 19; 22; 28; 0; 22; 28	258; **387**; 397; 391; 406; 416; 392; 411; 421; 419	78; **96**; 99; 97; 99; 94; 87; 96; 98; 98

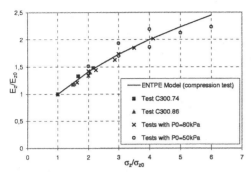

Figure 5. E_z/E_{z0} for compression triaxial tests (C type) (cf Table 1)

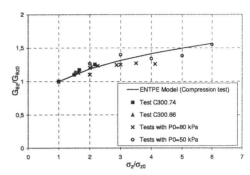

Figure 6. $G_{\theta z}/G_{\theta z0}$ for compression triaxial tests (C type) (cf Table 1)

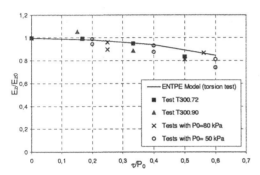

Figure 7. E_z/E_{z0} for pure torsion tests (T type) (cf Table 1)

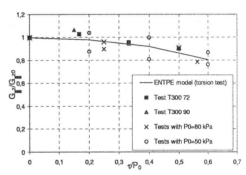

Figure 8. $G_{\theta z}/G_{\theta z0}$ for pure torsion tests (T type) (cf Table 1)

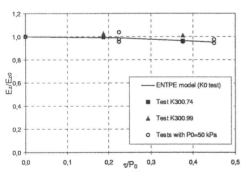

Figure 9. E_z/E_{z0} for anisotropic and torsion tests (K type) (cf Table 1)

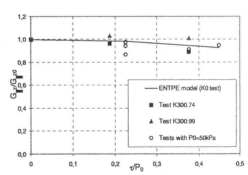

Figure 10. $G_{\theta z}/G_{\theta z0}$ for anisotropic and torsion tests (K type) (cf Table 1)

6 VISCOUS BEHAVIOUR MODELLING

6.1 *Experimental observations*

An example of viscous phenomenon for sand is observed during creep tests (Di Benedetto & al. 1999c).

When plotting the strain evolution with time a rapid variation, which decreases rapidly, is observed. Nevertheless, even after some thousands of seconds the strain is not totally stabilised and a small strain creep evolution is observed (Di Benedetto & al. 1999c, Tatsuoka & al. 1998). It is important to note that the creep direction is not linked with the stress level but seems to depend on the last inversion strain value(Di Benedetto & al. 1999c and 2000). It appears that the "small" cyclic loading, as well as creep, modify the response only locally, while the monotonic loading curve is closely rejoined after reloading.

Other triaxial tests with some rapid strain rate changes were made at Tokyo University (Tatsuoka & al. 1998), for which the main conclusions were:

223

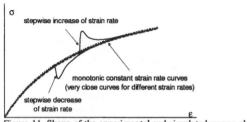

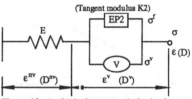

Figure 12. Analogical asymptotic body chosen for simulations presented in Figure 15. The viscous body "V" has a viscous evanescent behaviour. The tangent modulus K2 is not constant and is different when loading or unloading..

Figure 11. Shape of the experimental and simulated curves obtained for sand.

1) The monotonic stress strain curve seems to be unique and independent of the applied constant strain rate,

2) When applying a rapid change in the strain rate during loading, the stress value rapidly decreases (for slower rate) or increases (for faster rate), then rejoins the monotonic curve.

This specific behaviour (Figure 11) cannot be simulated with traditional models (viscoelastic, viscoplastic, etc.). The following paragraphs introduce a new concept of a model called "viscous evanescent", which was developed to model the experimental observations as described in 1) and 2). For simplicity's sake, only the mono-dimensional case is considered. A 3D expression can be obtained from equation 3.4 to 3.7.

6.2 *Simplified mono-dimensional case*

The chosen analogical asymptotic body is presented in Figure 12. As the aim of the proposed simulation is only to show that the developed formalism is able to model the specific behaviour as described in paragraph 6.1, only a mono-dimensional case is considered. In addition, the EP1 model is a spring with a constant modulus E (for general cases, the hypoelastic model, described in part 5, should be considered). EP2 is a non viscous body with a tangent modulus (obtained from Equation 6.6 when loading), which is different for loading or unloading paths.

A new approach is presented in treating the viscous body "V", introducing a new kind of behaviour called viscous evanescent (Ve). This has been developed to take into account the specific experimental observations.

6.3 *Description of the "viscous evanescent" behaviour*

Only a brief description of the model is presented in this paper. More details can be obtained in Di Benedetto & al. 1999c. At a constant viscous strain rate the viscous stress is given by:

$$\sigma^v = f(h2, D^v).R(\varepsilon^v) \qquad \text{if } D^v \text{ is constant} \qquad (6.1)$$

where f can be obtained from creep periods. The proposed expression is:

$$f = \eta_0 (D^v)^{1+b} \qquad (6.2)$$

and R is a monotonous decreasing function (Figure 14).

If the superposition of the stress path increments applied during the loading history is weighted by the function R, the viscous stress for any strain (or stain rate) evolution becomes :

$$\sigma^v = \int_0^t R(\varepsilon^v_{(t)} - \varepsilon^v_{(\tau)}) d(f(D^v_{(\tau)})) \qquad (6.3)$$

which can be rewritten:

$$\sigma^v(t) = f(D^v(t)) + \int_0^t f(D^v_{(\tau)}) R'(\varepsilon^v_{(t)} - \varepsilon^v_{(\tau)}) d\varepsilon^v_{(\tau)} \qquad (6.4)$$

$R'(X)$ is the derivative of $R(X)$.

6.4 *Simulations with the viscous evanescent law*

The viscous evanescent law is shown in Figure 2 with the body "V" having a viscous evanescent behaviour. The properties of the V type body are determined by Equations 6.2, and 6.3 or 6.4. The chosen expression for the function R is:

$$R(\varepsilon^v) = \exp(-\frac{\varepsilon^v}{\varepsilon^v_0}) \qquad (6.5)$$

where (ε^v_0) is a constant.

A better fit for the experimental data could probably be obtained with a less simple function, but our aim in this paper is to show that the main experimental trends are modelled. This explains why R(t) is defined by a simple exponential function, which needs only one constant (ε^v_0).

For each approach, it is then possible to perform

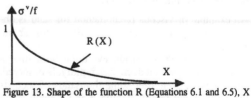

Figure 13. Shape of the function R (Equations 6.1 and 6.5), X is the viscous strain (ε^v)

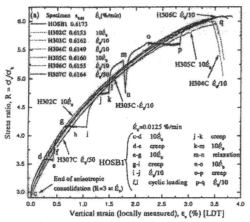

Figure 14. Monotonous PSC (Plain Strain Compression) tests (figure 2.10 in Tatsuoka et al. 1998)

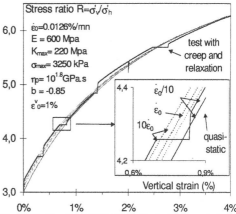

Figure 15. Simulated PSC (Plane Strain Compression) tests with the viscous evanescent model (same conditions as figure 14)

integration of the law, including viscous effects, for different stress paths.

Different monotonous PSC (Plain Strain Compression) tests at constant but different axial strain rates, performed at Tokyo University, did not show a clear influence by the strain rate on the stress-strain curve. Figure 14, which corresponds to figure 2.10 of Tatsuoka et al. 1998, confirms this last assumption for a strain rate ranging from 4.10^{-8} s^{-1} to 2.10^{-5} s^{-1}. While a monotonous strain rate has very little influence, consequent creep and relaxation do occur.

Simulations of the tests presented in Figure 14 are reported in Figure 15.

For this monodimensional calculus the "EP1" body (Figure 2) is a spring of rigidity "E". The body "EP2" is represented by a hyperbolic function:

$$\sigma = (\sigma_{max} \, \varepsilon^V) / (\frac{\sigma_{max}}{K_{max}} + \varepsilon^V) \qquad (6.6)$$

where K_{max} and σ_{max} are constants.

The chosen expressions correspond to a rough fit using few constants (only 3). Let us underline that our aim is to show the ability of the law to describe the qualitative observations. A better fit can be obtained in a further step while improving the different analytical expressions.

As no viscous stress (or strain) rate direction change occurs for the considered stress path, the cycling rule for the body EP2 needs not to be taken into account.

The viscous evolution is given by Equations (6.2, 6.4 and 6.5). The values of the 6 constants are indicated in Figure 15.

From the curves (Figure 15), one can conclude that the proposed law correctly simulates the general trends. Meanwhile, the introduction of a convolution product brings rather complicated numerical resolutions.

7 CONCLUSION

The hollow cylinder prototype "T4C StaDy" has revealed itself to be a powerful experimental tool for the investigation of sand behaviour from small to large strain domains.

A general formalism for a viscoplastic law has been presented. It is very general and can be applied to different types of modelling according to the choice of the non-viscous and viscous components.

Specific loading paths were applied with the hollow cylinder, "T4C StaDy". Small cyclic loading and creep periods were considered as they allow us to obtain experimental information on the non-viscous and viscous part of the law.

With the help of the "T4C StaDy" data and considering the general formalism, a new hypoelastic law was formulated for the non-viscous part. The complete anisotropic hypoelastic tensor can be obtained for any loading with or without rotation of the axes.

To complete the law, the viscous part was also formulated. This introduced a new type of model, called "viscous evanescent". In this paper only the monodimensional case was considered for the viscous deformation.

Some comparisons between experimental results from the hollow cylinder device and numerical simulations were proposed for dry Hostun RF sand. Non-viscous and viscous effects were considered. They showed that the law was able to translate correctly the experimental observations.

The authors wish to thank "Electricité de France" (EDF) for their support in this research program.

REFERENCES

Cazacliu, B. 1996. Comportement des sables en petites et moyennes déformations - réalisation d'un prototype d'essai de torsion compression confinement sur cylindre creux. *PhD, ECP/ENTPE, Paris.*

Cazacliu, B. & Di Benedetto, H. 1998a. Nouvel essai sur cylindre creux de sable, *Revue Française de Génie Civil* 2(27).

Cazacliu, B. & Di Benedetto, H. 1998b. Behaviour of sand in the small strain domain observed with a hollow cylinder apparatus, *European Conference on Earthquake Engineering.*

Darve, F. 1978. Une formulation incrémentale des lois rhéologiques. Application aux sols. *Thèse d'Etat, IM Grenoble.*

Di Benedetto, H. 1985. Viscous part for incremental non linear constitutive laws. *Proc. of the Fifth Int. Conf. On Numerical Methods in Geomechanics, Nagoya*: 429-435. Rotterdam:Balkema.

Di Benedetto, H. 1987. Modélisation du comportement des géomatériaux: application aux enrobés bitumineux et aux bitumes. *Thèse de Doctorat d'Etat, INP Grenoble.*

Di Benedetto, H. 1997. Effets visqueux et anisotropie des sables. *Proc. of the XIVth Int. Conference on Soil Mechanics and Foundation Engineering, Hamburg.* Rotterdam:Balkema.

Di Benedetto, H. & Tatsuoka, F. 1997a. Small strain behavior of geomaterials : modelling of strain rate effects. *Soils and Foundations* 37(2): 127-138.

Di Benedetto, H., Cazacliu, B., Boutin, C., Doanh, T. & Touret, J.P. 1997b. Comportement des sables avec rotation d'axes : nouvel appareil couvrant quatre décades de déformation. *Proc. of the XIVth Int. Conference on Soil Mechanics and Foundation Engineering, Hamburg.* Rotterdam: Balkema.

Di Benedetto, H., Ibraim, E. & Cazacliu, B. 1999a. Time dependent behavior of sand. In Jamiolkowski & al (ed.), *Proc. of IS on Prefailure Deformation Characteristics of Geomaterials, Torino.* Rotterdam: Balkema.

Di Benedetto, H., Cazacliu, B., Geoffroy, H. & Sauzéat, C. 1999b. Sand behavior at very small to medium strain-influence of stress. In Jamiolkowski & al (ed.), *Proc. of IS on Prefailure Deformation Characteristics of Geomaterials, Torino.* Rotterdam: Balkema.

Di Benedetto, H., Geoffroy, H. & Sauzéat, C. 1999c. Modelling viscous effects for sand and behaviour in the small strain domain. *Proc. of IS on Pre-failure Deformation Characteristics of Geomaterials, Torino.* Rotterdam: Balkema.

Di Benedetto, H., Geoffroy, H., Sauzéat, C. & Duttine A. 2000. *Etude du comportement cyclique des sables en petites et moyennes déformations, Rapport n°ND 2702 GC.* Ecole Nationale des Travaux Publics de l'Etat.

Di Prisco, C. & Imposimato, S. 1996. Time dependent mechanical behaviour of loose sands. *Mechanics of Cohesive-Frictional Materials* 1: 45-73.

Duttine, A. 2000. Etude du comportement cyclique des sables en petite et moyenne déformations. *DEA, ENTPE.*

Flavigny, E., Desrues, J. & Palayer, B. 1990. Note Technique – Le sable d'Hostun RF. *Revue Française de Géotechnique* 53:67-70.

Geoffroy, H., Di Benedetto, H. & Sauzéat, C. 2000. Etude du comportement du sable d'Hostun des petites aux moyennes déformations. *Colloque Physique et Mécanique des Matériaux Granulaires, ENPC, Marne la Vallée* 1:103-108.

Hardin, B.O. & Blandford, G. 1989. Elasticity of particulate materials. *Journal of Geotechnical Engineering, ASCE,* 115(6):788-805.

Hoque, E. 1996. Elastic deformation of sands in triaxial tests. *PhD, University of Tokyo,* 146 p.

Ibraim, E. 1998. Différents aspects du comportement des sables mis en évidence à partir d'essais triaxiaux. *Thèse de Doctorat, INSA Lyon.*

Jardine, R.J., Kuwano, R., Zdravkovic, L. & Thornton, C. 1999. Some fundamental aspects of the pre-failure behaviour of granular soils. In Jamiolkowski & al (ed.), *Keynote lecture in IS on Pre-failure Deformation Characteristics of Geomaterials, Torino.* Rotterdam:Balkema.

Lade, P. V., Yamamuro, J. A. & Bopp, P. A. 1997. Influence of Time Effects on Instability of Granular Materials. *Computers and Geotechnics* 20(3/4): 179-193.

Lo Presti, D., Sibuya, S. & Rix, G.J. 1999. Innovation in Soil Testing. In Jamiolkowski & al (ed.), *Keynote lecture in IS on Pre-failure Deformation Characteristics of Geomaterials, Torino.* Rotterdam:Balkema.

Murayama, S., Michihiro, K. & Sakagami, T. 1984. Creep characteristics of sands. *Soils and Foundations* 24(2):1-15.

Santucci de Magistris, F., Sato, T., Koseki, J. & Tatsuoka, F. 1998. Effects of strain rate and ageing on small strain behaviour of a compact silty sand. In Evangelista & Picarelli (ed.), *Proc. of 2nd IC on Hard Soils and Soft Rocks, Napoli.* Rotterdam: Balkema.

Sauzéat, C. 1997. Etude du comportement des sables des petites aux grandes déformations. *DEA, ENTPE.*

Tatsuoka, F., Teachavorasinsun, S., Dong, J., Kohata, Y. & Sato, T. 1994. Importance of measuring local strains in cyclic triaxial tests on granular materials, *Dynamic Geotechnical Testing II,* ASTM STP 1213, R. Ebelhar, V. Drnevich, B. Kutter (ed.), ASTM, Philadelphia: 288-302.

Tatsuoka, F., Jardine, R.J., Lo Presti, D., Di Benedetto, H. & Kohata, Y. 1997. Testing and characterizing pre-failure deformation of geomaterials. Keynote lecture in *XIVth Int. Conference on Soil Mechanics and Foundation Engineering, Hamburg.* Rotterdam:Balkema.

Tatsuoka, F., Santucci de Magistris, F., Hayano, K., Momoya, Y. & Koseki, J. 1998. Some new aspects of time effects on the stress-strain behaviour of stiff geomaterials. In Evangelista & Picarelli (ed.), Keynote Lecture for 2nd IC on Hard Soils and Soft Rocks, Napoli. Rotterdam: Balkema

Tatsuoka, F., Uchimura, T., Hayano, K., Di Benedetto, H., Koseki, J. & Siddiquee, M.S.A. 1999. Time-dependent deformation characteristics of stiff geomaterials in engineering practice. Keynote lecture in *IS on Pre-failure Deformation Characteristics of Geomaterials, Torino.* Rotterdam:Balkema.

Yasin, S. 1998. Strength and deformation characteristics of sands in plane strain shear tests. *PhD, University of Tokyo.*

Advanced Laboratory Stress-Strain Testing of Geomaterials, Tatsuoka, Shibuya & Kuwano (eds),
© *2001 Taylor & Francis, ISBN 90 2651 843 9*

Micaceous sands: stress-strain behaviour and influence of initial fabric

V.N. Georgiannou
National Technical University of Athens

ABSTRACT: The paper describes the stress-strain behaviour of micaceous sands. The presence of sand-size mica plates within a sand increases its void ratio, modifies its volume change characteristics and introduces a collapse potential. The effects depend on the quantity, distribution and orientation of the mica. The effect of the size of mica on the stress-strain response of granular mixes containing coarser and fine grains of mica was also derived from laboratory studies on a uniform and a well graded sand. It is shown that for the same size of mica the response of the host sand can be altered completely (i.e. from dilatant to contractant) due to the presence of mica, depending on the grading of the host sand. The effects of the mica can be appreciated if the packing in a clean sand and a sand with mica is compared. The test results imply that the importance of shape and location of additives in sand is not reflected in measures such as void ratio or granular void ratio.

1 INTRODUCTION

In this study the effects of platy mica particles on the behaviour of sand are investigated. A description of micas is given by Blyth & de Freitas (1984). In this study the muscovite mica was employed. Two sands were used: a uniform and a well graded sand.

The well graded sand used in these tests is the Jamuna sand and has the grading shown in Fig. 1. The grading was intended to be similar to natural micaceous sands found in Bangladesh in relation to the Jamuna bridge project. Commercial batches of quartz sand of different gradings were mixed up by trial and error until the required grading was obtained (Mundegar 1997). This sand was used as the host sand with which mica was mixed.

The grading of a uniform sand the HRS sand is also shown in Figure 1 and was used in a number of tests intended to identify the effect of varying the grading of the host sand on undrained shear.

The main type of mica used for testing was a commercially available muscovite type labelled MF60. Its grading is shown in Figure 1 and is of comparable grain size to the host sand. However, it should be noted that the aspect ratio for mica MF60 is 50.

In addition to testing sand and mica mixtures with mica grain size similar to the sand two different fines were employed to identify the effect of varying the relative size of the fines with respect to the host sand. These were mica SX powder, a silt size mica, and HPF4 silt (Zdravkovic 1996). The gradings for both silt sized materials are shown in Figure 1.

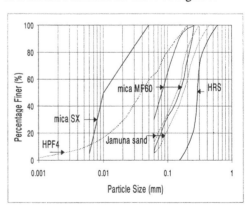

Figure 1. Particle size distributions of the materials used.

2 FORMATION OF MICACEOUS SANDS

Triaxial tests were performed on 38mm diameter cylindrical specimens with height to diameter ratio of 2:1. Specimens created by air pluviation were formed by placing in a funnel dry sand and mica mixtures, which had been previously weighed and thoroughly mixed within a graduated cylinder by turning it upside down and back again (approximately 10sec per movement) until a consistent texture could be observed by naked eye, and this required 4-5 turns of the cylinder. A tube of 1cm diameter was attached to the spout of the funnel with a length that brought the end of the tube 2cm above the top of the mould. The soil was then allowed to fill the mould falling from a constant height. Ishihara (1993) presented tests performed on a silty sand that showed néarly the same steady state line whether the specimens were prepared by water sedimentation or by dry deposition.

A suction of 20kPa was applied after placing the top cap on the specimen and was maintained throughout the saturation period. Saturation of the specimens was attained by flushing them with carbon dioxide for 30min, after which de-aired water was slowly percolated from the bottom through the top of the specimens. The volume of the water flushed through each specimen was approximately three times the volume of the specimen. Following water percolation a back pressure of around 300kPa was applied to the specimens while the effective stress of the specimens remained at 20kPa. The B value was measured to test specimen saturation and a minimum value of 0.97 was obtained.

Pairs of electrolytic level strain gauges of the type described by Burland & Symes (1982), were mounted diametrically opposite over a central gauge length of the specimen to measure local axial displacements. These transducers were mounted on the membrane after the specimens were formed and subjected to suction.

Finally, the specimens were anisotropically consolidated following initially a constant σ_r' drained stress path up to the line of constant stress ratio $\acute{o}_r'$ / $\acute{o}_v'$ =0.49, and then following the constant stress ratio line to an effective stress $s'=(\acute{o}_v'+\acute{o}_r')/2=75$kPa.

3 EXPERIMENTAL RESULTS

3.1 Effect of mica content variation

The mica content was varied in the laboratory on air pluviated specimens. The different soils resulting from the different gradations were then compared purely on the basis of using the same depositional method during specimen formation. As a result different void ratios resulted for the different gradations

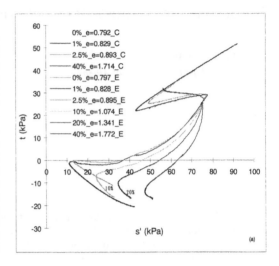

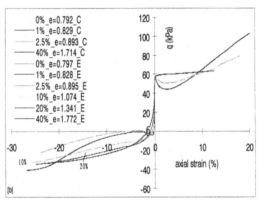

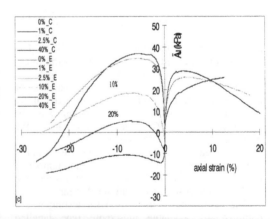

Figure 2. Undrained triaxial compression and extension with varying mica content; (a) effective stress paths, (b) stress-strain curves and (c) excess pore water pressure against axial strain curve.

the void ratio increasing with increasing mica content. The initial conditions of the tests are given in Table 1.

Table 1. Test data.

Test No.	Test type	Fine content (%)	Void ratio e
JS	C*	0	0.792
Jm1	C	1	0.829
Jm25	C	2.5	0.893
Jm40	C	40	1.714
JS	E*	0	0.797
Jm1e	E	1	0.828
Jm25e	E	2.5	0.895
Jm10e	E	10	1.074
Jm20e	E	20	1.341
Jm40e	E	40	1.772
JmSX	C	2.5	0.760
JHPF4	C	2.5	0.798
HRS	C	0	0.772
Hm1	C	1	0.769
Hm25	C	2.5	0.804
Hm5	C	5	0.891
Hsxm	C	2.5	0.746
HHPF4	C	2.5	0.761

C* = triaxial compression, E*=triaxial extension

Figure 2 shows (a) the stress paths, (b) the stress-strain curves and (c) the pore water pressures during undrained loading for a range of mica contents up to 40%. The specimens were all anisotropically consolidated along the same constant stress ratio line although this did not represent Ko conditions for the other mixtures, apart from the 2.5%, so that all stress paths started from a common point in the stress space.

In the compression space and for contents of mica up to 2.5% the specimens exhibit brittle response, with the deviatoric stress dropping after peak (Fig. 2(b)). Moreover, after approximately 3% axial strain dilative tendencies take over and all specimens follow what appears to be the sand failure line. On the contrary the 40% mica mixture does not show any brittleness (Fig. 2(a,b)) or any dilation tendencies (Fig. 2(c)).

In the extension side of the stress space a similar picture is emerging. For mica contents up to 2.5% brittleness is exhibited after peak strength to strains higher than 5% (Fig. 2(a,b)), followed by dilation after approximately 10% axial strain (Fig. 2(b,c)). For these small mica contents the rate of dilation appears to be very close for 0% and 1% mica content but the 2.5% mica content specimen appears to show a slightly reduced rate of dilation in both triaxial compression and extension (Fig. 2(c)).

At higher mica contents in extension brittleness is suppressed even at 10% (Fig. 2(a,b)). The graph of pore water pressure versus axial strain (Fig. 2(c)) shows that as mica content increases the contractive response of the specimens decreases, with minimum fluctuation of pore water pressure up to peak ob-

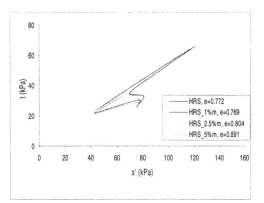

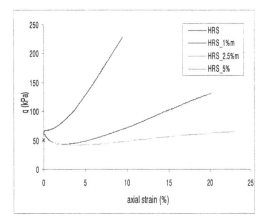

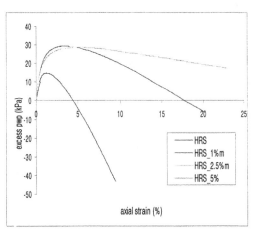

Figure 3. Undrained triaxial compression with varying mica content: effective stress paths, stress-strain curves and excess pore water pressure against axial strain curves.

served for the 40% mica content specimen. Therefore, the corresponding stress path never reaches the sand failure envelope (Fig. 2(a)).

The behaviour for low clay contents is less distinct and closer to the typical sand response. For mica contents up to 2.5% all specimens show brittle response with nearly identical effective stress paths up to the phase transformation line (Ishihara et al. 1975). In compression the 1% mica content specimen shows more brittleness than the 2.5% mica specimen but is still slightly less brittle than the clean sand although it has a higher void ratio.

However, it should be noted that the host sand is a well graded sand with a base gradation ranging between 0.06mm and 0.4mm. This makes the sand more susceptible to instability (brittleness and/or liquefaction) than a more uniform sand of similar void ratio.

Figure 3 describes the effect of mica on the subsequent behaviour of a uniform sand, the HRS sand, during shearing in triaxial compression. The sand was mixed with 1%, 2.5% and 5% sand size mica particles. Figure 3 shows the stress paths, the stress-strain curves and the excess pore water pressures during undrained loading. The response of the mixtures of HRS with mica is highly contractant and brittle. The presence of mica plates in this case transforms the behaviour of the host sand. Only 1% of mica introduces brittle behaviour to the sand which without any additives does not show any brittleness at all in compression. Moreover, for a mica content of only 5% any dilation tendencies exhibited by the sand appear to be suppressed.

Similar findings with regard to undrained shear of sands of different gradings have been reported by other researchers. Lade & Yamamunro (1997) noticed that between two different base gradations of Nevada sand (0.300-0.175mm) and (0.300-0.074mm) the latter exhibited complete liquefaction as opposed to temporary liquefaction exhibited by the more uniform sand. They inferred that the presence of sand between 0.175mm-0.074mm may have an effect of adding fines to a sand namely, increasing fines content increases liquefaction potential. The authors were comparing the sands at a loose state.

3.2 Effect of mica size variation

Apart from changing the grading of the sand the size of mica particles was changed from sand to silt size. In Figure 4 the stress paths and the corresponding stress strain curves are shown for Jamuna sand and mixtures of Jamuna sand with 2.5% of sand and silt (sx) size mica particles, respectively. It appears that the reduced size mica has a greater influence on the response of the host sand compared with the sand size mica. It increases the brittleness of the sand at least in triaxial compression thus making it more

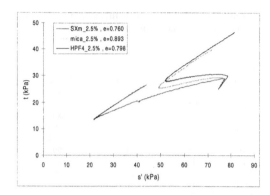

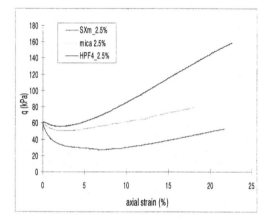

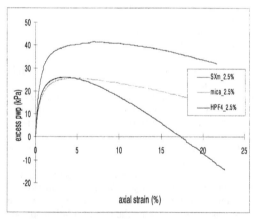

Figure 4. Mixtures of Jamuna sand and sand size mica, silt size mica (SX) and silt (HPF4).

susceptible to instability since higher brittleness is associated with higher excess pore water pressure, as shown in Figure 4. The test marked (HPF4) in the figure will be referred to later.

A similar effect can be observer in Figure 5 for the HRS. However, compared with the initial non brittle response of the sand any size of mica would create an unstable structure. The test on the silt specimen will also be referred to later.

3.3 Influence of initial fabric

It has been postulated by Hight et al. (1999) that the effects of the mica can be appreciated if the packing in a clean sand and a sand with mica is compared. The sand-size mica plates can bridge across adjacent sand grains preventing the overlying grain from nestling between as shown in Figure 6(a). This mechanism would be more effective in a uniform sand as suggested by the results of Figures 4 and 5. A study of thin sections prepared from specimens impregnated with chemical grout also supports the above argument.

The function of the platy mica particles was further investigated by tests on mixtures of the sands with silt grains. These tests have been included in Figures 4 and 5 for Jamuna sand and HRS respectively.

The presence of silt results in more stable behaviour by reducing contraction in Figure 4 while the response of the sand becomes purely dilatant in Figure 5. It appears that the silt grains fill in the voids between the sand grains, decrease the resulting void ratios and make the material response more stable (Fig. 6(b)).

The mixtures of the host sands with silt size mica (Figs 4,5) have smaller void ratios than the mixtures with the sand size mica yet they introduce more unstable response in triaxial compression. In this case the presence of silt size mica influences the response of the Jamuna sand dramatically. It is implied by the above that the importance of shape and location of additives in sand is not reflected in measures such as void ratio or granular void ratio.

3.4 Undrained Stiffness

The undrained secant stiffness of the various mixtures of mica and Jamuna sand is shown in Figures 7 and 8 for compression and extension respectively. The secant stiffness has been normalised with respect to effective mean normal consolidation pressure ($p_0'=(6_v'+26_r')/3$) and has been plotted against axial strain to a log scale. As the mica content increases so does the compressibility of the material and this is reflected by the corresponding stiffness values at small strains. In extension the 20% and 40% mica content specimens have nearly identical stiffness characteristics implying that mica particles take control over sand.

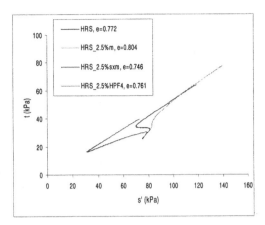

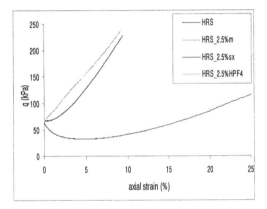

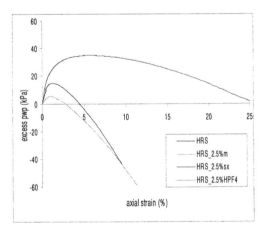

Figure 5. Mixtures of HRS and sand size mica (m), silt size mica (sx) and silt (HPF4).

However, for small contents of mica, up to 2.5% in compression and 1% in extension, the stiffness is similar to that of the sand. For the mixtures as for the sand specimens stiffness in extension is larger than stiffness in compression.

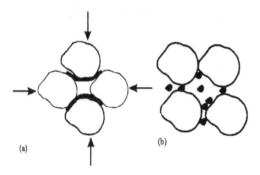

(a)   (b)

Figure 6. Importance of shape and location of additives in sand.

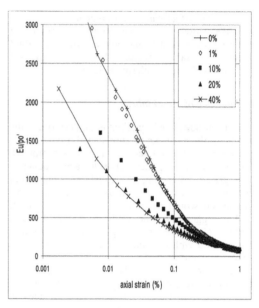

Figure 8. Variation of normalised secant stiffness for Jamuna sand and sand size mica mixtures. Triaxial extension.

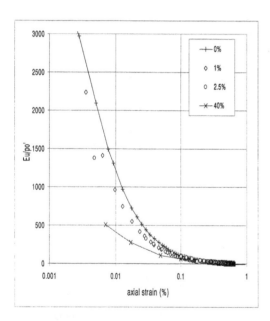

Figure 7. Variation of normalised secant stiffness for Jamuna sand and sand size mica mixtures. Triaxial compression.

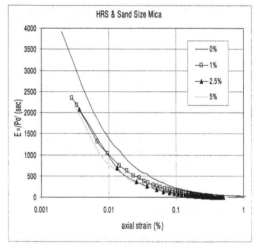

Figure 9. Variation of normalised secant stiffness in compression for HRS and sand size mica mixtures.

This reflects the anisotropy of these materials due to the method of formation and subsequent anisotropic consolidation.

The results for compression tests on the mixtures of mica and HRS are shown in Figure 9.

Compared with the results of Figure 7 even 1% mica content reduces the stiffness compared to clean sand and for mica contents up to 5% stiffnesses form a narrow band.

4 CONCLUSIONS

The importance of shape and location of additives in sand is not reflected in measures such as void ratio or granular void ratio. Moreover, the effect of the same additive depends also on the grading of the host sand. The addition of mica plates of silt and sand size was found to alter completely (i.e. from dilatant to contractant) the response of the host sand of uniform grading, even at small contents of mica (less than 2.5%).

However, for higher contents of mica brittleness is suppressed at contents of 20% and 5% for Jamuna sand and HRS, respectively. The stiffnesses measured in the tests are in good agreement with the observed shear behaviour.

5 REFERENCES

Blyth, F.G.H. & De Freitas M.H. 1984. *A Geology for Engineers*:325. London: Edward Arnolds

Burland, J.B. & Symes, M. 1982. A simple axial displacement gauge for use in the triaxial apparatus. *Geotechnique* 32(1):62-65.

Hight, D.W., Georgiannou, V.N., Martin, P.L. & Mundegar, A.K. 1999. Flow Slides in Micaceous Sands, Proc. Int. Symp. On Problematic Soils, IS TOHOKU'98, Sendai, Japan, (2):945-958.

Ishihara, K., Tatsuoka, F. & Yasuda, S. 1975. Undrained deformation and liquefaction of sand under cyclic stresses, *Soils and Fdns* 15(1): 28-44.

Ishihara, K. 1993. Liquefaction and flow failure during earthquakes. 35th Rankine Lecture. *Geotechnique* 43(3):351-415.

Lade, P.V. & Yamamuro, J.A. 1997. Effects of nonplastic fines on static liquefaction of sands, *Can. Geotechnique* 34:918-928.

Mundegar, A. K. 1997. An investigation into the effects of platy mica particles on the behaviour of sand, MSc Dissertation, Imperial College of Science Technology and Medicine, University of London.

Zdravkovic, L. 1996. The stress-strain-strength anisotropy of a granular medium under general stress conditions, PhD thesis, University of London.

Advanced Laboratory Stress-Strain Testing of Geomaterials, Tatsuoka, Shibuya & Kuwano (eds),
© 2001 Taylor & Francis, ISBN 90 2651 843 9

Effect of ageing on stiffness of loose Fraser River sand

J. A. Howie, T. Shozen and Y.P.Vaid
University of British Columbia

ABSTRACT: This paper presents the results of a laboratory study of the effect of time on the secant stiffness and stress-strain behaviour of loose Fraser River sand on four stress paths. The results show that the stiffness increased linearly with the logarithm of time when samples were allowed to age at constant stress ratio. Stiffness changed very rapidly over the first 1000 minutes after completion of consolidation, particularly for small increments of strain. The rate of stiffness increase varied with stress path. The modulus degradation curve became steeper with time for specimens aged at a particular stress ratio; with the effect more pronounced at higher stress ratios. The ratio of volumetric to shear strain during shear also varied with ageing time and stress path. Isotropically consolidated and aged samples displayed much lower influence of time on stiffness and volume change behaviour than samples aged at higher stress ratios.

1 INTRODUCTION

Recent laboratory work at the University of British Columbia has provided new insight into the effect of ageing on the stress-strain behaviour of sands. Ageing is shown to have a very large effect on the interpreted stiffness, with the effect becoming greater as the incremental strain range of interest drops below 0.2% and the consolidation stress ratio increases to 2.0 (i.e. K_o=0.5) and above. The need for a consistent approach to ageing of laboratory specimens used to study stress-strain behaviour is demonstrated.

2 RECENT TRENDS IN THE UNDERSTANDING OF SOIL STIFFNESS

Soil stiffness is commonly defined as the secant to a non-linear stress strain curve over the strain-range of interest. Figure 1 from Atkinson (2000) summarizes the current understanding of the variation of soil stiffness. Three strain ranges are recognized (Atkinson and Sallfors, 1991):

a) Very small strain (<0.001%) – stiffness is relatively constant;

b) Small strain (0.001%-1%) – stiffness reduces (or attenuates) with strain;

c) Large strain (>1%) – soil is approaching failure and stiffness is small.

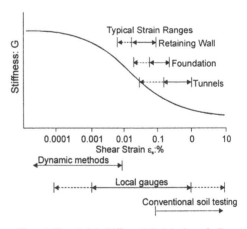

Figure 1. Characteristic Stiffness-strain behaviour of soil with typical strain ranges for laboratory tests and structures (after Atkinson & Sallfors, 1999 and Mair 1993)

Jardine (1992) identified the strain range of $0.001<\varepsilon_{axial}<0.1\%$ as one where strain is non-linear but unload-reload loops show completely recoverable behaviour.

Tests capable of allowing determination of stiffness in the three ranges are also shown in Figure 1. In the very small strain range, seismic wave velocities are conventionally used both in the field and in the laboratory. Triaxial testing with careful measurement of deformations using external monitoring devices can resolve strains down to about 0.01%

(Atkinson and Sallfors, 1991). Recent advances in strain monitoring systems such as on-sample strain measurements are closing the gap between these two measurement methods. This paper will present data on stiffness from advanced triaxial tests where axial (ε_a) and volumetric strains (ε_v) were measured to a resolution better than 0.01% and 0.005%, respectively.

3 LABORATORY PROGRAM

3.1 Material Tested

Tests were carried out on Fraser River sand: a grey, uniform, semi-angular medium-grained sand, with D_{50} and D_{10} of 0.271 mm and 0.161 mm, respectively, a uniformity coefficient of 1.88 and a Specific Gravity of 2.72. Particles less than 0.1 mm in diameter were removed. The maximum and minimum void ratios were measured to be 0.989 and 0.627 respectively.

3.2 Sample Preparation and Test Procedure

Samples were prepared by water pluviation, a technique that has proved to give very uniform samples (Vaid and Negussey, 1984). This procedure mimics alluvial deposition and ensures full saturation. Rigorously consistent laboratory procedures led to very repeatable sample densities. Most samples were prepared in their loosest state. To allow removal of the

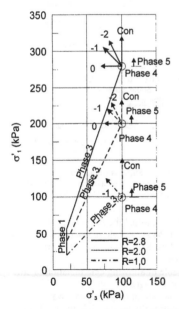

Figure 2. Applied stress paths

metal former, a 12 to 15 kPa vacuum was applied to the sample. At this point, sample relative densities ranged from 13 to 17%. Medium dense samples were prepared by applying low frequency vibration during pluviation and again after placement of the loading cap. Prior to the start of consolidation, samples were at an isotropic confining pressure of 20 kPa under a back-pressure of 200 kPa.

The stress paths investigated and the definition of Test Phases 1 to 5 are shown in Figure 2. All tests were drained. Samples were consolidated first to the prescribed confining pressure (most commonly $\sigma'_3=100$kPa) along paths of constant stress ratio, $R=\sigma'_1/\sigma'_3$, of 1.0, 2.0 and 2.8 (Phase 3). Initial loading from $\sigma'_3=20$kPa to the prescribed stress ratio followed the conventional triaxial path of $\Delta\sigma'_3=0$ (Phase 1). Phase 2, an initial stage of ageing, was eliminated after the first few tests. Upon completion of consolidation, samples were allowed to age at constant stress ratio for times ranging from 1 to 10,000 minutes. Problems of test instability were experienced for the ageing period of 10,000 minutes and so only two tests were attempted.

Samples were then sheared along one of four stress paths:

a) Conventional triaxial path, "Con" in Figure 2;

b) Constant p'= $\dfrac{\sigma'_1 + 2\sigma'_3}{3}$, i.e. $\dfrac{\Delta\sigma'_1}{\Delta\sigma'_3} = -2$, "-2" in Figure 2;

c) Constant t = $\left(\dfrac{\sigma'_1 + \sigma'_3}{2}\right)$, i.e. $\dfrac{\Delta\sigma'_1}{\Delta\sigma'_3} = -1$, "-1" in Figure 2;

d) Constant σ'_1, .e. $\Delta\sigma'_1 = 0$, "0" in Figure 2.

Tests were carried out under load control, with stress increased in increments of 1 to 1.5 kPa applied every 15 seconds. Careful attention was paid to minimization of bedding errors, correction for membrane penetration and for membrane stiffness (Kuerbis and Vaid, 1990).

4 DEFORMATION DURING AGEING

Figure 3 shows ε_v vs. ε_a during ageing at constant stress ratio (at σ'_3 =100 kPa) for loose samples. The ratio $\varepsilon_v/\varepsilon_a$ depends only on stress ratio, with $\varepsilon_v/\varepsilon_a$ = 0.55, 1.0 and 3.0 to 3.5 for R=2.8, 2.0 and 1.0, respectively. These ratios were found to be essentially independent of relative density, D_r, and maximum consolidation cell pressures, σ_3 =50 to 150 kPa. $\varepsilon_v/\varepsilon_a=1.0$ is equivalent to one dimensional compression deformation indicating that K_o is just less than 0.5 for this soil. Maximum creep strains of only 0.28% axial and 0.13% volumetric were observed over 1000 minutes in very loose samples at σ'_3 =100

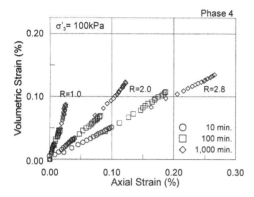

Figure 3. Relationship between axial and volumetric strains

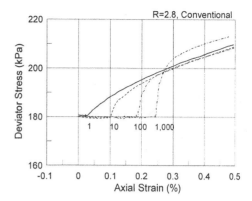

Figure 5. Stress-strain curves during Phase 4 and 5

kPa and constant R=2.8. The magnitude of creep strain decreased as D_r increased and consolidation stress ratio decreased.

5 STRESS-STRAIN BEHAVIOUR DURING SHEAR

The effects of ageing at R=2.8 on the initial portions of the stress strain curves obtained from loading along the conventional triaxial path are illustrated in

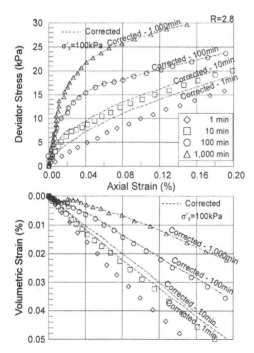

Figure 4. Stress-strain responses with residual creep correction

Figure 4. Corrected curves show the effect of removing deformation which would have occurred if no stress increments had been applied, i.e. creep deformation which would have occurred under no change in stress. The correction is small for 10 minutes ageing and insignificant for 100 minutes and longer. At the resolution of strain possible in these tests, the initial, apparently linear, portion of the curve increases in length as the ageing period increases. The secant stiffness increases with ageing for the strain range shown.

The volumetric strain at a given axial strain as a result of an increase in shear stress is reduced by periods of ageing. The stiffening effect of time is only apparent over the very early portions of the overall stress strain curve as shown in Figure 5. This figure shows deviator stress, σ_d vs. axial strain for Phases 4 and 5 for ageing times of up to 1000 minutes. All stress strain curves coincide beyond a deviator stress increase of about 20 kPa. The curve for 1000 minutes lies above the others because of slightly higher relative density.

A similar increase in the initial slope of the stress-strain curve with time and a corresponding reduction in the amount of contractive volumetric strain was observed along the other stress paths followed. Figure 6 shows the effect of ageing on the '0' path at R=2.8. Samples became more dilatant in the initial stages of shear as the ageing time increased. Similar effects were observed at other consolidation stress ratios.

6 INTERPRETATION OF SOIL STIFFNESS

Interpretation of soil stiffness from element tests requires an assumption of a soil model. The conventional approach is to interpret the tests assuming the soil to be an isotropic elastic solid. If the soil fitted this model, then it should be possible to define the

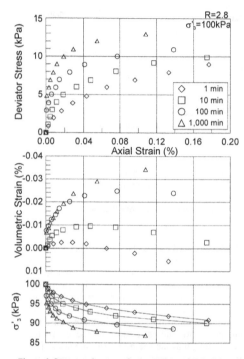

Figure 6. Stress-strain curves for stress slope of 0, R=2.8

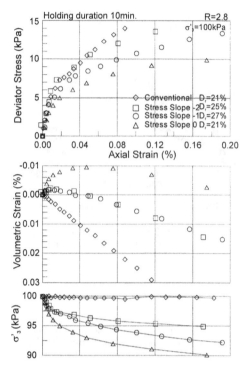

Figure 7. Stress-strain curves at R=2.8 with holding duration of 10 min. on various stress paths

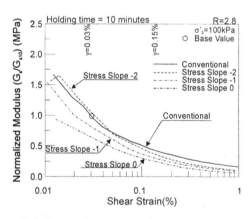

Figure 8. Attenuation curves for various stress paths

stress-strain behaviour in terms of two independent elastic parameters such as Young's modulus, E, and Poisson's ratio, ν, or Shear Modulus, G, and Bulk modulus, K. Under this framework, all elastic deformation parameters can be calculated based on knowledge of two. For example, shear modulus can be calculated using the expression:

$$G = \frac{E}{2(1+\nu)} \tag{1}$$

Poisson's ratio is commonly assumed to be around 0.1 to 0.2 for small strains. For very loose Fraser River sand, Poisson's ratio evolved with stress ratio and time. Poisson's ratio was ν=0.1 for initial loading from R=1.0, ν=0.2 for R=2.0 and ν=0.36 for R=2.8. At R=2.8, ν changed from 0.36 to 0.48 as ageing time increased. E and ν can be measured easily in conventional triaxial tests, but the interpretation of these parameters is more difficult for tests on other stress paths as both vertical and lateral stress are changing. Figure 7 shows the effect of stress path on the initial stages of the stress-strain curves and on ε_v vs ε_a.

To ease the comparison of modulus for loading along different stress paths, the stress-strain curves have been interpreted in terms of G_s with G_s taken to be the secant to the curve of maximum shear stress

$(\sigma_1-\sigma_3)/2$ against shear strain, $\gamma = (\varepsilon_1-\varepsilon_3)$. The effect of strain increment level was studied by comparing secant moduli at shear strains of 0.03% and 0.15%, $G_{s0.03}$ and $G_{s0.15}$, respectively. Figure 8 illustrates how G_s varies with stress path and strain level for R=2.8 and a sample aged for only 10 minutes. The moduli have been normalized by $G_{s0.03}$ on the con-

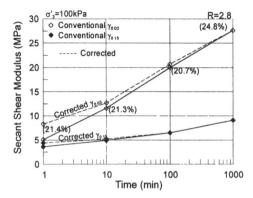

Figure 9. Secant shear modulus and after correction

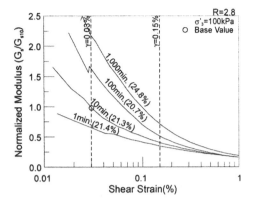

Figure 10. Attenuation curves for R=2.8 on conventional path

conventional path. Stiffness is higher for the 'Con' and '-2' stress paths below strains of 0.1%. At larger strains, the 'Con' path response is stiffer.

The effect of time on $G_{s0.03}$ and $G_{s0.15}$ is illustrated in Figure 9 for R=2.8. The effect of correcting the stress-strain curve for continuing creep is shown. The increase in $G_{s0.03}$ with ageing is greater than for $G_{s0.15}$. $G_{s0.03}$ increased to about 3.5 times its initial value in 1000 minutes while $G_{s0.15}$ approximately doubled. The variation of stiffness with time for strain increments up to 1% is shown for the same samples in Figure 10. The values have been normalized to $G_{s0.03}$ at 10 minutes. The rate of modulus attenuation with strain increases dramatically with ageing time.

Testing on samples at D_r of about 50% showed a very similar magnitude of increase in shear modulus with time (Shozen, 2000) but, as the initial moduli were higher for higher D_r, the percentage increase over time was much less. Nevertheless, Figure 11 shows that at R=2.8, ageing for 1000 minutes resulted in a higher stiffness at $\gamma < 0.1\%$ than in a denser but younger sample. Stiffness attenuated more severely with strain in the looser sample.

Similar effects were noted in tests aged at R=2.0 and 1.0. The effect of time on $G_{s0.03}$ and $G_{s0.15}$ for R=1.0 is shown in Figure 12. Comparison of Figures 9 and 12 shows that the initial values of $G_{s0.03}$ and $G_{s0.15}$ were much greater for R=1.0 than for R=2.8. Under isotropic ageing, increases in stiffness were not observed until samples had been aged for longer than 100 minutes.

7 DISCUSSION

7.1 Effects of time

The above results show that ageing time can have a large effect on the stiffness of loose sand samples,

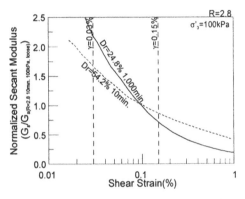

Figure 11. Comparison between the effect of time and of relative density

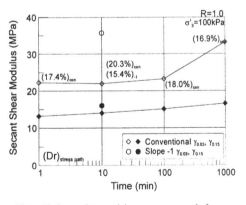

Figure 12. Secant shear modulus on two stress ratio for R=1.0

with the increase proportionately greater at higher initial stress ratios. Ageing decreases the tendency towards initial contraction of the samples under increases in shear stress. These effects do not seem to have been systematically studied. Anderson and Stokoe (1978) observed an increase with time in the small strain modulus of Ottawa sand measured in a Resonant Column device. They used the stiffness measured after ageing for 1000 minutes under isotropic confining stress, G_{1000}, as the base values for their observations. They found the stiffness increased by about 1% of the base value, G_{1000}, for each log cycle of time. The samples appear to have been fairly dense, having a void ratio of 0.5. Jamiolkowski and Manassero (1995), drawing on the work of Anderson and Stokoe, Mesri at. al. (1990); and Schmertmann (1991), suggested that the effect of ageing on small strain shear modulus, G_o, interpreted from resonant column tests or seismic wave velocities, may be approximated by the expression

$$: \quad \frac{G_o(t)}{G_o(t_p)} = 1 + N_G \log\left(\frac{t}{t_p}\right) \quad (2)$$

where $G_o(t)$ is G_o at time $t > t_p$; $G_o(t_p)$ is G_o at time t_p;
   t is any time $> t_p$;
   t_p is the time to the end of primary compression (EOP); and
   N_G is the slope of a plot of G_o versus the log of time expressed as a fraction of $G_o(t_p)$.

Mesri et al. (1990) suggested that for 6 sands investigated, N_G varied from 1 to 3 % and that N_G increased as the soil became finer. Fahey (1998) suggested that this rate of increase was not sufficiently high to explain the difference between the measured stiffnesses of undisturbed and reconstituted samples. The work presented here, with G_s obtained over much larger strain increments than G_o, suggests that if 1 minute is taken as EOP, N_G depends on stress ratio and stress path and could be as high as 0.75. N_G is much smaller for samples aged under isotropic stress conditions.

Daramola (1980) also studied the effects of ageing on the stiffness of sand. He tested dense Ham River sand under the conventional triaxial stress path after ageing under an isotropic stress of 350 kPa for 0 to 152 days. The stiffness was observed to be a function of D_r when samples were not aged for a long time. However, after ageing, time was observed to have great influence on stiffness and D_r was not the main factor controlling stress-strain response. Secant stiffnesses at strains less than 0.5% increased by 100% over three log cycles of time.

From the above, it is clear that the measured secant stiffness of loose Fraser River sand is strongly affected by sample age and stress ratio during ageing

in addition to the more commonly considered factors such as D_r; magnitude of strain increment; stress ratio; and stress path.

7.2   Effect of stress level

All tests presented above were aged at an effective confining pressure of 100 kPa. As stress ratio varied, the mean normal stress, p', was 133 kPa and 160 kPa for R=2.0 and 2.8, respectively. The stress dependence of soil stiffness is commonly modelled by expressions of the type (Janbu, 1963):

$$G_s = k_G p_a \left(\frac{p'}{p_a}\right)^n \quad (3)$$

where p' is the current mean normal effective stress, p_a is a reference pressure to make the equation non-dimensional, k_G is a dimensionless parameter which varies with the strain level of interest and n is a material parameter. For sands, n is typically close to 0.5 (Hardin, 1978). Atkinson (2000) noted that the variation of small strain stiffness of soils could be represented by an expression of the form:

$$G_o = A F(e) p_a \left(\frac{p'}{p_a}\right)^n R_o^m \quad (4)$$

where A, n and m are material parameters, R_o is the over-consolidation ratio, F(e) is a function of void ratio, and p' is as before. Comparing equations (3) and (4), k_G is an empirical parameter which takes account of material type, density and degree of over-consolidation in addition to strain level. For the tests reported in this paper for one type of normally consolidated sand at similar values of D_r, normalization of the measured stiffness by (p'/p_a) allows comparison of stiffnesses measured at different initial confining pressures and stress ratios. In addition, the effect of stress path can be accounted for by defining the initial shear stress level, τ, relative to the shear stress at failure along the given stress path, τ_f where the stress path meets the failure line at a stress ratio of about 4.0. Different stress paths result in values of τ/τ_f ranging from about 0.33 to 0.67 for consolidation and ageing at R=2.0 and from 0.6 to 0.86 for R=2.8.

Figure 13 is a plot of $G/(p'/p_a)$ vs. τ/τ_f for tests on very loose samples of Fraser River sand. It is clear that the effect of ageing on stiffness is proportionately larger for ageing at higher stress ratios or as the initial shear stress gets closer to its limiting value. For ageing to 1000 minutes, the trend is for stiffness normalized by mean normal stress to reduce with shear stress level. Ageing for 1000 minutes results in an increase in $G_{s0.03}$ of about 50% for ageing at R=1.0, an increase of 75% to 120% at R=2.0 and an

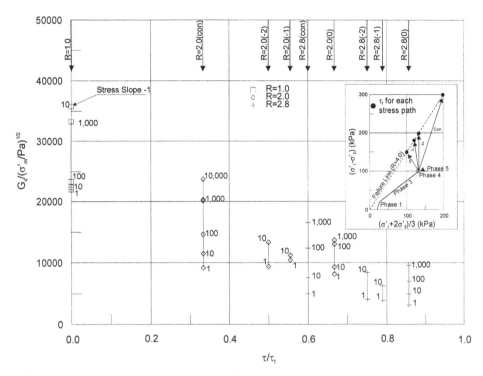

Figure 13. Normalized secant shear modulus for all conditions

increase of about 220% at R=2.8. The effect of ageing is to reduce the effect of initial stress ratio on stiffness.

From Figure 10, the effect of ageing is greater at small strain increment levels. If this trend were to continue with reducing strain increment level, ageing would be expected to be a significant factor in the effect of stress ratio on the small strain shear modulus, G_o, conventionally determined by measurement of shear wave velocity, V_s. This suggests the need for caution in the present trend towards using V_s or G_o as an indicator of in situ soil state (Robertson and Fear, 1995).

7.3 Effect of stress path

The effect of stress path on the measured stiffness varies depending on the level of the initial stress ratio. All stress paths studied involved increases in stress ratio. The conventional path also results in an increase in p'. The '-1' and '0' stress paths cause reductions in p' and the '-2' path consists of an increase in shear stress alone. Of the four stress paths studied, the strongest dilation at strains <0.1% was observed on the '0' path for samples which were aged for a longer period. Figure 14 is a comparison

of ε_v vs ε_a for loose and medium dense samples at R=2.8. The much greater influence of ageing on the initial deformation of the looser samples is clearly illustrated, with the dilation continuing for a greater proportion of the subsequent shearing for older samples. The denser samples dilated much more strongly than the loose ones beyond axial strains of 0.1%. This suggests that the initial deformations were dominated by the unloading of p' for the less stiff, young samples. Dilation due to shear dominated for denser samples. Increased stress ratio quickly caused contraction of the loose, young samples but loose samples displayed less tendency towards contraction as ageing time increased. The implication of this finding for undrained tests requires further study.

8 IMPLICATIONS AND CONCLUSION

Ageing up to 1000 minutes after completion of consolidation has an important effect on stiffness of loose to medium dense Fraser River sand. Limited tests beyond this period suggest that the effect continues. The increase in stiffness appeared to be constant with the logarithm of time. This agrees with

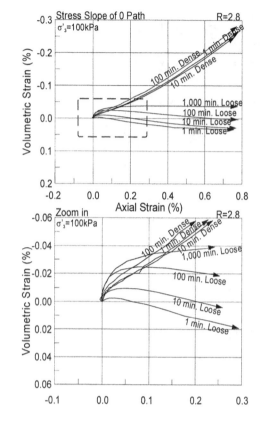

Figure 14. Initial dilative volumetric strain on stress-slope of 0 path for R=2.8

previous laboratory observations. The effect was greater as the value of constant stress ratio during ageing became larger. As the stress ratio during ageing was increased, the proportion of stiffness due to ageing became larger. The proportion of stiffness due to ageing also increased as the soil became looser. Very loose aged samples can be stiffer than denser but younger sands.

The rate of attenuation of stiffness with strain was greater for samples which had been aged for a longer period. Ageing had little effect on secant stiffness beyond shear strains of about 0.5%. In fact, the effect of ageing up to 1000 minutes was overcome early in the shearing process with the stress –strain curves thereafter coinciding with those of the unaged specimens. Although no small strain measurements were undertaken, the results suggest that very loose, aged sands may have high small strain moduli but the stiffness may drop rapidly as the soil strains. It also implies that small strain stiffness as interpreted from shear wave velocity should be interpreted in

conjunction with other data such as geological considerations when it is being used to assess in situ state.

The implications for laboratory test programs are also considerable. Consistency of laboratory test results would be improved by careful attention to the age of the samples at initiation of shear. This study suggests that allowing samples to sit overnight under consolidation stresses prior to testing could eliminate variability in the early stages of stress-strain curves caused by variations in sample age. It also suggests that the in situ small strain behaviour of sands cannot be reliably assessed from tests on reconstituted samples and that matching shear wave velocities of field soil and laboratory samples will not be a reliable indicator of identical initial state.

The effect of stress path on stiffness also appears to vary with stress ratio during ageing and the strain increment over which stiffness is measured. For all stress paths and stress ratio, ageing decreased the tendency for initial contraction during shear. This may have importance in consideration of liquefaction of very loose sands.

ACKNOWLEDGEMENTS

This research was supported by a grant from the Natural Science and Engineering Research Council of Canada. Technical assistance of Harald Schremp, Scott Jackson and John Wong is gratefully acknowledged.

REFERENCES

Anderson, D.G. and Stokoe, K.H. 1978. Shear modulus, a time-dependent soil property. *Dynamic Geotechnical Testing. ASTM Special Technical Publication 654*, American Society for Testing and Materials. 66-90.

Atkinson, J.H. 2000. Non-linear soil stiffness in routine design. *Géotechnique* 50(5):487-508.

Atkinson, J. H. and Sallfors, G. 1991. Experimental determination of stress-strain-time characteristics in laboratory and in situ tests, *Proc. 10th European Conf. On SMFE*, Florence, Italy, Balkema, Rotterdam. 3:915-956,

Burland, J. 1989. Ninth Laurits Bjerrum Memorial Lecture: "Small is Beautiful"-the stiffness of soil at small strains. *Canadian Geotechnical Journal*, 26, 499-516.

Daramola, O., 1980. Effect of consolidation age on stiffness of sand. *Geotechnique*, 30(2):213-216

Hardin, B.O. 1978. The nature of stress-strain behaviour for soils. *Proceedings of ASCE Geotechnical Division Specialty Conference on Earthquake Engineering and Soil Dynamics, Pasadena.* 1,3-39.

Fahey, M. 1998. Deformation and in situ stress measurement. Geotechnical Site Characterization, Robertson and Mayne (eds). *Proceedings of the First International Conference on Site Characterization-ISC'98.* Atlanta, Georgia. A.A. Balkema, Rotterdam.

Ishihara, K. 1996. *Soil Behaviour in Earthquake Engineering.* Clarendon Press, Oxford.

Jamiolkowski, M. and Manassero, M. 1995. The role of in-situ testing in geotechnical engineering – thoughts about the future. *Proceedings of the International Conference on Advances in Site Investigation Practice,* London, U.K. 929-951. Thomas Telford.

Jamiolkowski. M., Lo Presti, D.C.F. and Froio, F. 1998. Design parameters of granular soils from in situ tests. *Geotechnical Hazards,* Maric, Lisac and Szavits-Nossan (eds). 65-94. Balkema, Rotterdam

Janbu, N. 1963. Soil compressibility as determined by oedomter and triaxial tests. *Proceedings of the fourth European Conference on Soil Mechanics and Foundation Engineering.*

Jardine, R.J., 1992. Some observations of the kinematic nature of soil stiffness. *Soils and Foundations,* **32**(2):111-124.

Kuerbis, R.H. and Vaid, Y.P. 1990. Corrections for membrane strength in the triaxial test. *Geotechnical Testing Journal,* **13**(4):361-369.

Mair, R.J. 1993. Developments in geotechnical engineering research: applications to tunnels and deep excavations. Unwin Memorial Lecture 1992. *Proceedings of Institution of Civil Engineers Civil Engineering,* **3**:27-41.

Mejia, C., Vaid, Y.P. and Negussey, D. 1988. Time dependent behaviour of sand. *Proceedings of First International Conference on Rheology and Soil Mechanics,* Coventry, U.K. Keedwell (ed.), 312-326.

Mesri, G., Feng, T.W. and Benak, J.M. 1990. Postdensification penetration resistance of clean sands. *Journal of Geotechnical Engineering,* ASCE, **116**(7): 1095-1115.

Mitchell, J.K. and Solymar, Z.V. 1984. Time-dependent strength gain in freshly deposited or densified sand. *Journal of Geotechnical Engineering,* ASCE, **110**(11): 1559-1576.

Robertson, P.K., and Fear, C.E. 1995. Liquefaction of sands and its evaluation. *Earthquake Geotechnical Engineering, Ishihara (ed.)., Balkema, Rotterdam. 1253-1289.*

Schmertmann, J.H. 1970. Static Cone to compute static settlement over sand. *Journal of Geotechnical Engineering Division,* ASCE, **96**(SM3): 1011-1043.

Schmertmann, J.H. 1991. The mechanical aging of soils. 25th Karl Terzaghi Lecture, *Journal of Geotechnical Engineering,* ASCE, **117**(9): 1288 - 1330.

Shozen, T., (2001) Deformation under the constant stress state and its effect on strain-strain behaviour of Fraser River Sand, *M.A.Sc. Thesis, Dept of Civil Engineering, The University of British Columbia*

Tatsuoka,F., Santucci de Magistris,F., Hayano,K., Momoya,Y. and Koseki,J. 1998. Some new aspects of time effects on the stress-strain behaviour of stiff geomaterials, *Keynote Lecture for 2nd International Conference on Hard Soils Soft Rocks,* Evangelista and Picarelli (eds), Napoli, Balkema, Vol.2.

Vaid, Y. P. and Negussey, D., (1984), A critical assessment of membrane penetration in the triaxial test., *Geotechnical Testing Journal,* Vol.7, No.2, pp.70-76

Withers,N.J., Howie,J.A., Hughes,J.M.O. and Robertson,P.K. 1989. Performance and Analysis of Cone Pressuremeter Tests in Sands. *Géotechnique,* **39**(3): 433-454.

Advanced Laboratory Stress-Strain Testing of Geomaterials, Tatsuoka, Shibuya & Kuwano (eds),
© 2001 Taylor & Francis, ISBN 90 2651 843 9

Deformation characteristics of recompressed volcanic cohesive soil

M. Katagiri
Nikken Sekkei Nakase Geotechical Institute, Kawasaki, Japan

K. Saitoh
Chuo University, Tokyo, Japan

ABSTRACT: This paper describes the deformation characteristics of recompressed volcanic cohesive soil obtained from cyclic triaxial tests, and discusses the effects of saturation and drainage on its deformation characteristics. The result is that the equivalent Young's Modulus of the recompressed volcanic cohesive soil is located within the lower limit of the range of previous studies for marine clays. However, the relationship between damping ratio and axial strain is twice as large as the previous data. The equivalent Young's Modulus and dumping ratio are dependent on the saturation of the specimen and the drainage conditions.

1 INTRODUCTION

As more than 40 % of the land in Japan is covered by volcanic materials, volcanic cohesive soils are commonly used as banking materials. This kind of soil is classified as a special material in geotechnical engineering because the strength ratio of recompressed samples to undisturbed ones is extremely small. Volcanic cohesive soils have high locality, that is, their properties are dependent on their locality.

Recently, a seismic response analysis for an important earth structure constructed from volcanic cohesive soils has been performed and its deformation behaviour assessed (New Aomori Airport, Kudo et al., 1989). However, the deformation characteristics of recompressed volcanic cohesive soil at the small strain level have not yet been fully clarified. Generally, earth structures made from such volcanic cohesive soils are in an unsaturated condition. It is important to clarify the characteristics of unsaturated ground in practice. The effect of saturation on the deformation behaviour of such recompressed volcanic cohesive soils, however, has never been studied.

In this paper, in order to study the deformation characteristics of a recompressed volcanic cohesive soil with a relatively high saturation (more than 90 %), cyclic triaxial tests to determine the deformation properties were performed. The effects of saturation and drainage conditions on the specimens' deformation characteristics were also investigated.

2 PHYSICAL PROPERTIES OF THE MATERIAL

The material tested for this test series was volcanic cohesive soil sampled at Gunma Prefecture, north of Tokyo. Its physical properties are listed in Table 1.

Table 1 Index properties of Volcanic Cohesive soil

w_n	ρ_s	$F_c(<75\mu m)$	w_L	w_P	Ip
%	g/cm^3	%	%	%	
95.2	2.803	63.2	97.5	64.5	33

Figure 1 shows the particle size distributions of volcanic cohesive soil tested. Toyoura sand, a standard sandy material in Japan, is also plotted in the same figure. This volcanic cohesive soil is classified as a fine-grained soil.

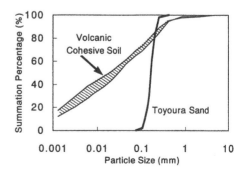

Figure 1 Particles size distribution of Volcanic Cohesive Soil

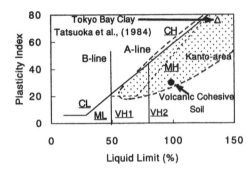

Figure 2 Plasticity chart plotting Volcanic Cohesive Soil

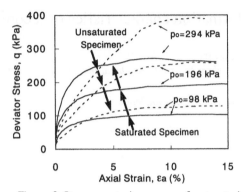

Figure 3 Stress – strain curves of saturated
and unsaturated specimens

Figure 2 is a plasticity chart for the volcanic cohesive soil tested. This material is classified as a
high plasticity volcanic soil. It is located in the
Kanto area, and is situated in the lower limit of the
range. In this figure, the sample tested by Tatsuoka
et al. (1984) is also plotted, and it is also classified
as a high plasticity cohesive soil.

3 PREPARATION OF TRIAXIAL SPECIMENS AND SETTING METHOD

The specimens for the monotonic and cyclic triaxial
tests were prepared by the static one-dimensional
precompression method. The procedure was as follows. Volcanic cohesive soil particles passed
through a 4.75 mm sieve were put into a preconsolidation container 800 x 250 mm in section. A piston
with a drainage filter was inserted into the container.
A pressure of 147 kPa was applied to the piston for 2
hours. Triaxial specimens of 50 mm in diameter and
100 mm in height were vertically trimmed from the
consolidated soil block.

The saturated specimens were prepared by the dry
setting method with the vacuuming and subsequent
back pressurisation proposed by Ampadu and Tatsuoka (1993). In this series, minimum back pressure
was applied to –78.4 kPa under a maximum effective stress of 19.6 kPa. At a back pressure of –78.4
kPa and a confining pressure of –58.8 kPa, de-aired
water was run from bottom to top for 2 hours under
a constant hydraulic gradient of about 7 kPa, to drive
out any entrapped air from the specimen and from
the drainage lines. A final back pressure of 196 kPa
was applied for this series. An average increase in
back pressure of about 5 kPa/min was chosen as being adequate to ensure stable readings of changes in
height and volume of the specimen. All the specimens prepared by this procedure were fully saturated
(B-value > 0.96). That is, the calculated saturation

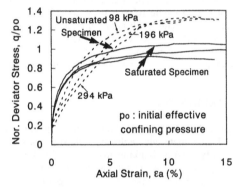

Figure 4 Normalised deviator stress and axial
strain curves of saturated and unsaturated specimens

degree of the saturated specimens ranged from 99 to
101 %.

The unsaturated specimens, on the other hand,
were set on a pedestal in a triaxial cell. A confining
pressure of 196 kPa and no back pressure were applied. Measurement of the B-value before the
undrained compression process was not carried out
since the specimen would have damaged. The volume change of the specimen during the consolidation
process was calculated by an axial displacement,
having assumed that the specimen deformed isotropically. The degree of saturation of the unsaturated specimens after isotropic consolidation ranged
from 90 to 94 %.

4 STATIC SHEAR BEHAVIOUR

The stress-strain curves of the saturated and unsaturated specimens are shown in Fig. 3. A great difference in the initial slope of the stress-strain curve and
the maximum deviator stress on the same confining

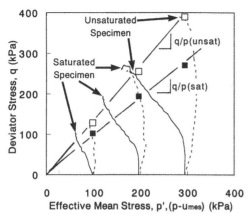

Figure 5 Effective stress paths and failure criteria

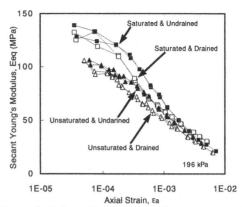

Figure 6 Deformation modulus of saturated and unsaturated specimens under drained and undrained conditions

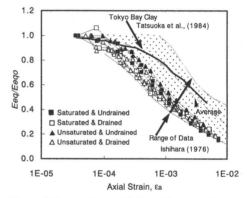

Figure 7 Comparison of deformation modulus

pressure are shown in the saturated and unsaturated specimens.

Figure 4 shows the relationships between axial strain and normalised deviator stress on the saturated and unsaturated specimens. Here, the normalised deviator stress is the deviator stress divided by the initial effective confining pressure. The normalised stress-strain curves of the saturated specimens converge roughly at a given relationship, in particular, the relationships in the lower strain level less than 2 % were almost piled up on the others. This means that the stress-strain curves of saturated volcanic cohesive soil can be normalised by effective confining pressure.

The relationships between axail strain and normalised deviator stress of the unsaturated specimens, however, are separated from each other up to 7 % of axial strain, and converge after that. In the unsaturated specimens, migration of pore water may occur during the compression process. The degree of migration may be dependent on initial effective confining pressure. After 7 % of axial strain, the normalised deviator stresses converged at about 1.3. At this stage, the degree of migration may be almost the same.

Figure 5 shows the effective stress paths of saturated specimens expressed by solid lines, and the relationships between effective confining pressure and shear strength in both the saturated and unsaturated specimens marked by ■ and □, respectively. The q/p of saturated specimens is 0.9 and is smaller than that of 1.3 in the unsaturated specimens.

The paths of deviator stress and mean stress minus measured excess pore pressure in the unsaturated specimens are also drawn as broken lines in the same figure. The measured excess pore water pressures of the unsaturated specimens were smaller than those of the saturated specimens. Although the true effective mean stress could not be determined

without measurement of the pore air pressure of the unsaturated specimen in the undrained compression tests, the effective stress of the unsaturated specimen at failure was thought to be larger than that of saturated one. Therefore, deviator stress at failure is thought to increase.

5 SHEAR BEHAVIOUR AT SMALL STRAIN LEVELS

5.1 Deformation behaviour

Figure 6 shows the secant Young's modulus E_{sec} and the axial strain ε_a curves of the saturated and unsaturated specimens obtained from cyclic loading tests under undrained and drained conditions. The isotropic confining pressure of each test was 196 kPa. The E_{sec}-values for the strain range of less than 0.0001 in all specimens are considered to be constant, and the maximum magnitudes of E_{sec0} are almost 140 MPa and 110 MPa for the saturated and unsaturated specimens, respectively. The E_{sec}-value decreases as axial strain level increases. Moreover, there is no difference in the E_{sec}-value in both the specimens for a strain range of several thousandths. The relationships between E_{sec} and ε_a for the saturated specimens marked by ■ and □ are located above

those for the unsaturated ones in the whole strain level. The reason for this is thought to be the reduced volume of pore air in the unsaturated specimen at the same axial loading.

The relationships between E_{sec} and ε_a under undrained conditions marked by ■ and ▲ are located above those under drained conditions. As drainage from the specimen under drained conditions is permitted during cyclic loading, the deformation of the specimen is thought to be larger than that at the same deviator stress under undrained conditions. Tatsuoka et al. (1995) carried out drained and undrained cyclic loading tests on saturated Pleistocene clay, and turned up the same results as in this test series.

Figure 7 shows the relationship between the normalised deformation modulus, E_{sec}/E_{sec0} and the axial strain ε_a. Here, the E_{sec0} is E_{sec} at the minimum strain obtained from the cyclic tests. The data for several cohesive soils expressed by Ishihara (1976) and Tatsuoka et al. (1984) are also shown in the same figure. The curves of the recompressed volcanic cohe-

Figure 8 Comparison of damping ratio

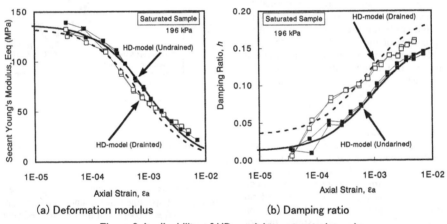

(a) Deformation modulus   (b) Damping ratio

Figure 9 Applicability of HD-model to saturated specimens

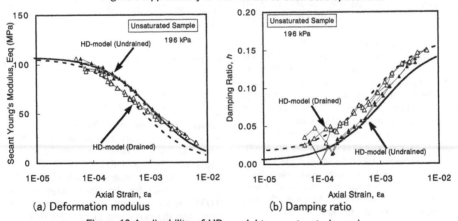

(a) Deformation modulus   (b) Damping ratio

Figure 10 Applicability of HD-model to unsaturated specimens

sive soil are located within the range of marine clays expressed by Ishihara (1976), and are situated near its lower limit, especially in the relatively large strain range.

The deformation modulus of clay is a function of consolidation time, overconsolidation ratio, void ratio and current confining pressure, as stated by Hardin and Black (1968). In this study, the consolidation time was 24 hours under normally consolidated conditions. These conditions were almost the same as the marine clays. The void ratio of the volcanic cohesive soil specimens, however, was approximately 3, and was thought to be slightly larger than those of marine clays. Therefore, the relationships between E_{eq}/E_{eq0} and ε_a of volcanic cohesive soils are thought to be located near the lower limit of the range for marine clays.

Figure 8 shows the relationship between the damping ratio h and the axial strain ε_a. The relationships expressed by Ishihara (1976) and Tatsuoka et al. (1984) are also shown in the same figure. For a strain range from less than one ten-thousandth to several thousandths, the damping ratio of the recompressed volcanic cohesive soil is approximately twice as large as the average value of the cohesive soils referred to by Ishihara (1976).

The relationships between h and ε_a on the saturated specimen under drained condition marked by the open symbols □ are, remarkably, located above those of the other three. The relationship of the unsaturated specimen under drained conditions marked by △ is situated slightly above the specimen under undrained conditions, ▲. The maximum difference between undrained and drained conditions for both the specimens is 0.03 at axial strain of 0.0002. The difference, however, decreases as axial strain increases. The reason why the damping ratio under drained conditions is larger than that under undrained conditions is thought to be related to the resistance to drainage of the specimen.

5.2 Modelling of $\varepsilon_a - E_{sec}$, h relations

Now we will try to formulate the above deformation characteristics using the hyperbolic stress-strain curve proposed by Hardin and Drnevich (1972). This model, called here the HD-model, can indicate the Young Modulus as the slope of the stress-strain curve, and can compute the coefficient of hysteresis damping from the ratio of energy lost in one cycle loading to its work.

Using the measured data, the relationship between ε_a and $1/E_{sec}$ was plotted, and the coefficients of the $\varepsilon_a - E_{sec}$ curve were determined by a linear function of the relationship between ε_a and $1/E_{sec}$. Also, the relationship between E_{sec} and h was plotted, and the coefficients of the $\varepsilon_a - h$ curve were decided by the linearity in the plotted relation.

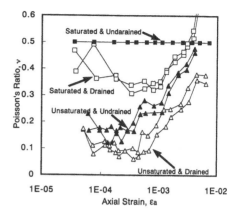

Figure 11 Comparison of Poisson's ratio

Figure 9(a) shows the measured $\varepsilon_a - E_{sec}$ curves of the saturated specimen under drained and undrained conditions, and the relationships fitted by the HD-model. Fig. 9(b) indicates the measured and fitted $\varepsilon_a - h$ curves. Fig. 10(a) and (b) show the same relations for the unsaturated specimen. The E_{sec0} in both the figures are the magnitudes of E_{sec} at an axial strain of 0.00001.

The fitted relationships between $\varepsilon_a - E_{sec}$ and h for the saturated specimen under undrained conditions marked by ■, are almost consistent with the measured relationships in the whole strain range. In the other three cases, the fitted relationships agree with the measured ones in a strain level range of less than 0.001. In an axial strain range of more than 0.001, the fitted $\varepsilon_a - E_{sec}$ relationships underestimate in all cases. The fitted $\varepsilon_a - h$ relationship depends on the saturation of the specimen and the drainage conditions. That is, the fitted curve of the saturated specimen under drained conditions overestimates, that of the unsaturated specimen under drained conditions agrees with the measured data, and that of the unsaturated specimen under undrained conditions underestimates. However, these differences between fitted and measured relations are small, 0.02 at most. Therefore, the modelling of the deformation characteristics of recompressed volcanic cohesive soil can use the HD-model. In the limited range of test conditions conducted, it is concluded that the HD-model can practically explain the deformation behaviour of saturated and unsaturated specimens under drained and undrained conditions.

5.3 Estimation of Poisson's ratio

Assuming a material to be an isotropic body, its shear modulus G is expressed simply by Young's Modulus E and Poisson's ratio v. The shear Modulus G is a deformation modulus without volume change, and does not depend on the saturation and drainage conditions. In undrained compression test on a saturated isotropic material, the ν-value becomes 0.5, and the G is calculated by the E and the ν.

To assess the horizontal deformation characteristics of saturated and unsaturated specimens during the drained or undrained compression process, the ν-values obtained from the measured E_{sec} and the reference G_r using equation (1) are compared.

$$\nu = \frac{E_{sec}}{2G_r} - 1 \qquad (1)$$

Here, the reference G_r is the value determined by $\nu = 0.5$ and the measured E_{sec} of the saturated specimen at each strain level under the undrained compression test.

Figure 11 shows the relationships between axial strain and Poisson's ratio for four combinations of saturation and drainage conditions. In the combination of saturated specimen under undrained conditions, the ν-values in the whole strain range are assumed to be 0.5. In the saturated specimen under drained conditions, the ν-value at a strain less than 0.0001 is approximately 0.45, and as axial strain increases, the value decreases to about 0.3 and keeps constant. Above an axial strain of 0.001, the ν-value increases with a larger axial strain. The reduction of the ν-value is caused by the volume change of the specimen due to drainage from it. In the other two unsaturated specimen cases, the ν-value decreases, keeps constant and then increases, as the axial strain increases.

The ν-values at an axial strain of 0.0001 are 0.4, 0.18 and 0.1 for the saturated specimen under drained, the unsaturated under undrained, and the unsaturated under drained conditions, respectively. The ν-values for the unsaturated specimens are smaller than those of the saturated ones under the same drained conditions. The horizontal deformation is also dependent on the compressibility of pore air in the specimen.

In the above discussion, the recompressed volcanic cohesive soil is assumed to be an isotropic material. In practice, soil constructed by static compaction shows an anisotropic behaviour. In this discussion, however, to investigate the effect of saturation and drainage conditions on horizontal deformation, isotropy is assumed. The obtained results are roughly thought to the effect of saturation and drainage conditions on horizontal deformation characteristics.

6 CONCLUDING REMARKS

From cyclic loading tests on saturated and unsaturated volcanic cohesive soil under drained and undrained conditions, the following conclusions can be drawn;

(1) The relationships between axial strain – secant Young's modulus and the damping ratio depend on the saturation and drainage conditions in a small strain range.

(2) The deformation modulus and axial strain relations influenced by saturation and drained conditions are located within the range of the previous studies for marine clays, and are situated near its lower limit. However, the relationship between the damping ratio and axial strain are twice as large as shown in the previous data.

(3) The Hardin-Drnevich model, a hyperbolic stress-strain relationship can express the deformation characteristics of volcanic cohesive soil with different combinations of saturation and drainage.

There are many kinds of volcanic cohesive soils worldwide. In this study, however, only one sample was used. In the future, using many materials with different characteristics will be necessary to confirm the deformation behaviour obtained from this study.

REFERENCES

Ampadu, S.K. & F. Tatsuoka, 1993. Effect of setting method on the behaviour of clays in triaxial compression from saturation to undrained shear, S&F. Vol.33, 2:14-34.

Hardin, B. O. & W. L. Black, 1969. Vibration modulus of normally consolidated clay (closure). ASCE Vol. 95, SM6: 1531-1537.

Hardin, B.O. & V.P. Drnevich, 1972. Shear modulus and damping in soils; Design equations and curves, ASCE. Vol.98, SM7:667-692.

Ishihara, K., 1976. Basic soil dynamics. Kajima Publishing Co. :1996-202 (in Japanese).

Kudo, N., S. Hara & T. Chiba, 1989. Chap.4 Representative construction works – Construction of New Aomori Airport –. TUCHI-TO-KISO, Vol.37, 3:75-79 (in Japanese).

Tatsuoka, F., S. Yamada & T. Satoh, 1984. Design and manufacturing of shear testing apparatuses for soils No. 6 – Methods of controlling and measuring stress and loads 3 – Chishitsu-to-chosa, No. 1984-1:56-62 (in Japanese).

Tatsuoka, F., D.C.F. Lo Presti & Y. Kohata, 1995. Deformation characteristics of soils and soft rocks under monotonic and cyclic loads and their relationships. SOA-Report, Proc. of 3rd Int. Conf. On Recent Advances in Geotechnical Earthquake Engineering and Soil dynamics, St Luis (Prakash eds.), Vol.2, 851-879.

Advanced Laboratory Stress-Strain Testing of Geomaterials, Tatsuoka, Shibuya & Kuwano (eds),
© *2001 Taylor & Francis, ISBN 90 2651 843 9*

Effects of ageing on stress-strain behaviour of cement-mixed sand

L.Kongsukprasert, R.Kuwano, and F.Tatsuoka
Department of civil engineering, University of Tokyo

ABSTRACT: The effects of ageing with shear stress on the post-ageing stress-strain behaviour of a cement-mixed sand and the associated development of yield locus were experimentally investigated. The post-ageing stress-strain behaviour was substantially different from that of unaged specimens under otherwise the same test conditions. A high stiffness zone developed, with quasi-elastic behaviour in the vicinity of the ageing stress point and a yield locus as the boundary. A part of the yield locus was located outside the ultimate failure plane for some stress paths with decreasing p'. The volumetric strain behaviour after yielding of aged specimens was controlled by the location of yielding point with respect to the potentially dilative zone and the loading stress path.

1 INTRODUCTION

Cement-mixed soils have been used in many onshore ground improvement projects, but basically limited to secondary structures, such as road base and embankment fills. However, some recent laboratory stress-strain tests on cement-treated soils showed a feasibility of constructing critical structures allowing very small deformation, such as bridge abutments, by using cement-mixed soil (e.g., Barbosa-Cruz and Tatsuoka, 1999; Sugai et a l., 2001). Small-scale shaking table tests have also been performed to evaluate the seismic stability of the bridge abutments made of cement-mixed sand supporting the full weight of bridge girder (Tateyama et al., 2000).

For reliable design of such critical structures, accurate evaluation of the deformation of the concerned cement-mixed soil structure during and after construction is essential. The previous laboratory tests referred to above showed that it is important to take into account positive effects of curing (or ageing with development of cementation) with shear stress on the subsequent stress-strain behaviour, particularly that at small strains. That is, the stress-strain behaviour of cement-mixed soil that has been subjected to ageing with shear stress for a long duration in the field would be totally different from that of those that have been cured under the atmospheric pressure (as in the ordinary practice of laboratory testing).

In view of the above, the effects of ageing with shear stress on the stress-strain behaviour of a typi-cal cement-mixed sand were investigated by means of an advanced triaxial testing system. In this paper, post-ageing stress-strain behaviour, including that at small strains and associated development of yield locus around the ageing point in stress space, is reported.

2 MATERIAL AND TEST PROCEDURES

A natural sand originated from Aomori Prefecture, Japan, was mixed with a small amount of Portland cement and water and then compacted at an initial water content w_i= 22 % and a cement/sand ratio in weight c/s= 4.3 % to a nominal dry density ρ_d = 1.23 gf/cm^3 (n.b., the actual compacted dry densities were different from this value). The sand had a specific gravity of 2.801 and a uniformity coefficient of 3.0 with a maximum particle diameter of 2 mm having a 3.6 % of fines particles with diameters less than 0.075 mm. Each specimen was compacted manually in 5 layers in a mold with controlled thickness of each compacted layer of 30 mm. Rectangular prismatic specimens with dimensions of 60 x 99 x 150 (mm) were produced instead of cylindrical ones, so that lateral strains free from bedding errors at the side surface of specimen taking place when the effective confining pressure changes can be reliably obtained by measuring locally with a set of laterally placed LDTs (as described later). Each specimen was stored in a mold at a constant water content (= w_i) under the atmospheric pressure for seven days before removing from the mold. The specimens were then wrapped with a piece of plastic wrapping sheet and the curing was continued also at w_i under the atmospheric pressure for the other four days, i.e., the total initial curing period before setting in the triaxial cell was eleven days. The value of wi, c/s and ρ_d of each corresponding specimen are listed in Table 1.

Table 1. The specification of the specimens.

Test No.	w_i (%)	c/s (%)	ρ_d (%)
A11A1	22.440	4.378	1.231
A11A2	22.749	4.364	1.238
A11A3	22.350	4.354	1.231
A11A4	21.706	4.331	1.239
A11A5	22.540	4.354	1.229
A11APSC	22.378	4.349	1.231
C11APSC*	>21.07	>4.0	<1.243
LC11APSC	22.178	4.322	1.235

* The exact initial water content of sand alone was not obtained for this specimen. However, the same batch of sand sample and the same portions design were used for all specimens.

A triaxial testing system used was the one that had been developed to perform automated loading along various stress paths, creep tests during otherwise monotonic loading at a constant axial strain rate and small-amplitude cyclic loading tests at arbitrary stress states and to change arbitrarily the strain rate and the direction of loading (Santucci de Magistris et al., 1999). Axial strains were accurately evaluated by using a pair of longitudinal LDTs (120 mm-long) and lateral strains were by using three pairs of lateral LDTs (70 mm-long) (Figure 1). Uniform deformations in the horizontal planes were assumed to obtain local lateral strains.

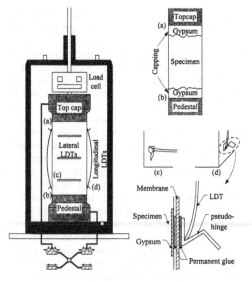

Figure 1. The triaxial test apparatus. a)&b) capping; c)&d) the attachment supporting LDTs.

Each specimen was water saturated before consolidation. Isotropic and anisotropic compression and shear loading was performed at a constant axial strain rate of 0.03%/min. An unload/reload cycle with small strain amplitude was applied at many stress states to evaluate the elasto-plastic property at each stress state. The series of triaxial test was performed to investigate the effects of ageing with shear stress on the subsequent stress-strain behaviour along different stress paths. Six different loading stress paths following the state of ageing were used to explore the development of yield locus around the ageing stress point. The stress paths employed within this study were illustrated in Figure 2. In this paper, "ageing" or "re-curing" are defined as "drained creep at an anisotropic stress state with σ_v' = 900 kPa and σ_h =200 kPa."

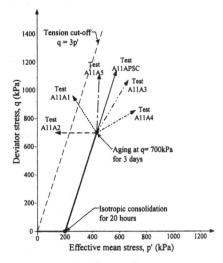

Figure 2. The six different stress paths employed to study the development of the yield locus.

3 DEVELOPMENT OF A HIGH STIFFNESS ZONE BY AGEING

A set of three triaxial tests was performed. Each specimen was first isotropically compressed to σ_v' = σ_h = 200 kPa. At this stress state, the specimens were aged for 20 hours in tests A11APSC and C11APSC and for 92 hours in test LC11APSC. After aged, i.e. consolidated, loading was started with a constant axial strain rate at a constant confining pressure σ_h' = 200 kPa. The specimens in tests A11APSC was second aged (or re-curing) at q = 700 kPa (σ_v' = 900 kPa and σ_h = 200kPa) for three days before the restart of shearing until the ultimate failure state, while a pair of specimens in tests C11APSC and LC11APSC were sheared without an intermission of ageing.

Figures 3, 4 show the results from the test sets performed to evaluate the effects of ageing at an anisotropic stress state for three days, without changing the stress path before and after ageing (i.e., constant $\sigma_h' = 200$ kPa). In the case of continuous loading (tests C11APSC and LC11APSC), there is no sharp change in the tangent modulus of stiffness throughout the test. With aged specimens (test A11APSC), on the other hand, a sharp change in the tangent modulus of stiffness took place three times; 1) when the creep stage started; 2) when loading restarted at a constant axial strain rate that was significantly larger that at the end of creep stage; and 3) at the end of the high stiffness zone where large-scale yielding started taking place.

It may be seen that in test A11APSC, a high stiffness zone developed as can be seen from the behaviour after loading was restarted after ageing (or recuring) with shear stress; this behaviour can be observed as the second sharp change in tangent modulus of stiffness. At small strains immediately after the restart of loading, the stress-strain behaviour was nearly quasi-elastic. Then, small scale yielding started taking place at an increasing rate with the increase in the deviator stress. Finally, large scale yielding and an associated peak deviator stress were observed as the end of high stiffness zone.

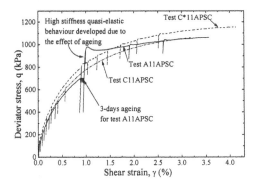

Figure 3. The effect of ageing on shear strain behaviour.

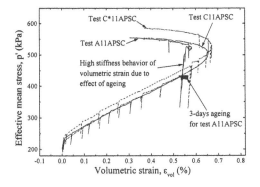

Figure 4. Effect of ageing on volumetric strain behaviour

A comparison of the results from tests C11APSC LC11APSC shows that ageing for some period at an isotropic stress state also develops a high stiffness zone around the isotropic ageing stress state. However, the size of high stiffness zone is smaller and the yielding point can be defined much less clearly than it is in the case of ageing at an anisotropic stress state.

It may be seen from Figure 3, 4 that for the same total curing time at the ultimate failure state, the peak strength of the specimen aged at an anisotropic stress state (test A11APSC) is slightly smaller than that of the corresponding unaged specimen (test LC11APSC). This result would be possibly due to an inevitable variation in the initial conditions of the specimens (as observable differences in the stress-strain relations at the beginning stage in tests A11APSC and C11APSC). On the other hand, the peak strength of the aged specimen was not remarkably different from that of the corresponding unaged specimen having a shorter total curing period at the ultimate failure state. It is likely therefore that the peak strength is not significantly affected by such ageing at an anistropic stress state.

3.1 Yielding and Large-scale yield point

In this paper, "large-scale yielding" or, later, "yielding" is used for the behaviour of the sharp change in the stiffness at the end of the high stiffness zone.

As small-scale yielding has already taken place at an increasing rate before reaching the large scale yielding point, it is not simple to accurately define each large- scale yielding point. In the present study, the large-scale yielding point was defined as the point of maximum curvature along an appropriate arc length of the relationship between the stress path length and the strain increment, plotted in a full-log figure, around the respective large-scale yield point. The method to obtain stress path lengths and strain increments is described in Appendix.

The ultimate strength $q_{ultimate}$ was defined as the deviator stress at large strains, which may be smaller than the peak strength attained at the large-scale yielding point, at a smaller strain.

4 YIELD LOCUS AND EFFECTS OF STRESS PATH ON THE STRESS-STRAIN BEHAVIOUR

Six specimens with the same initial curing history before the start of triaxial loading at $\sigma_h' = 200$ kPa were sheared along six different stress paths, as shown in Figure 2, after ageing with shear stress.

4.1 Yield locus

The term 'yielding' will herein be referred to large scale yielding. Each yield point at q_{yield} and ultimate failure state at $q_{ultimate}$ are indicated along the respective stress path and listed in Table 2. The yield locus developed around the ageing point and the ultimate failure surfaces were constructed based on the test results, as shown schematically in Figure 5. The potentially dilative zone that was inferred based on the test results was also identified.

Table 2. Yield and ultimate strength from each stress path

Test No.	Yield strength (kPa)		Ultimate strength (kPa)	
	q_{yield}	p'_{yield}	$q_{ultimate}$	$p'_{ultimate}$
A11A1	780.458	306.809	-	-
A11A2	no yielding was observed			
A11A3	969.768	601.556	1655.0	953.042
A11A4	795.932	605.634	1499.853	896.909
A11A5	980.752	445.480	932.260	443.057
A11APSC	979.971	527.144	1062.783	553.517

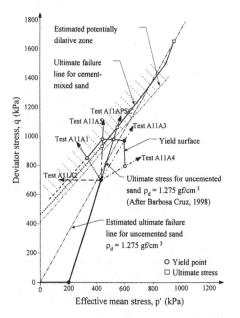

Figure 5. The development of yield locus around the stress point of ageing.

It may be seen from Figures 5 that a part of the yield locus is located above the ultimate failure line. That is, in tests A11A1, A11A2 and A11A5, having the stress paths with nearly constant p' or decreasing p', the yield point was first attained then followed by a reduction in the deviator stress before approaching the ultimate stress state. In these tests, therefore, the value of q_{yield} is equal to the respective maximum deviator stress q_{max} and larger than $q_{ultimate}$. In test A11A2, the yield point was not observed, because the stress state reached the tension cut-off line,

where $\sigma_h = 0$, before approaching the yield point. In the other tests (A11A3, A11A4 & A11APSC), the yield point was located below the ultimate failure line with the respective q_{yield} value being lower than the corresponding q_{max} value, which was in turn lower than $q_{ultimate}$. In some of these tests, immediately after yielding, the q value decreased temporarily with a substantial drop in the tangent modulus of stiffness.

4.2 Stress-strain behaviour before and after yielding

Figures 6, 7 and 8 are the results from test A11A5, A11A3, and A11APSC respectively, show three typical relationships between the stress path length and the shear and volumetric strain increments from the restart of loading. The stiff stress-strain behaviour was observed for a range from the restart of loading and the yielding point in all the tests. The pre-yielding volumetric deformation characteristics were observed complicated as in the followings

1) In tests A11A1, A11A2 & A11A5, the yield points were located above the ultimate failure line. The volumetric strain increments were negative (dilative behaviour) before yielding, followed by larger dilative behaviour after yielding; these phenomena can be, for example, as seen in Figure 6.

2) In tests A11A3, A11A4 & A11APSC, the yield points were located below the ultimate failure line. Noticeably contractive behaviour was observed before yielding. In tests A11A3 & A11A4, in which p increased at a rate dp/dq larger than 1/3, post-ageing volumetric strain behaviour was contractive (Fig. 7). In test A11APSC, in which σ'_h was kept constant, immediately after the yield point, the shear stress dropped rapidly with small contractive volume strain increments, followed by a significant rate of volume expansion due to positive dilatancy (Fig. 8).

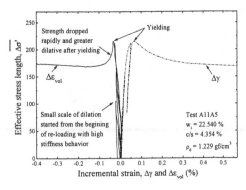

Figure 6. Post-ageing dilative behaviour (test A11A5).

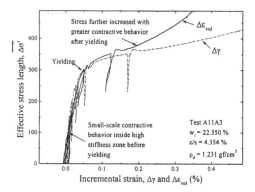

Figure 7. Post-ageing contractive behaviour (test A11A3).

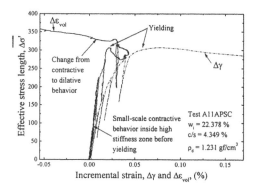

Figure 8. Post-ageing contractive-then-dilative behaviour (test A11APSC)

The aforementioned shear deformation and volumetric deformation properties of the specimens aged with shear stress are interpreted below.

4.3 Shear deformation

The stress-strain characteristics of the cement-mixed sand with and without the effects of ageing with shear stress are compared schematically in Figure 9. It is illustrated that the overall pre-peak stress-strain behaviour of sand is improved by cementation with the ultimate strength of cement-mixed sand being substantially greater than that of uncemented sand under otherwise the same test conditions. With cement-mixed sand, the modulus of stiffness after loading was restarted following ageing with shear stress is very high, which is followed by noticeably stress overshooting with large-scale yielding at the end of high stiffness zone.

The shaded area indicates the effects of cementation developed by ageing with shear stress in comparison to the behaviour of an unaged one having the same total curing time at the moment of ultimate failure. The dotted area indicates the effects in comparison to the behaviour of an unaged one having the

same curing time at the start of shearing. As the aged specimen is sheared beyond the yielding point, the effect of ageing is gradually erased.

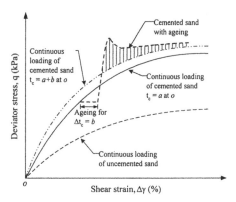

Figure 9. Illustration of effects of ageing.

4.4 Volumetric deformation

By ignoring viscous deformation, a given total volumetric strain increment $\Delta\varepsilon_{vol}$ is decomposed as follows;

$$\Delta\varepsilon_{vol} = \Delta\varepsilon_{vol}^e + \Delta\varepsilon_{vol}^p \qquad (1)$$

where $\Delta\varepsilon^e_{vol}$ is the elastic volumetric strain increment, which is positive and negative for positive and negative increments of pressure level, which is approximately equal to $\Delta p'$; $\Delta\varepsilon^p_{vol}$ is the plastic volumetric strain increment, given as;

$$\Delta\varepsilon_{vol}^p = \Delta\left(\varepsilon_{vol}^p\right)_{shear} + \Delta\left(\varepsilon_{vol}^p\right)_{compression} \qquad (2)$$

where $\Delta(\varepsilon^p{}_{vol})_{shear}$ is the plastic volumetric strain increment caused by shearing (i.e., dilatancy component), which is positive and negative when the stress ratio is lower and higher than the critical value (i.e., the boundary between potentially contractive and dilative zones); and $\Delta(\varepsilon^p{}_{vol})_{compression}$ is the positive plastic volumetric strain increment by a pressure increase. From the test results, the post yielding volumetric strain behaviour can be categorized into 1) dilative behaviour; 2) contractive behaviour; 3) combining contractive-then-dilative behaviour.

Figure 10. shows three different corresponding stress paths prepared to explain the above three different post-yielding volumetric strain behaviours.

4.4.1 Post-ageing dilative behaviour

The stress-strain relationship in test A11A5 (Fig. 6) is herein analyzed, in which p' increased at a small rate and the yield point is located above the ultimate failure line. Before the yielding point, the behaviour was stiff and nearly elastic and a small volume expansion was observed; i.e., a negative

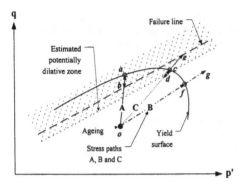

Figure 10. Schematically illustration of three different stress paths corresponding to three different post-ageing stress-strain behaviours.

elastic strain $\Delta\varepsilon^e_{vol}$ was observed despite a positive $\Delta p'$. This behaviour can be explained based on the following cross-anisotropic elasticity model.

With uncemented sands and gravels, the vertical elastic Young's modulus E_v^e is a rather unique function of σ'_v, while the horizontal elastic Young's modulus E_h^e is a rather unique function of σ'_h. For cemented geomaterials, the following equations would be relevant:

$$E_v^e = E_{v1}\left(\frac{\sigma'_v + c_v}{\sigma_1}\right)^m ; E_h^e = E_{h1}\left(\frac{\sigma'_h + c_h}{\sigma_1}\right)^m \qquad (3)$$

when c_v and c_h are positive constants; E_{v1} is the E_v^e value when $\sigma_v + c_v$ is equal to σ_1; and E_{h1} is the E_h^e value when $\sigma_h + c_h$ is equal to σ_1. According to Eq. 3, the elastic volumetric strain increment is obtained as:

$$\Delta\varepsilon^e_{vol} = \Delta\varepsilon^e_v + 2\Delta\varepsilon^e_h \qquad (4)$$

$$\Delta\varepsilon^e_v = \frac{1}{E_v^e}\Delta\sigma'_v - \frac{2v^e_{hv}}{E_h^e}\Delta\sigma'_h$$

$$\Delta\varepsilon^e_h = \frac{1-v^e_{hh}}{E_h^e}\Delta\sigma'_h - \frac{v^e_{vh}}{E_v^e}\Delta\sigma'_v$$

It is possible therefore to have $\Delta\varepsilon^e_{vol} < 0$ for $dp' = 0$ when E_v^e is larger than E_h^e at anisotropic stress conditions with $\sigma'_v > \sigma'_h$.

During loading after the yielding point, the q value dropped sharply associated with a significant rate of positive dilatancy (volume expansion.) Considering stress path A in Figure 10, loading was restarted at the ageing point o and point a is the yield point which was located above the ultimate failure line. As yield took place in a zone above the ultimate failure line, where the plastic volumetric strain characteristics upon yielding were dilative, a large negative value of $\Delta(\varepsilon^p_{vol})_{shear}$ therefore, took place followed by the stress state dropped sharply towards point b, located on the ultimate failure line. Com-

pared with this value, the value of $\Delta(\varepsilon^p_{vol})_{compression}$ that took place associated with yielding should be insignificant.

The test results show that after the start of large-scale yielding, the stress-strain relation tended to rejoin the original relation of the un-aged specimen having the same total curing period, as schematically shown in Figure 11. Moreover, the test results also show that the cementation effect that existed before ageing with shear stress was not totally lost upon yielding, but only the component that was gained during ageing with shear stress was gradually lost as the shear strain increases after ageing.

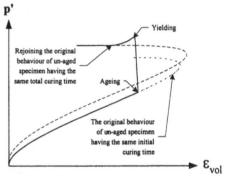

Figure 11. Schematically illustration of rejoining with post-ageing dilative behaviour.

4.4.2 Post-ageing contractive behaviour

In test A11A3, the yield point was located below the potentially dilative zone. The pre-yielding behaviour was stiff and nearly elastic. As p' increased at a high rate of $dp'/dq = 0.62$, a small positive value of $\Delta\varepsilon_{vol} \approx \Delta\varepsilon^e_{vol}$ was observed before reaching the yield point (Fig. 7). Stress path B (Fig. 10) illustrates loading path in which the yield point is located largely below the failure plane and also below the potentially dilative zone. As the yielding took place below the potential dilative zone and at a relatively high value of p' (compared with the previous case), the volumetric strain behaviour was, therefore, continuing contractive at a higher degree than that in the pre-yielding. Similarly to the previous case, upon yielding, the post-yielding volumetric strain behaviour of the aged specimen trended to rejoin the original behaviour of the un-aged one having the same total curing time that can also be observed in this case. During loading from the yield point until the rejoining point to the original stress-strain curve, the aged specimen exhibited contractive behaviour, which was mostly plastic, as illustrated in Figure 12.

With cement-mixed soils, with the increase in the pressure level, the strength component due to cementation becomes less important in comparison with the frictional strength component. For this reason, at pressures higher than some critical value, the yield strength associated with the yielding of bonded inter-particle contacts becomes lower than the ultimate failure strength, which is largely controlled by the frictional strength component. In this case, the ultimate strength is equal to the peak strength.

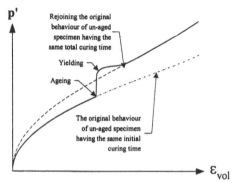

Figure 12. Schematically illustration of rejoining with post-ageing contractive behaviour.

4.4.3 Post-ageing combined contractive-then-dilative behaviour

The behaviour after the restart of loading from the ageing state in test A11APSC is shown in Figure 8. In this test, $dp'/dq = 1/3$, which was in between the values in the two cases described above, the schematic corresponding stress path C is shown in Figure 10. For this reason, the pre-yielding behaviour was slightly contractive, due mostly to positive elastic volumetric strains.

The post-yielding volumetric strain behaviour was very complicated, as seen from Figure 8, which could be interpreted as follows.
1) After the yielding started, the shear stress temporarily dropped and the associated plastic volumetric strain increment was positive. This is due to the following two mechanisms:
   a) Upon yielding, the stress-strain relation of the aged specimen tended to rejoin that of the corresponding un-aged specimen.
   b) As seen from Figure 4, during loading for a stress range from the restart of loading until the yielding point for the aged specimen, the un-aged specimen exhibited a relatively large positive volumetric strain increment (mostly plastic).
2) As schematically illustrated in Figure 10, the yield point was located already in the potentially dilative zone, while it was close to the boundary between the potentially contractive and dilative zones. That is, at a stress level below of the yield point of the aged specimen, the un-aged specimen started dilating (see Fig. 4). For this reason, the volumetric strain behaviour of the aged specimen

changed from a contractive one to a dilative one as the shear stress restarted increasing as a result of rejoining the original stress-strain behaviour of un-aged specimen.
3) By further loading until the ultimate failure state, the positive dilative behaviour became essentially the same as that of the corresponding un-aged specimen, exhibiting a high rate of positive dilatancy.

5. CONCLUSIONS

The following conclusions can be derived from the test results presented above:
1) The stress-strain behaviour after ageing with shear stress of a cement-mixed specimen was substantially different from that of the corresponding un-aged specimen under otherwise the same test conditions.
2) A yield locus developed around the stress point of ageing, in which the stress-strain behaviour was very stiff, in particular it was nearly quasi-elastic at small strains immediately after the loading was restarted from the ageing state. A part of yield locus for some certain stress paths with large decreasing rates of p' was located above the ultimate failure line.
3) The dominant component of the volumetric strain taking place before yielding was quasi-elastic, which was basically negative and positive for stress paths with decreasing and increasing p.
4) The dominant component of the volumetric strain taking place after yielding was plastic, which was dilative, or contractive, or contractive-then-dilative, depending on the position of the yield point relative to the potentially dilative zone and the loading stress path direction.

APPENDIX : DEFINING A YIELD POINT

In order to exactly define the yield point along a given stress path, usually the stress-strain relationship with the stress in terms of q or p' or σ'_v or σ'_h is referred to. In the present study, however, along some stress paths, stress and strain components as above are kept constant or nearly constant, which made difficult to define a yielding point based on the usual plotting method. For the reason above, the following plotting method was introduced, in which the effective stress path length, $\overline{\Delta\sigma'}$, starting from the stress point where the specimen was aged, was plotted against the strain increments.

Referring to Figure 13, the stress path length $\overline{\Delta\sigma'}_b$ for loading from point a until point b, can be obtained as:

$$\overline{\Delta\sigma'}_b = \int_a^b d\sigma' = \sum_a \sqrt{(d\sigma'_v)^2 + (d\sigma'_h)^2} \qquad (A\text{-}1)$$

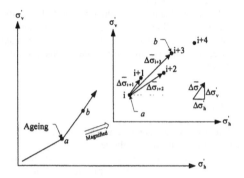

Figure 13. The stress path length

Despite that all the stress paths for shearing after ageing employed in this study were nominally straight, with actual data, each individual data point more-or-less deviated from the target straight stress path. Consequently, the exact calculation of the stress length by the above equation integrating all scattering errors resulted into a false stress path length. In order to alleviate this type of error, the stress path length was calculated as:

$$\overline{\Delta\sigma'}_b = \sqrt{\{(\sigma'_v)_b - (\sigma'_v)_a\}^2 + \{(\sigma'_h)_b - (\sigma'_h)_a\}^2} \quad (A\text{-}2)$$

The shear and volumetric and shear strain increments for loading from point a (i.e., the starting point of the stress psyh length) until point b were obtained as:

$$\Delta\gamma_b = \{(\varepsilon_v)_b - (\varepsilon_v)_a\} - \{(\varepsilon_h)_b - (\varepsilon_h)_a\}$$

$$\Delta\gamma_b = \gamma_b - \gamma_a \qquad (A\text{-}3)$$

$$(\Delta\varepsilon_{vol})_b = \{(\varepsilon_v)_b - (\varepsilon_v)_a\} + 2\{(\varepsilon_h)_b - (\varepsilon_h)_a\}$$

$$(\Delta\varepsilon_{vol})_b = (\varepsilon_{vol})_b - (\varepsilon_{vol})_a \qquad (A\text{-}4)$$

REFFERENCES

Barbosa-Cruz E., Sato, Y., Tatsuoka, F. and Sugo K. (1997). "Effects of curing conditions on small behaviour of cement-treated sand," Proc. 52th Annual Conf. of JSCE, 3-(B), pp. 426-427.

Goto, S., Tatsuoka, F., Shibuya, S., Kim, Y.S. and Sato, T. (1991). "A simple gauge for local small strain measurements in the laboratory," Soils and Foundations, Vol. 31, No. 1, pp. 169-180.

Hayano, K., Sato, T. and Tatsuoka, F. (1997). "Deformation characteristics of a sedimentary soft rock from triaxial compression tests using rectangular prism specimens," Geotechnique, Vol. 47, No. 3, pp. 439-449.

Hoque, E. (1996) "Elastic deformation of sands in triaxial tests", Doctor thesis, Univ. of Tokyo.

Kohata, Y., Tatsuoka F., Wong, L., Jiang, G.L., Hoque, E. and Kodaka T. (1997), "Modelling of nonlinear deformations for stiff geomaterials," Geotechnique Symposium on Prefailure deformation of Geomaterials, London.

Kongsukprasert, L. (2000), "Effects of ageing on stress-strain behaviour of cement-mixed sand," Master thesis, Univ. of Tokyo.

Santucci de Magistris,F., Koseki,J., Amaya,M., Hamaya,S., Sato,T. and Tatsuoka,F. (1999). "A triaxial testing system to evaluate stress-strain behaviour of soils for wide range of strain and strain rate," Geotechnical Testing Journal, ASTM, Vol.22, No.1, pp.44-60.

Sugai,M., Tatsuoka,F., Kuwabara,M. and Sugo,K. (2000). "Strength and deformation characteristics of cement-mixed soft clay," IS Yokohama.

Tatsuoka,F., Jardine,R.J., Lo Presti,D., Di Benedetto,H. and Kodaka,T. (1999), "Characterising the Pre-Failure Deformation Properties of Geomaterials," Theme Lecture for the Plenary Session No.1, Proc. of XIV IC on SMFE, Hamburg, September 1997, Volume 4, pp.2129-2164.

Advanced Laboratory Stress-Strain Testing of Geomaterials, Tatsuoka, Shibuya & Kuwano (eds),
© 2001 Taylor & Francis, ISBN 90 2651 843 9

Dependency of horizontal and vertical subgrade reaction coefficients on loading width

J. Koseki
Institute of Industrial Science, Univ. of Tokyo, Japan

Y. Kurachi
Shiraishi *Corporation*, Japan

T. Ogata
Japan Highway Public Corporation, Japan

ABSTRACT: Comparisons are made on horizontal and vertical subgrade reaction coefficients that are numerically obtained with modeling soil properties based on relevant triaxial test results on undisturbed or reconstituted samples of softrock, gravel, and clayey gravel. It is demonstrated that the dependency of the subgrade reaction coefficients on the loading width is affected by both the non-linearity and the pressure-level-dependency in the deformation properties of subsoil. Effects of the non-linearity appear in both the horizontal and the vertical subgrade reaction coefficients, when compared at the same displacement of the loading plate. On the other hand, effects of the pressure-level-dependency predominantly appear in the vertical ones. In the cases with subsoil of gravel and clayey gravel, combined effects of these factors result into apparent independency of the vertical subgrade reaction coefficients on the loading width.

1 INTRODUCTION

In practice, coefficients of subgrade reaction have been conventionally employed in designing foundations in Japan (e.g., JRA, 1996). Corrections have been made for the effects of the loading width on the coefficients of subgrade reaction, based on experimental data collected for relatively small-scale foundations (e.g., Yoshida and Yoshinaka, 1972). It has not been, however, confirmed if similar correction is applicable to large-scale foundations. In addition, the nonlinear nature of the soil deformation characteristics has not been considered rationally in evaluating the coefficient of subgrade reaction.

In the present study, in order to investigate the scale effects on coefficients of horizontal and vertical subgrade reaction, finite element analyses were conducted on three types of level grounds consisting of softrock (sedimentary soft mudstone), gravel, and clayey gravel, respectively. Nonlinear and pressure-level-dependent deformation properties of subsoil were modeled based on relevant triaxial test results that were performed on undisturbed or reconstituted samples.

2 NUMERICAL MODELS

A series of non-linear elastic analyses by means of a three dimensional finite element method was performed to simulate behaviors of horizontally loaded well foundations. The model employed for the

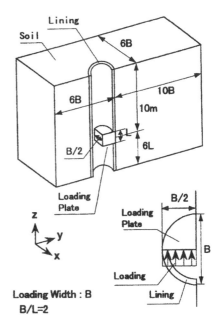

Fig. 1 3-D Model of horizontally loaded well foundation

analyses is shown in Fig. 1, which consists of 4728 elements and 6107 nodes. Horizontal loads were applied partly to a section of the foundation at a depth of 10 m. The loading plate was assumed to be

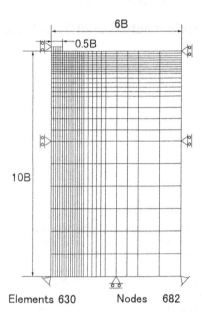

Elements 630   Nodes   682

Fig. 2 2-D Model of vertically loaded spread foundation

infinitely stiff, while the other parts of the foundation lining were modeled with finite stiffness by using elastic solid elements with Young's modulus of 1600 MPa and Poisson's ratio of 0.13, which were determined based on relevant compression test results on cored samples (Ogata et al., 1999). The width B of the loading plate was changed between 1 and 5m. Referring to Desai (1972), the model size was set proportionally to the value of B.

In order to simulate behaviors of vertically loaded spread foundations, another series of analyses by means of a two dimensional finite element method was performed on a model shown in Fig. 2. In this series, vertical loads were applied to the whole spread foundation, assuming that the stiffness of the foundation was infinitely high. The width B of the foundation was changed between 1 and 20 m, and the adjustment of the model size was made in a similar manner as in the former series of analyses.

The level ground in these models was assumed to be uniform. Analyses were made on three types of subsoil, consisting of softrock, gravel, and clayey gravel. The deformation properties of softrock were assigned based on triaxial test results conducted by Wang (1996) on undisturbed specimens retrieved from a sedimentary soft mudstone deposit. Those of gravel and clayey gravel were assigned based on large-scale triaxial test results by Jiang (1996) and Koseki et al. (1999), respectively, on reconstituted specimens of compacted gravel with fines content of about 2 % and on undisturbed specimens retrieved

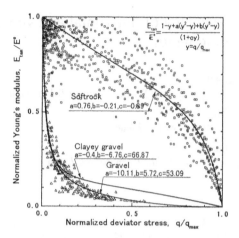

Fig. 3   Relationships between normalized deviator stress and normalized Young's modulus

from a Pleistocene gravel deposit with fines content of about 10 to 20 %.

Referring to Fig. 3, non-linearity in the deformation properties of these subsoil materials in terms of the tangential Young's modulus E_{tan} was formulated as;

$$E_{tan}/E^e = \{1 - y + a(y^2 - y) + b(y^3 - y)\}/(1 + cy), \quad y = q/q_{max} \quad (1)$$

where E^e is an undamaged Young's moduli that was evaluated by applying small-amplitude cyclic axial loading on the specimen under isotropic stress states; q is the deviator stress (= $\sigma_1 - \sigma_3$); q_{max} is the maximum deviator stress; and a, b and c are parameters assigned for the respective soil as indicated in the figure. Values of q_{max} and E^e were evaluated from the minor principal stress σ_3 as shown in Figs. 4 and 5. It should be noted that possible effects of the inherent and induced anisotropy are neglected in the present study.

Considering the large amount of scattering in the measured values of the tangential Poisson's ratio ν_{tan}, it was assumed to be constant, for simplicity, in the present analyses. Values of ν_{tan} was set to 0.2 for the softrock, 0.3 for the gravel and 0.35 for the clayey gravel.

Applicability of the numerical modeling adopted in the present study has been verified by Ogata et al. (1999) through a simulation of in-situ horizontal loading tests conducted in a deposit of clayey gravel using loading plates of different sizes, as shown in Fig. 6.

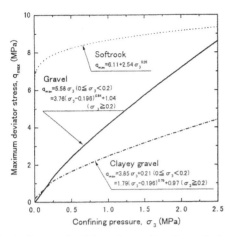

Fig. 4 Pressure-level dependency of maximum deviator stress

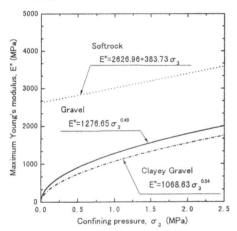

Fig. 5 Pressure-level dependency of undamaged Young's modulus under isotropic stress states

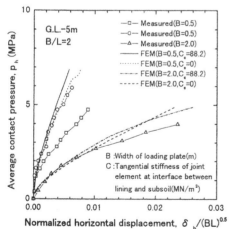

Fig. 6 Comparison of measured and calculated load-displacement relationships (Ogata et al., 1999)

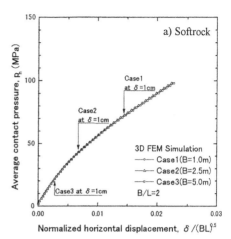

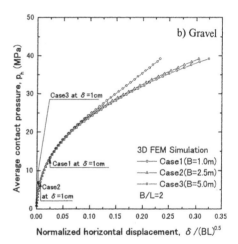

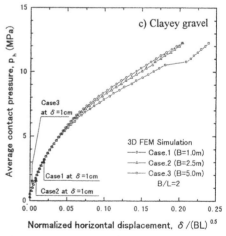

Fig. 7 Load-displacement relationships of horizontally loaded well foundation

3 HORIZONTAL SUBGRADE REACTION

Load-displacement relationships computed for the horizontally loaded well foundation are shown in Fig. 7. The contact pressure p was obtained by dividing the total horizontal load applied to the loading plate with the its contact area BL, where L is a height of the horizontally-loaded section set equal to $0.5B$. The horizontal displacement δ of the loading plate was normalized with the value of $(BL)^{0.5}$, which is an equivalent width of the loading plate converted to a square section while keeping the same contact area. It is seen that, after such normalization, the load-displacement relationships for respective subsoil type became similar among the different sizes of loading plate.

In order to compare the results among different subsoil types, the coefficient of subgrade reaction k, obtained as a secant stiffness from the above load-displacement relationships (i.e., $k=p/\delta$), was normalized with the reference coefficient k_0 at $B=B_0$ = 1m. When compared at the same normalized displacement of $\delta/(BL)^{0.5}= 1$ % as shown in Fig. 8a, values of the normalized horizontal subgrade reaction coefficient k/k_0 decreased with the increase in the normalized loading width B/B_0. Irrespective of the subsoil types, their relationships could be approximated to $k/k_0= (B/B_0)^{-1}$, which corresponds to the theoretical relationship obtained with an assumption that subsoil is linear-elastic material exhibiting no pressure-level-dependency.

On the other hand, when compared at the same horizontal displacement of $\delta= 1$cm as shown in Fig. 8b, the dependency of the k/k_0 values on the B/B_0 values became smaller. Their relationships were affected by different degree of non-linearity in the deformation properties of subsoil, since the strain level in subsoil at the same displacement of the loading plate became smaller with the increase in the loading width. For reference, the state at $\delta= 1$ cm for the respective size of loading plate is indicated by a vertical arrow in Fig. 7.

4 VERTICAL SUBGRADE REACTION

Figure 9 shows load-displacement relationships computed for the vertically loaded spread foundation. The same normalization as employed for the horizontally loaded well foundation was made to prepare these figures, whereas the relationships were not similar among the different sizes of spread foundation. It should be noted that, since the non-linearity in the deformation properties of the softrock was relatively small (refer to Fig. 3), the

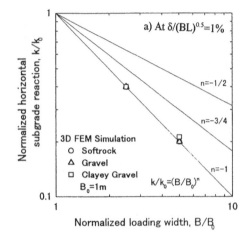

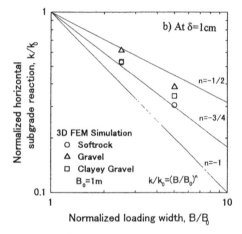

Fig. 8   Relationships between normalized loading width and normalized subgrade reaction coefficient of horizontally loaded well foundation

corresponding load-displacement relationships (Fig. 9a) did not exhibit yielding in the range of the present computation. Rather, they were slightly upward-concave due to the pressure-level-dependency of E^e; i.e., with the increase in the contact pressure p, the value of E^e increased accordingly, resulting into increase in the value of E_{tan}, even under a reduced value of E_{tan}/E^e that was to a lesser extent.

Figure 10a shows the vertical subgrade reaction coefficients compared at the same normalized displacement of $\delta/B= 1$ % among different subsoil, where the dependency of the k/k_0 values on the B/B_0 values became smaller than that for the horizontal subgrade reaction coefficients (refer to Fig. 8a). Their relationships were affected by different degree of pressure-level-dependency in the deformation properties of subsoil, since the affected zone in the

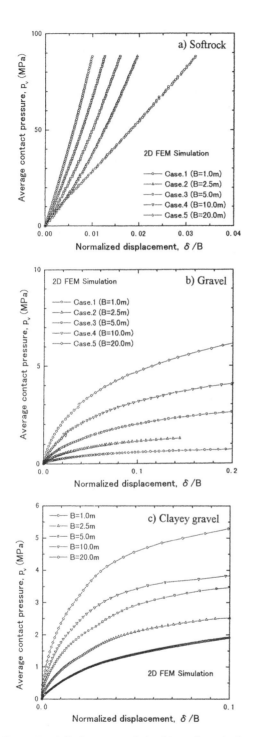

Fig. 9 Load-displacement relationships of vertically loaded spread foundation

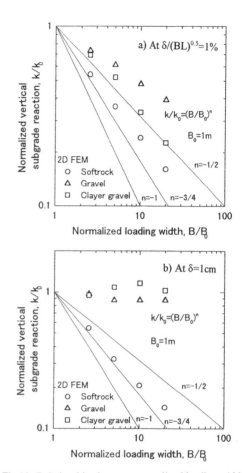

Fig.10 Relationships between normalized loading width and normalized subgrade reaction coefficient of vertically loaded spread foundation

subsoil became deeper with the increase in the loading width, as schematically shown in Fig. 11.

In addition, when the vertical subgrade reaction coefficients were compared at the same vertical displacement of $\delta = 1$cm as shown in Fig. 10b, values of k/k_0 for the gravel and clayey gravel grounds were almost unchanged irrespective of the loading width. Such an apparent independency on loading width was obtained due to combined effects of the non-linearity and the pressure-level-dependency in the deformation properties of subsoil.

The different degrees of dependency on loading width between the horizontal and vertical subgrade reaction coefficients are caused by their different extents on the effects of pressure-level-dependency in the deformation properties of subsoil. As schematically shown in Fig. 12 on the horizontal subgrade reaction coefficients, a larger loading plate

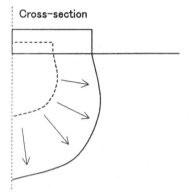

Cross-section

Fig.11 Expansion of affected area in subsoil for vertically loaded spread foundation with increase in loading width

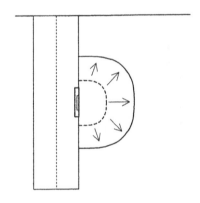

a) Cross-section along center line

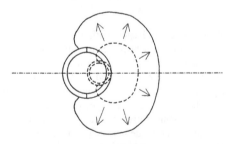

b) Plan view at the depth of loading plate

Fig.12 Expansion of affected area in subsoil for horizontally loaded well foundation with increase in loading width

mobilizes a larger affected zone in the subsoil, whereas effects of the deeper and stiffer portion and that of the shallower and softer portion of the newly mobilized zone are canceled out to each other. As discussed by Ogata et al. (1999), the horizontal reaction from the subsoil will be reduced if the loading plate is located relatively close to the subsoil surface. Therefore, the dependency of the horizontal subgrade reaction coefficient on the loading width will become more pronounced when compared at a shallower depth than in the present study. Such combined effects of the loading depth and the loading width may be seen in Fig. 6 that the results from in-situ horizontal loading tests conducted at a depth of 5 m were dependent on the loading width.

5 CONCLUSIONS

Through the comparisons of numerical results that were modeled based on relevant laboratory stress-strain tests, it was demonstrated that the dependency of the subgrade reaction coefficients on the loading width is affected by both the non-linearity and the pressure-level-dependency in the deformation properties of subsoil. Effects of the non-linearity could be seen in both the horizontal and the vertical subgrade reaction coefficients, when compared at the same displacement of the loading plate. On the other hand, effects of the pressure-level-dependency could be predominantly seen in the vertical ones. In the cases with subsoil of gravel and clayey gravel, combined effects of these factors resulted into apparent independency of the vertical subgrade reaction coefficients on the loading width.

REFERENCES

Desai,C.S. 1972: Theory and applications of the finite element method in geotechnical engineering, *Proc. Application of the FEM in Geotech. Eng.*, Vol.1, pp.3-90.

Japan Road Association 1996: Design specifications of highway bridges -part 4: foundation-, pp.237-239 (in Japanese).

Jiang, G.L. 1996: Small strain behavior and strength characteristics of gravel by large triaxial tests, *Doctor of Engineering Thesis, University of Tokyo* (in Japanese).

Koseki, J., Balakrishnaiyer, K. and Tatsuoka, F. 1999: Large scale triaxial tests on elastic properties of undisturbed gravel containing fines, *Proc. of 2nd International Symposium on Pre-failure Deformation Characteristics of Geomaterials,Torinio*, Vol.1, pp.423-430.

Ogata, T., Kurachi, Y., Oishi, M., Ouchi, M., Maeda, Y. and Koseki, J. 1999: In-situ horizontal loading tests on gravelly subsoil and their numerical simulations, *Proc. of 2nd International Symposium on Pre-failure Deformation Characteristics of Geomaterials, Torino*, Vol.1, pp. 379-386.

Yoshida, I. and Yoshinaka, R. 1972: A method to estimate modulus of horizontal subgrade reaction for a pile, *Soils and Foundations*, Vol.12, No. 3, pp.1-17.

Wang, L. 1996: Study on field deformation characteristics of sedimentary softrocks by triaxial tests, *Doctor of Engineering Thesis, University of Tokyo* (in Japanese).

Advanced Laboratory Stress-Strain Testing of Geomaterials, Tatsuoka, Shibuya & Kuwano (eds),
© 2001 Taylor & Francis, ISBN 90 2651 843 9

Low-strain stiffness and material damping ratio coupling in soils

C.G. Lai
Studio Geotecnico Italiano, Milano, Italy

O. Pallara, D.C.F. Lo Presti & E. Turco
Politecnico di Torino, Italy

ABSTRACT: Experimental evidence shows that soils subjected to dynamic excitations have both the ability to store and to dissipate strain energy. The phenomenon of energy dissipation takes place even at very small strain levels, below the linear cyclic threshold shear strain. From a phenomenological point of view this type of material behaviour can conveniently be described by the linear theory of viscoelasticity. An important result predicted by this theory is the functional dependence between the velocity of the propagation of body waves and the material damping ratio. Hence, in contrast to usual practice, these parameters should be measured simultaneously and at the same frequency of excitation. In this article the authors present a new experimental procedure to be conducted in laboratory with resonant column apparatus where the shear wave velocity and the shear damping ratio are measured simultaneously. Since these parameters are determined at specific frequencies of excitation, the method is also well suited to investigate the frequency dependence laws of these important soil parameters.

1 INTRODUCTION

During the past thirty years, a considerable amount of research has been carried out in an effort to better understand the mechanical response of soils to dynamic excitations. These studies were performed using a variety of laboratory techniques (e.g., resonant column tests, cyclic torsional shear tests, cyclic direct simple shear tests, and cyclic triaxial tests), that allowed researchers to investigate the influence of variables including strain amplitude and frequency of excitation on soil behavior. The results obtained from this work have helped in identifying the most important variables and factors affecting the dynamic behavior of soils.

These variables and factors can be broadly divided into two categories according to their origin: *external variables* and *intrinsic variables*. The external variables correspond to externally applied actions and include the stress/strain path, stress/strain magnitude, stress/strain rate, and stress/strain duration depending on the nature of the applied action (i.e., stress-controlled versus strain-controlled tests). The intrinsic variables correspond to the inherent characteristics of soil deposits and include the soil type, the size of the soil particles, and the state parameters. These parameters include the geostatic effective stress tensor (which is a measure of the current state of in-situ effective stress), some measure of the arrangement of the soil particles (e.g., the fab-

ric tensor), and some measure of the stress-strain history (e.g., the yield surface or, at least, the pre-consolidation pressure). Figure 1 summarizes the relationships between causes and effects in the response of soils to dynamic excitations.

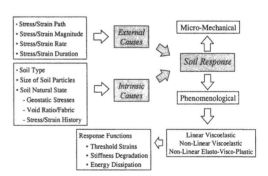

Figure 1. Cause-effect relationships in soil response to dynamic excitations (from Lai, 1998).

Soil behaviour may be studied by using either a phenomenological or a micromechanical approach. The main focus of the phenomenological approach is to understand the relationship between causes and effects from a macroscopic point of view, without attempting an explanation of the observed phenomena at the microscopic level.

This microscopic interpretation is the objective of the micromechanical approach, which is implemented by using the framework of either continuum or discrete mechanics (Lai, 1998). The approach used in this paper to model the dynamic behavior of soils at a low strain level is phenomenological, and coincides with that of classical, linear-isotropic viscoelasticity.

2 DYNAMIC BEHAVIOUR OF SOILS

2.1 Experimental observations

Among the external variables affecting soil response to dynamic excitations, experimental evidence shows that the one playing the most important role is the magnitude of the applied strain, or more precisely, the magnitude of the induced deviatoric strain tensor.

It is commonly accepted that the stress-strain relationship be essentially linear, even though inelastic, only at very small strains. This region of linear behavior is characterized by shear strain values within the range $0 < \gamma \leq \gamma_t^l$ where the upper bound γ_t^l has been labeled the *linear cyclic threshold shear strain* by Vucetic (1994).

Within this region no stiffness degradation is observed in the soil response, however the hysteretic loop, in the stress-strain plane, exhibits a non-null area (see as an example Lo Presti et al., 1997). The phenomenon of energy dissipation at very small strain levels is caused by the existence, in a strain-controlled test, of a time-lag between a driving cyclic strain and the driven cyclic stress (the word hysteresis comes from the ancient Greek and means "lag" or "delay"). This time lag is responsible for energy losses over a finite period of time, which is typical, from a phenomenological point of view, of a viscoelastic behavior. Current experimental data do not support, for $\gamma \leq \gamma_t^l$, the existence of instantaneous energy dissipation phenomena, which would be more typical of an elastoplastic behavior.

According to many authors, γ_t^l is in practice equal to about 10^{-3} %. Yet γ_t^l depends on both external and intrinsic factors as shown experimentally by many authors (see for instance Tatsuoka et al., 1997).

The linear cyclic threshold shear strain γ_t^l has been defined by considering simple shear strain paths. The soil response to both static and dynamic loading is strain/stress-path dependent and hence different values of γ_t^l would be obtained if different strain/stress-paths were used. However the relevance of this concept and its implications in understanding the dynamic behavior of soils would be unchanged.

Another important feature of dynamic soil behavior at very small strains observed experimentally is the invariance of important soil properties, such as stiffness and the material damping ratio, as the number of cycles progresses. Indirect evidence of this

fact is that the shape of the hysteretic loop is not observed to change with the increase in the number of cycles (see as an example EPRI, 1991 and Ishihara, 1996).

2.2 Constitutive modeling

From the previous section it appears clear that for $0 < \gamma \leq \gamma_t^l$ soils subjected to dynamic excitations exhibit both the ability to store strain energy (which is typical of an elastic behavior), and to dissipate strain energy over a finite period of time (typical of a viscous behavior). Constitutive models able to accurately describe from a phenomenological point of view these aspects of material behaviour are those given by the theory of linear viscoelasticity (Christensen, 1971).

In hypoelastic materials the current state of stress is completely determined, via an injective correspondence, by the current state of strain. Conversely, viscoelastic materials have the distinctive feature that the current state of stress is a function not only of the current state of strain but also of all past states of strain characterizing the strain-history of the material under study.

A complete description of a linear viscoelastic constitutive model for an isotropic material requires specification of two material functions (Christensen, 1971). In the time domain they are either the shear and bulk *relaxation functions*, $G_S(t)$ and $G_B(t)$, or the shear and bulk *creep functions*, $J_S(t)$ and $J_B(t)$. Stress and strain history are then linked together in a constitutive relationship by Boltzmann's integral equation (Pipkin, 1986).

When the prescribed strain or stress history is a harmonic function of time, the constitutive relation assumes, in the frequency domain ω, a very simple algebraic form:

$$s_{ij}(t) = 2G_S^*(\omega) \cdot e_{0ij} \cdot \exp(i\omega t) \tag{1}$$

$$\sigma_{kk}(t) = 3G_B^*(\omega) \cdot \varepsilon_{0kk} \cdot \exp(i\omega t) \tag{2}$$

where $s_{ij} = \sigma_{ij} - \frac{1}{3}\delta_{ij}\sigma_{kk}$ and $e_{ij} = \varepsilon_{ij} - \frac{1}{3}\delta_{ij}\varepsilon_{kk}$ are respectively the components of the deviatoric stress and strain tensors, e_{0ij} and ε_{0kk} are the amplitudes of the deviatoric and volumetric strains, and δ_{ij} is the Kronecker symbol (summation convention is implied on repeated indices). Finally, $G_S^*(\omega)$ and $G_B^*(\omega)$ are two material functions denoted as the *complex shear* and *bulk moduli* respectively.

The argument of a complex modulus is a measure of the amount by which the stress lags behind a prescribed harmonic strain. It is also possible to show that this quantity is proportional to the amount of energy dissipated by the material undergoing harmonic oscillations (Pipkin, 1986). The reciprocal of the complex moduli $G_S^*(\omega)$ and $G_B^*(\omega)$ are called *complex compliances*, and are denoted by $J_S^*(\omega)$ and $J_B^*(\omega)$.

3 PHASE VELOCITY AND ATTENUATION IN LINEAR VISCOELASTIC MEDIA

A mechanical disturbance propagating at its own phase velocity in a linear viscoelastic medium undergoes spatial attenuation due to the phenomena of energy dissipation taking place along the direction of propagation. The phase velocity and attenuation coefficient of P and S body waves can be considered in a sense as an alternative form assumed in viscoelastic media by the constitutive parameters. In fact, these kinematical quantities are related to the two material functions $G_S^*(\omega)$ and $G_B^*(\omega)$ by the following expressions (Fung, 1965):

$$V_P(\omega) = \left[\Re\left(\sqrt{\frac{\rho}{G_B^* + \frac{4}{3}G_S^*}} \right) \right]^{-1} \qquad (3)$$

$$\alpha_P(\omega) = \omega \cdot \Im\left(\sqrt{\frac{\rho}{G_B^* + \frac{4}{3}G_S^*}} \right) \qquad (4)$$

$$V_S(\omega) = \left[\Re\left(\sqrt{\frac{\rho}{G_S^*}} \right) \right]^{-1} \qquad (5)$$

$$\alpha_S(\omega) = \omega \cdot \Im\left(\sqrt{\frac{\rho}{G_S^*}} \right) \qquad (6)$$

where the symbols $\Re(\cdots)$ and $\Im(\cdots)$ denote the real and imaginary parts of a complex number, respectively.

Equations 3 through 6 can be obtained from direct application of the *elastic-viscoelastic correspondence principle* (Christensen, 1971). According to this principle, the solution of a harmonic boundary value problem in linear viscoelasticity can be obtained from the solution of the corresponding elastic problem, by extending the validity of the latter solution to complex values of the field variables. It should be remarked, however, that application of the elastic-viscoelastic correspondence principle is restricted to problems where the boundary conditions (i.e. specified stresses and displacements at the boundary of the body) are *time-invariant*.

The mechanics of wave propagation in linear viscoelastic media is completely described by the phase velocities, V_P and V_S, and by the attenuation coefficients, α_P and α_S. Whereas V_P and V_S give a measure of the speed at which *irrotational* and *equivoluminal* disturbances propagate in a viscoelastic medium and are hence related to its *stiffness*, α_P and α_S give a measure of the spatial attenuation of these waves as they propagate through the medium and are hence related to its *dissipative properties*. In fact, the *material damping ratio* D_χ and the attenuation coefficient α_χ are univocally related by the following relationship:

$$D_\chi(\omega) = \left[\frac{\dfrac{\alpha_\chi \cdot V_\chi}{\omega}}{1 - \left(\dfrac{\alpha_\chi \cdot V_\chi}{\omega} \right)^2} \right] \qquad (7)$$

where $\chi = S, P$. It can also be shown that the material damping ratio D_χ can be calculated directly from the complex moduli via the following expression (Lai, 1998):

$$D_\chi(\omega) = \frac{1}{2} \cdot \tan\left[\arg\left(G_\chi^* \right) \right] = \frac{\Im\left(G_\chi^* \right)}{2 \cdot \Re\left(G_\chi^* \right)} \qquad (8)$$

where $\arg[\cdots]$ denotes the argument of a complex number and $G_P^*(\omega) = (G_B^* + \frac{4}{3}G_S^*)$ is the *complex constrained modulus*. Equation 8 can be used to obtain an expression for computing the phase velocity of body waves V_χ as a function of the material damping ratio D_χ. From Equations 3 and 5 the following relation can be immediately derived (Lai, 1998):

$$V_\chi(\omega) = V_\chi^e \cdot \sqrt{\frac{2(1 + 4D_\chi^2)}{1 + \sqrt{1 + 4D_\chi^2}}} \qquad (9)$$

where V_χ^e is the velocity of propagation of body waves in an elastic medium ($\chi = S, P$). Equation 9 is evidence that in viscoelastic media, the velocity of propagation (and hence the stiffness) of body waves depends upon the material damping ratio D_χ and, in general, is a frequency dependent function as a consequence of *material dispersion*, a phenomenon affecting all dissipative media.

An important corollary of the functional dependence of V_χ upon D_χ is that an experimental procedure for the measurement of these two parameters should determine them *simultaneously*. Yet, it is current practice in experimental geotechnics to determine V_χ and D_χ using methods and procedures that are completely independent from each other. Whereas V_χ is most commonly obtained from a back-calculation after measuring the *resonant fre-*

quency corresponding to the fundamental mode of vibration of a soil sample subjected to cyclic loading, the material damping ratio D_χ is determined using, for instance, either the *steady-state vibration technique* or the *amplitude decay method* (Ishihara, 1996; Kramer, 1996).

The next section will illustrate the principles of an experimental procedure to be conducted in laboratory with the resonant column test, where the shear wave velocity V_S and material damping ratio D_S of a soil specimen are measured simultaneously and at the same frequency of excitation. Since these parameters can be determined at different frequencies of excitation, the proposed method is also well suited to investigate the frequency dependence laws of these important dynamic soil parameters.

A final remark before concluding this section. The two functions $V_\chi(\omega)$ and $D_\chi(\omega)$ in Equation 9 cannot be prescribed arbitrarily since they are not independent from each other. In fact, it is possible to prove that in order to fulfill the fundamental *principle of causality*, a mechanical disturbance propagating in a linear viscoelastic medium must have phase velocity and attenuation satisfying *Kramers-Krönig equation*, which simply states that the real and the imaginary part of the body wave complex wavenumber are the *Hilbert transforms* of each other (Aki and Richards, 1980).

The phase velocity $V_\chi(\omega)$ and the material damping ratio $D_\chi(\omega)$ pairs measured independently from each other as occurs when using techniques common in soil mechanics do *not* satisfy, in general, the Kramers-Krönig relation and hence the causality constraint.

4 SIMULTANEOUS MEASUREMENT OF SHEAR WAVE VELOCITY AND SHEAR DAMPING RATIO

In the resonant column test, a solid or hollow circular cylindrical soil specimen is subjected to harmonic excitation by an electromagnetic driving system (Drnevich, 1985). The soil specimen can be excited in either the torsional or the longitudinal modes of vibration. The study presented in this paper refers to a stress-controlled resonant column test set in the torsional mode of oscillation. The frequency and amplitude of the harmonic excitation is controlled by an electromagnetic driving system.

Figure 2 shows a simplified scheme of a fixed-free resonant column apparatus, which is fixed at the base and free to rotate at the top, where a driving torque $T_0 e^{i\omega t}$ is applied. The parameter that is measured experimentally is the shear complex modulus $G_S^*(\omega)$.

Introducing a system of cylindrical coordinates $\{r, \vartheta, z\}$ the equation of motion governing the vibrations of an elastic cylinder is:

$$\nabla^2 u_\vartheta = \frac{1}{V_s^2} \cdot \frac{\partial^2 u_\vartheta}{\partial t^2} \qquad (10)$$

where $\nabla^2 = \dfrac{\partial^2}{\partial r^2} + \dfrac{1}{r}\dfrac{\partial}{\partial r} + \dfrac{1}{r^2}\dfrac{\partial^2}{\partial \vartheta^2} + \dfrac{\partial^2}{\partial z^2}$ denotes the *Laplacian operator* in cylindrical coordinates, $V_s = \sqrt{G_s/\rho}$ is the elastic shear wave velocity, G_s is the elastic shear modulus, and ρ the mass density of the soil specimen. Finally, $u_\vartheta(r, z, t)$ is the displacement component in the direction ϑ .

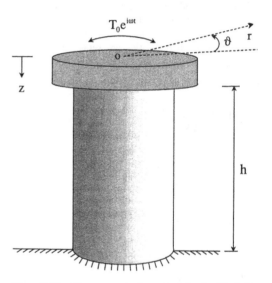

Figure 2. Fixed-free resonant column apparatus (modified after Ishihara, 1996).

If at $z = 0$ the cylinder is subjected to a specified harmonic torque $T_0 e^{i\omega t}$, the solution to Equation 10 may be sought in the form:

$$u_\vartheta(r, z, t) = \xi(r) \cdot \phi(z) \cdot e^{i\omega t} \qquad (11)$$

where ω is the angular frequency of oscillation. If Eq. 11 is substituted into Eq. 10, this second order partial differential equation gets transformed into two ordinary differential equations in the unknown functions $\xi(r)$ and $\phi(z)$ which can be easily solved. The result is:

$$u_\vartheta(r, z, t) = \left[A\sin\left(\frac{\Omega z}{h}\right) + B\cos\left(\frac{\Omega z}{h}\right) \right] \cdot r \cdot e^{i\omega t} \qquad (12)$$

where $\Omega^2 = \left(\dfrac{\rho h^2 \omega^2}{G_s}\right)$ and h is the height of the cylinder. A and B are two constants to be determined from the boundary conditions:

$$u_\vartheta(r, h, t) = 0 \qquad (13)$$

$$\int_S \tau_{\vartheta z}(r, 0, t)\,ds = M(0, t)$$
$$= T_0 e^{i\omega t} - I_0 \ddot{\vartheta}(0, t) \qquad (14)$$

where I_0 is the rotational moment of inertia about the z axis of the cap of the resonant column placed on top of the specimen, and $\vartheta(z, t)$ is the angle of twist, defined by $\vartheta(z, t) = -u_\vartheta(r, z, t)/r$. Finally, $S = \pi R^2$ is the cross sectional area of the cylinder having a radius R.

Since the only non-zero stress component is $\tau_{\vartheta z} = G_s \dfrac{\partial u_\vartheta}{\partial z}$, the twisting moment $M(z, t)$ is computed by integrating $\tau_{\vartheta z}(r, z, t)$ over the cross sectional area to yield:

$$M(z, t) = \frac{\pi R^4}{2} \cdot \frac{\rho \omega^2 h}{\Omega} \cdot$$
$$\cdot \left[A \cdot \cos\left(\frac{\Omega z}{h}\right) - B \cdot \sin\left(\frac{\Omega z}{h}\right) \right] \cdot e^{i \cdot \omega t} \qquad (15)$$

By applying the boundary conditions expressed by Eqs. 13-14, it is found that the ratio between the driving torque $T_0 e^{i\omega t}$ and the angle of twist at the top of the specimen $\vartheta(0, t) = -\dfrac{u_\vartheta}{r} = \phi(0) \cdot e^{i\omega t}$ is given by (Lai, 1998):

$$\frac{T_0 e^{i\omega t}}{\vartheta(0, t)} = \frac{\pi R^4}{2} \cdot \frac{\rho \omega^2 h}{\Omega} \cdot \cot(\Omega) - I_0 \omega^2 \qquad (16)$$

The solution of the corresponding viscoelastic boundary-value problem is obtained by applying the elastic-viscoelastic correspondence principle. The result is:

$$\frac{T_0 e^{i\omega t}}{\vartheta(0, t)} = \frac{\pi R^4}{2} \cdot \frac{\rho \omega^2 h}{\Omega^*} \cdot \cot(\Omega^*) - I_0 \omega^2 \qquad (17)$$

where $\Omega^*(\omega) = \sqrt{\dfrac{\rho \omega^2 h^2}{G_s^*(\omega)}}$ and $G_s^*(\omega)$ is the complex shear modulus. An inspection of Eq. 17 suggests that experimental measurement of the angle of twist at the top of the specimen $\vartheta(0, t)$ will allow the complex modulus $G_s^*(\omega)$ to be determined once the amplitude of the applied torque T_0 and the geometry of

the specimen and of the resonant column apparatus are known.

It should be remarked that since the driving torque $T_0 e^{i\omega t}$ and the angle of twist $\vartheta(0, t)$ will be in general out of phase, $\phi(0)$ is in general a complex number. If this analysis is carried out for a sufficiently large range of frequencies, it is then possible to determine experimentally the frequency dependence law of the complex shear modulus $G_s^*(\omega)$. Finally, the shear wave velocity $V_s(\omega)$ and the shear damping ratio $D_s(\omega)$ of the soil specimen can be obtained from $G_s^*(\omega)$ by means of Eq. 5 and Eq. 8 respectively.

5 EXPERIMENTAL MEASUREMENTS

5.1 Testing apparatus

To assess the validity of the technique for the simultaneous measurement of shear wave velocity and shear damping ratio a series of dynamic laboratory tests were performed.

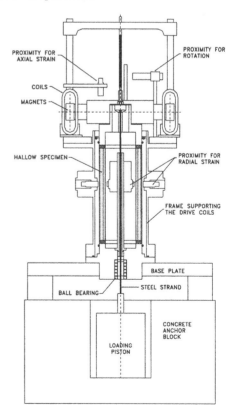

Figure 3. General scheme of the torsional shear/resonant column apparatus (from Lo Presti et al., 1993).

The equipment used is a torsional shear/resonant column apparatus whose main characteristics are briefly summarized as follows (see Fig. 3). The apparatus is equipped with an electric motor (made by SBEL, Arizona) having 8 coils and 4 magnets.

It is capable of a maximum torque of about 1 Nm and can operate under load-controlled mode. The motor is driven by an arbitrary function generator coupled with a power amplifier. The function generator is used to specify the waveform of the torque applied to the specimen. The power amplifier instead allows the amplitude of the applied torque to be controlled. The latter is computed by means of a torque-voltage calibration curve (Lo Presti et al., 1993). Both the arbitrary function generator and the data acquisition system are computer-controlled (Pallara, 1995).

The apparatus can host either solid or hollow cylindrical specimens having an internal diameter of 30 mm, an external diameter of 50 mm and a height of 100 mm. The specimen is fixed at the base and free to rotate at the top. The rotational moment of inertia I_0 of the cap of the apparatus placed on top of the specimen has been determined from a calibration procedure. Its value is $3.85 \cdot 10^{-6}$ Mgm^2 (Pallara, 1995).

The specimens can be consolidated anisotropically with a consolidation stress ratio $K_c = \sigma'_{hc} / \sigma'_{vc} \leq 1$. The tests performed for this research have been conducted on isotropically consolidated, solid specimens.

An accelerometer is used to determine the resonance frequency when using the apparatus for conventional resonant column tests. Shear strains at the top of the specimens are measured by means of a pair of proximity transducers having a resolution of 0.1 µm which corresponds to a shear strain of about 10^{-4}%.

The transducer targets are fixed to the drive system at a radial distance of about 30 mm, and hence they rotate with the sample.

The proximity transducers are firmly attached to the frame supporting the coils and monitor the overall rotational movement (see Fig. 4). The relative distance between the proximity transducers and their targets can be set or re-set during a test by moving the transducers along the horizontal direction by means of a system which can be operated from outside the cell. Acquisition of the applied voltage and of the output of the two proximity transducers was carried out by means of three simultaneously triggered digital multimeters. Thus, no delay exists between torque and rotation measurements.

The apparatus is also equipped with sensors for the measurements of longitudinal and radial strains.

5.2 Soil samples and testing program

With the equipment described in the previous section a series of tests have been performed on *undisturbed* soil specimens coming from two sites in Italy:

- Pisa
- Castelnuovo di Garfagnana

The material coming from Pisa was retrieved from a depth of about 20 m and is classified as a slightly overconsolidated (OCR<2) high plasticity *silty clay*.

The sample from Castelnuovo di Garfagnana was retrieved from depths between 25.5 and 26.1 m and is classified as *silt with clayey sand*.

Table 1 summarizes the main geotechnical characteristics of the specimens, while Figure 5 shows the grain size distribution curves of the two soils.

Table 1. Characteristics of soil specimens retrieved at the sites.

Feature	Pisa	Castelnuovo di Garfagnana
Sampler	Laval	Shelby
Depth (m)	20.00-20.20	25.50-26.10
G_S	2.731	2.709
γ (t/m^3)	1.658	2.173
γ_S (t/m^3)	1.048	1.928
w_N (%)	59.65	12.69
w_L (%)	81.62	20.40
w_P (%)	29.56	15.67
PI (%)	52.06	4.73
LI (%)	99.58	98.37
CaCO$_3$ (%)	5.8	21.2

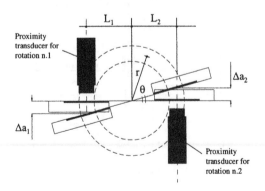

Figure 4. Scheme of the apparatus for the measurement of shear strains (from Pallara, 1995).

The undisturbed soil specimen retrieved at Pisa was isotropically reconsolidated to an effective confining pressure of 167 kPa.

After reconsolidation the specimen was subjected to three different tests:

- resonant column test (*new technique*)
- resonant column test (*conventional - RC*)
- cyclic torsional shear test (*conventional - CTS*)

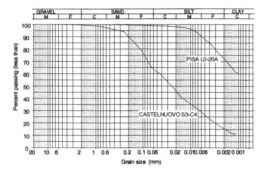

Figure 5. Grain size distribution curve of soil specimens retrieved at Pisa and Castelnuovo di Garfagnana.

The resonant column test with the new technique allowed the simultaneous measurement of V_S and D_S. The test was conducted by applying a cyclic torque at different frequencies of excitation from 5 to about 500 Hz. At each frequency the torque amplitude was modified in such a way as to keep constant the rotation angle of the top of the specimen to a value of about 10^{-4} rad. The measurements performed with this new technique have also been interpreted as conventional CTS tests.

Conventional resonant column and cyclic torsional shear tests were then run at different strain levels to allow a comparison of the results obtained with the new procedure. The resonant frequency for the specimen was equal to 39.9 Hz. The cyclic torsional shear test was performed under load-controlled mode using a torque with a triangular time-varying law at a constant frequency of 0.1 Hz.

The undisturbed soil sample retrieved at Castelnuovo di Garfagnana was isotropically reconsolidated to an effective confining pressure of 331 kPa. Following reconsolidation, the specimen was then subjected to the same testing program as the Pisa sample. The resonant frequency of the specimen was equal to 87 Hz. In contrast to the resonant column test carried out with the new technique on the Pisa sample, the test on the specimen coming from Castelnuovo di Garfagnana was run at a constant torque.

All the tests were conducted under *undrained conditions* and with a 24 hour resting period with opened drainage to allow for possible pore pressure dissipation.

5.3 Experimental results

Figure 6 shows the time-histories of the applied torque and of the rotation of the top of the Pisa specimen at 20 Hz obtained during the resonant column test run according to the proposed method. The time-histories in the figure have been fitted with numerical sinusoids for obtaining the phase and amplitude of the experimental signals.

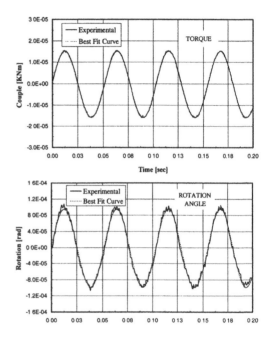

Figure 6. Time-history of applied torque and rotation at top of Pisa specimen at 20 Hz (from Turco, 2000).

Figure 7 shows the elliptically-shaped hysteretic loops obtained with the Pisa specimen at 10, 31, 40 and 70 Hz. From the plots it was noted that, while the shear strain remains constant with frequency and approximately equal to $1.7 \cdot 10^{-5}$, the shear stress varies.

It was also observed that at the resonant frequency, which is about 40 Hz, the hysteretic loop rotates of about 180°. This happens because at the resonance frequency response (i.e. the rotation of the top of the specimen) and excitation (i.e. the applied torque) are out of phase of 180° even in the case of *zero* damping.

This phenomenon of *phase-shift* between response and excitation resulting in a rotation of the elliptic hysteretic loop also occurs at higher resonance frequencies as can be predicted theoretically.

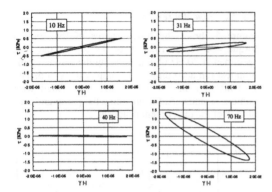

Figure 7. Elliptic hysteretic loops obtained with Pisa specimen at 10, 31, 40 and 70 Hz. (from Turco, 2000).

The values of the shear wave velocity and the shear damping ratio obtained with the new procedure for the Pisa soil at various excitation frequencies are illustrated in Figures 8-9.

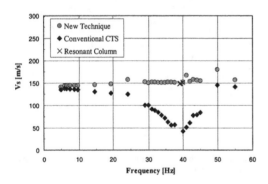

Figure 8. New and conventional measurement of shear wave velocity for Pisa soil (from Turco, 2000).

The shear wave velocity $V_S(\omega)$ and shear damping ratio $D_S(\omega)$ were determined from the solution of Equation 17 for the unknown $G_S^*(\omega)$ using the *Newton-Raphson method* (applied for complex values of the variables).

The values of T_0 and $\phi(0)$ required for the solution of Equation 17 were obtained from the experimental data. Once $G_S^*(\omega)$ was calculated, $V_S(\omega)$ and $D_S(\omega)$ were finally computed by means of Equations 5 and 8 respectively.

In Figure 8 the values of $V_S(\omega)$ and $D_S(\omega)$ obtained with the new procedure have been compared with those obtained with conventional methods. For the latter the shear wave velocity was computed from the slope of the major axis of the elliptic hys-

teretic loop, whereas the material damping ratio was obtained from the area enclosed by the hysteretic loop. The latter is proportional to the amount of energy per unit volume dissipated by the specimen during a cycle of harmonic loading.

The comparison of the values of $V_S(\omega)$ obtained with the new and conventional procedure is satisfactory. The results given by the two methods are different at frequencies close to resonance where the procedure based on the slope of the hysteretic loop tend to be unreliable.

Figure 8 also shows an excellent agreement between the shear wave velocity obtained with the conventional and with the new interpretation of the resonant column test.

With regards to the material damping ratio, Figure 8 shows that the values of $D_S(\omega)$ determined with the new procedure are in good agreement with those obtained with the method based on the area enclosed by the hysteretic loop at low frequencies only (below 10 Hz).

At frequencies above 10 Hz data from both the conventional and the new procedure are scattered denoting a certain degree of uncertainty in the experimental measurements.

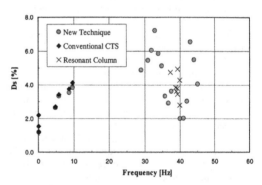

Figure 9. New and conventional measurement of shear damping ratio for Pisa soil (from Turco, 2000).

The last comparison of the values of $V_S(\omega)$ and $D_S(\omega)$ obtained with new and conventional procedures is illustrated in Figures 10-11, where these soil parameters have been determined for the Castelnuovo di Garfagnana soil.

Like the Pisa soil, the result of this comparison is satisfactory for frequencies below the resonant frequency, which in this case is 87 Hz. There is also a good agreement between shear wave velocities obtained with the new procedure and those determined with the conventional resonant column test.

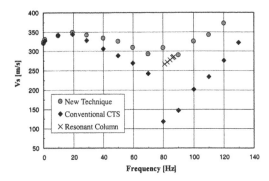

Figure 10. New and conventional measurement of shear wave velocity for Castelnuovo di Garfagnana soil (from Turco, 2000).

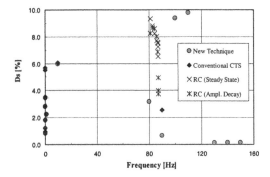

Figure 11. New and conventional measurement of shear damping ratio for Castelnuovo di Garfagnana soil (from Turco, 2000).

Finally, the values of the damping ratio are even more scattered than those of the Pisa soil, mainly because with the Castelnuovo di Garfagnana soil the shear strain was not held constant with frequency during the test. As a result, the agreement between conventional and new methods is good only at frequencies below 10 Hz.

6 CONCLUDING REMARKS

The mechanical behavior of soils subjected to dynamic excitations can accurately be described at very low strain levels (i.e. for $0 < \gamma \leq \gamma_t^l$) by means of the linear theory of viscoelasticity. This theory is able to accurately simulate, from a phenomenological point of view, two important features exhibited by soils undergoing harmonic oscillations: their ability to store, and at the same time, to dissipate strain energy over a finite period of time.

An important consequence of reformulating the low-strain dynamic properties of soils within the framework of the theory of viscoelasticity is that the velocity of propagation of body waves, in particular V_S, and the material damping ratio are not two independent parameters. Their functional coupling is a direct consequence of material dispersion, a phenomenon by which, in viscoelastic media, the velocity of propagation of mechanical disturbances has to be frequency dependent in order to satisfy the fundamental principle of causality.

As a result of their mutual dependence, $V_S(\omega)$ and $D_S(\omega)$ in soils should be measured simultaneously, even though in current geotechnical engineering testing practice, these two parameters are determined separately.

In this paper a novel experimental technique, to be conducted in laboratory with the resonant column test, has been proposed whereby shear wave velocity and the damping ratio are measured simultaneously and at the same frequency of excitation. An attractive aspect of the new procedure is its use for determining the frequency dependence laws of $V_S(\omega)$ and $D_S(\omega)$.

The new method, which offers a re-interpretation of the resonant column test, has been applied to an experimental campaign conducted on two different Italian soils. A preliminary comparison of the results obtained with the novel procedure and those determined with conventional methods is very encouraging. Values of shear wave velocity measured with the new method and for both types of soil are accurate up to the resonant frequency and beyond. As regards the shear damping ratio, results are reasonable up to a frequency of about 10 Hz. Beyond this frequency both conventional and new measurement procedures were affected by inaccuracies.

However, despite these inaccuracies, from a qualitative point of view the results obtained from these tests concerning the frequency dependence laws of $V_S(\omega)$ and $D_S(\omega)$ are in agreement with experimental data from the literature (Lo Presti et al., 1997; Shibuya et al., 1995; Stokoe et al., 1995; Tatsuoka et al., 1997; D'Onofrio et al., 1999).

In particular, whereas shear wave velocity is affected by frequency only moderately (over a frequency range from 1 to about 100 Hz), shear damping ratio exhibits a much stronger frequency dependence.

7 REFERENCES

Achenbach, J.D. 1984. *Wave Propagation in Elastic Solids*. North-Holland, Amsterdam, Netherlands, pp. 425.

Aki, K., and Richards, P.G. 1980. *Quantitative Seismology: Theory and Methods*. W.H. Freeman and Company, San Francisco, 932 pp.

Christensen, R.M. 1971. *Theory of Viscoelasticity - An Introduction*. Ed. Academic Press, 245 pp.

D'Onofrio, A., Silvestri, F. and Vinale, F. 1999. Strain Rate Dependent Behaviour of a Natural Stiff Clay. *Soils and Foundations*. 39(2):69-82.

Drnevich, V.P. 1985. Recent Developments in Resonant Column Testing., *Proceedings of Richart Commemorative Lectures*, Sponsored by Geotechnical Engineering Division, in Conjunction with ASCE Convention, Detroit, Michigan, October 23.

Electric Power Research Institute. 1991. *Proceedings: NSF/EPRI Workshop on Dynamic Soil Properties and Site Characterization*. Report NP-7337, Vol. 1, Research Project 810-14.

Fung, Y.C. 1965. *Foundations of Solid Mechanics*. Prentice-Hall, New Jersey, pp.525.

Ishihara, K. 1996. *Soil Behaviour in Earthquake Geotechnics*. Oxford Science Publications, Oxford, UK, pp. 350.

Kramer, S.L. 1996. *Geotechnical Earthquake Engineering*. Prentice-Hall, New Jersey, pp.653.

Lai, C.G., Lo Presti, D.C.F., Pallara O., and Rix, G.J. 1999. Misura Simultanea del Modulo di Taglio e dello Smorzamento Intrinseco dei Terreni a Piccole Deformazioni. *Proceedings 9th Italian National Conference on Earthquake Engineering*, ANIDIS99, Torino, Italy, September 20-23 (in Italian).

Lai, C.G. 1998. Simultaneous Inversion of Rayleigh Phase Velocity and Attenuation for Near-Surface Site Characterization. Ph.D. Dissertation, The Georgia Institute of Technology, Atlanta, Georgia, 370 pp.

Lo Presti D.C.F., Pallara O, Lancellotta R., Armandi M. and Maniscalco R. 1993. Monotonic and Cyclic Loading Behaviour of Two Sands at Small Strains, *Geotechnical Testing Journal*, Vol. 16, No 4, pp 409-424.

Lo Presti, D.C.F, Pallara, O., and Cavallaro, A. 1997. Damping Ratio of Soils from Laboratory and In-Situ Tests. Proceedings of discussion special technical session on Earthquake Geotechnical Engineering, Balkema, pp. 391-400. *14th International Conference on Soil Mechanics and Foundation Engineering*, Hamburg, Germany, 6-12, September, 1997.

Pallara, O. 1995. Comportamento Sforzi-Deformazioni di Due Sabbie Soggette a Sollecitazioni Monotone e Cicliche. Ph.D Dissertation, Department of Structural Engineering, Politecnico di Torino (in Italian).

Pipkin, A.C. 1986. *Lectures on Viscoelasticity Theory*. 2nd Edition, Springer-Verlag, Berlin, pp.188.

Shibuya, S., Mitachi, T., Fukuda, F., and Degoshi, T. 1995. Strain Rate Effects on Shear Modulus and Damping of Normally Consolidated Clay. *Geotechnical Testing Journal*, 18(3), 365-375.

Stokoe, K.H. II, Hwang S.K., Lee, J.N.K. and Andrus, R.D. 1995. Effects of Various Parameters on the Stiffness and Damping of Soils at Small to Medium Strains. Keynote Lecture 2, *IS Hokkaido*. 2: 785-816. Balkema.

Tatsuoka, F., Jardine R., Lo Presti, D.C.F., Di Benedetto, H. and Kodaka, T. 1997. Testing and Characterising Pre-Failure Deformation Properties of Geomaterials. Theme Lecture Session 1, Balkema, Vol. 4, pp. 2129-2164, *XIV ICSMFE*, Hamburg, Germany September 1997.

Turco, E. 2000. Determinazione Simultanea del Modulo di Taglio e del Rapporto di Smorzamento da Prove di Colonna Risonante. Master Thesis. Department of Structural and Geotechnical Engineering. Politecnico di Torino. pp. 93 (in Italian).

Vucetic, M. 1994. Cyclic Threshold Shear Strains in Soils. *Journal of Geotechnical Engineering*, ASCE, Vol.120, No.12, pp.2208-2228.

Advanced Laboratory Stress-Strain Testing of Geomaterials, Tatsuoka, Shibuya & Kuwano (eds),
© *2001 Taylor & Francis, ISBN 90 2651 843 9*

A solid cylinder torsional shear device

J.Manzanas & V.Cuellar
Laboratorio de Geotecnia (CEDEX), Madrid, Spain

ABSTRACT: A new torsional simple shear device, able to perform both monotonic and cyclic loading on solid cylindrical soil specimens, has been developed at the "Laboratorio de Geotecnia" of CEDEX (Spain). This apparatus was designed with a view to obtaining, with great simplicity and for a wide range of shear strain, accurately and reliably tenso-deformational characteristics of soils. To this end the device is able to perform tests on solid cylindrical specimens (both reconstituted and undisturbed) of standard dimensions 38 mm diameter and 76 mm high. The shear strain measurement system, consisting of two pairs of LVDT's, allows to obtain a continuous record of shear strains from 10^{-4}% to 3%. Torque is applied by an electromagnetic loading system which can grow up to 1.2 Nm and is measured and registered by means of a torsional load cell with a resolution of $5 \cdot 10^{-5}$ Nm. In this paper the apparatus is described and some experimental results are presented.

1 INTRODUCTION

During the last years it has been noted the importance of knowing the actual soil behaviour at strain levels below of values as small as 0.01% (Simpson et al. 1979, Jardine et al. 1986, Burland 1989). To this end, in addition to study the influence of strain rate and repeated loading on soil stiffness (Alarcon-Guzman et al. 1989, Shibuya et al. 1991, Stokoe et al. 1994, Tatsuoka et al. 1995, Shibuya et al. 1995, Lo Presti et al. 1997) several devices for torsional shear tests have been developed (Isenhower & Stokoe 1981, Ampadu & Tatsuoka 1993, Lo Presti et al. 1993, d'Onofrio et al. 1999). Within this research line, the Laboratorio de Geotecnia has designed a new apparatus with a double objective: performing torsional shear tests with a wide range of shear strain control and measurement, and carrying them out with simplicity, accuracy and reliability. To accomplish these requirements, a device which performs tests on solid cylindrical specimens was designed. In this way, tests on soils with bigger particle size and less "undisturbed" that those on hollow cylindrical specimens can be performed.

To verify the appropriate performance of this new equip-ment, some tests on Toyoura sand were carried out.

2 DESCRIPTION OF THE APPARATUS

2.1 *Mechanical Layout*

In Figure 1 the basic layout of the equipment is shown. The tested solid cylindrical specimen (A), 38 mm diameter and 76 mm high, is placed between two porous sintered bronze disks 5 mm thick (B) and jacketed with a latex membrane (E) for confining. Each bronze disk has six embedded radial stainless steel ribs to prevent slippage with the specimen surface. In the case of rigid materials, those bronze disks are replaced by other without steel ribs and fixed to the sample surface by means of a special hydraulic-hardening porous material able to follow the saturation and consolidation processes.

2.2 *Torsional loading system*

To avoid the problems connected with mechanical links, the torque loading on the specimen during torsional shearing is applied by means of an electromagnetic system. It consists of four permanent Neodymium magnets (K) located inside four couples of coils (L). These coils are fixed to the supporting frame (G) while the magnets are attached to the movable driving system (M). This configuration generates a maximum torque of 1.2 Nm, and allows a maximum relative vertical displacement of ±5 mm.

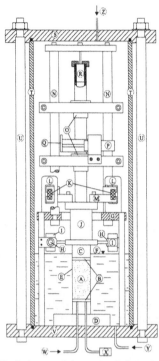

Figure 1. Basic layout of the torsional shear device: (A) specimen, (B) porous sintered bronze stones, (C) top cap, (D) base pedestal, (E) latex membrane, (F) glycol bath, (G) supporting frame, (H) small shear strain LVDT, (I) LVDT reference plate, (J) torque cell, (K) magnets, (L) coils, (M) drive system, (N) fixed columns, (O) bearings, (P) medium shear strain LVDT, (Q) LVDT reference plate, (R) axial strain LVDT, (S) top plate, (T) confining chamber, (U) tie rods, (V) base plate, (W) backpressure, (X) pressure cell, (Y) glycol   tank, (Z) air pressure

2.3 Pressure system

Confining pressure (Z) is supplied by an air pressure pump through a regulating pressure tank to ensure steady supply and regulated by a manual valve. The sample within the con-fining pressure is covered with glycol (F) to avoid the air migration through the membrane into the specimen. In this way confining pressures up to 1 MPa can be applied. Backpressure (W) is independently applied by means of a water-mercury system, reaching a maximum value of 1 MPa.

2.4 The measuring system

2.4.1 Measurement of stresses
It is well known that the use of torque-voltage calibration curves presents inherent difficulties associated with:
- demagnetization of the magnets with usage (Lo Presti et al. 1993)
- temperature effects on the coils

Figure 2. General view of the device

- the relative location of the magnets within the coils (Ray 1983).

For this reason, it was decided to measure the torsional loading by means of an internal torque cell (J) located between the drive system (M) and the specimen top cap (C). This torque cell is a DMA of 6 Nm full range, 2.0 mV/V sensitivity and $5 \cdot 10^{-5}$ Nm resolution (i.e. on a specimen of standard dimensions, the maximum torque of 1.2 Nm produces a shear stress of 90 kPa, being the resolution of 0.004 Pa).

The pore pressure is measured by a HBM transducer of 10 bar range, 2 mV/V rated sensitivity, while the confining pres-sure is measured by a manometer with 0.1 bar precision scale.

2.4.2 Measurement of strains
For the purpose of covering the shear strain range from 10^{-4}% to 3% two pairs of LVDT's (Linear Variable Differential Transformer) with different ranges and sensitivities were used: ±1 mm range and 69 mV/V/mm sensitivity for small strains (H) and ±7.5 mm range and 56 mV/V/mm sensitivity for medium strains (P). To avoid bending errors in the measurement, the LVDT's of each couple were placed diametrically opposed.

The selected placement of the small strain LVDT's directly on the specimen top cap (C) has the advantage of avoiding mechanical errors and those associated to torque cell twist. On the other hand, the placement of the LVDT's in this location and the use of springs in them, introduces an additional torque not measured by the torque cell. Such torque exclusively depends on the angular position of the top cap. Hence, it is possible to quantify this effect by running several loading-unloading tests without specimen, the driving system suspended from the supporting frame, and monitoring the applied torque as a function of the angular position measured by the small strain LVDT's. In this way, it was possible to verify the repetitivity of the calibration curve and incorporate this correction in the data treatment.

The small strain LVDT's are able to monitor the maximum angular displacement up to ±1.67 degrees (±0.029 rad) with a resolution of less than 0.0001 degrees ($< 1.7 \cdot 10^{-6}$ rad). In the case of specimens 38 mm diameter and 76 mm high, and considering an equivalent average radius 0.82 times the external radius of the sample, as proposed by Chen & Stokoe (1979), those angular displacements are equivalent to a shear strain of ±0.60% and $3.6 \cdot 10^{-5}$% respectively.

Another pair of LVDT's of greater measuring range are used to cover the medium strains (P). Their characteristics are: angular displacement from -1.67 degrees to +6.8 degrees, which are equivalent to a maximum shear strain of -0.60% to +2.43% (3.03% full range).

Axial displacement is measured by a LVDT transducer (R) installed within the rotation shaft, with a nominal linear range of ±5 mm and 100 mV/V/mm sensitivity.

2.5 *Control and acquisition system*

The data acquisition and control is performed by electronic boards integrated in a PC and specific software that allows the inclusion of the calibration curves of all the transducers. This system makes it possible to correct the transducers non-linearity both in data acquisition and test control.

Shearing can be controlled both in torque or in angular displacement. These peculiarities and the capabilities of the function generation allows making, both monotonic and cyclic, stress and strain rate controlled tests.

Maximum control rate is 0.15 Nm/s (11.33 kPa/s) and 1.0 degrees/s (0.36%/s) and the minimum is $2 \cdot 10^{-5}$ Nm/s (1.5 Pa/s) and $3 \cdot 10^{-5}$ degrees/s (10^{-5}%/s).

The control system on the registered error signal is a closed-loop PID type (Proportional Integral Derivative). The PID parameters are established by the user, as well as the filters used with the signal of each transducer.

3 CALIBRATION

3.1 *Angular displacement*

In order to avoid cumulative errors, the calibration of the transducer used for the angular displacement measurement was carried out with them installed in

the same position they were to have during the test, calculating directly the rotated angle. This operation was undertaken by the use of a laser beam placed on top of the shaft of the equipment (R) inciding on an horizontal rule divided in millimeters placed at 15 m from the shaft.

3.2 *Axial displacement*

For axial displacement calibration, a digital micrometer of 1 mm sensitivity and 3 mm uncertainty was employed.

3.3 *Torque*

The torque cell was calibrated by means of a Lorenz Meβtechnik torque calibrating cell, of 10 Nm full range and 0.50 mV/V sensitivity.

3.4 *Pressure*

Pressure transducers were calibrated with a precision pressure controller/calibrator Druck DPI 510 of 210 bar full-scale and ±0.01% accuracy.

4 TEST RESULTS

For checking this new device capabilities, the widely tested Japanese standard sand called Toyoura sand was used. Toyoura sand is an uniform fine sand with mainly sub-angular particle shape. Its index properties are: $D_{50} = 0.17$ mm, $U_c = 1.3$, $G_s = 2.650$, $e_{min} = 0.611$ and $e_{max} = 0.985$.

4.1 *Specimen preparation and testing*

Specimen was prepared by wet tamping method (Tatsuoka et al. 1986). The material with a water content of about 8% was placed in five equal portions into the split mould, compacting each one with six tamping blows. A 0.5 kg weight tamper with a constant free fall of 20 cm was employed. In this way, a specimen density of 1.47 g/cm^3 was obtained (e = 0.803). Afterwards, the specimen was jacketed with a latex membrane and frozen, and 24 hours later it was placed into the device. Then it was saturated and isotropically consolidated under an effective pressure p' = 100 kPa.

After 24 hours of consolidating time, an undrained monotonic torsional shear test was carried out. Test was performed under shear strain rate control, at a constant rate value of 0.003 degrees/s (0.001%/s). Initially, strain rate was controlled with the small strain LVDT's measures and, before these become out of range, strain rate control was automatically transferred to the medium strain LVDT's measures.

4.2 Results

Figures 3 and 4 illustrate the results obtained from the monotonic torsional shear test. Figure 3.a shows the entire stress-strain curve derived from the test, and Figure 3.b shows the initial part of that curve. Figure 3.b illustrates the different resolution achieved with both small and medium strain LVDT measurement systems.

In Figure 4, the shear secant modulus G is plotted against shear strain. It must be emphasized that shear strain values as small as about 10^{-4}% are measured with this new equipment, with an excellent agreement with values reported in literature.

5 CONCLUSIONS

A new torsional shear apparatus is presented and described. It is designed for testing solid cylindrical specimens with a diameter of 38 mm and a height of 76 mm. Torsional loading is applied via an electromagnetic loading system able to reach a maximum torque of 1.2 Nm. Angular displacements are measured by a double system of LVDT's capable of measuring shear strains from about 10^{-4}% to 3%. In order to check the capacity of the equipment, monotonic loading tests have been carried out on Toyoura sand.

REFERENCES

Alarcon-Guzman, A., Chameau, J.L., Leonards, G.A. & Frost, J.D. 1989. Shear modulus and cyclic undrained behaviour of sands. *Soils and Foundations* 29(4): 105-119.
Ampadu, S.K. & Tatsuoka, F. 1993. A hollow cylinder torsional simple shear apparatus capable of a wide range of shear strain measurement. *Geotechnical Testing Journal* 16(1): 3-17.
Burland, J.B. 1989. Small is beautiful-the stiffness of soils at small strains. Ninth Laurits Bjerrum Memorial Lecture. *Canadian Geotechnical Journal* 26(4): 499-516.
Chen, A.T.F. & Stokoe, K.H.II 1979. Interpretation of strain-dependent modulus and damping from torsional soil tests. Report N° USGS-GD-79-002, U.S. Geological Survey, Menlo Park, California.
D'Onofrio, A., Silvestri, F. & Vinale, F. 1999. A new torsional shear device. *Geotechnical Testing Journal* 22(2): 101-111.
Isenhower, W.M. & Stokoe, K.H.II. 1981. Strain rate dependent shear modulus of San Francisco bay mud. *Proc. Int. Conf. Recent Adv. Geotech. Earthquake Eng. and Soil Dyn., St. Louis, Missouri* 2:597-602.
Jardine, R.J., Potts, D.M., Fourie, A.B. & Burland, J.B. 1986. Studies on the influence of non-linear stress-strain characteristics in soil-structure interaction. *Geotechnique* 36(3): 377-396.
Lo Presti, D.C.F., Jamiolkowski, M., Pallara, O., Cavallaro, A. & Pedroni, S. 1997. Shear modulus and damping of soils. *Geotechnique* 47(3): 603-617.
Lo Presti, D.C.F., Pallara, O., Lancellota, R., Armandi, M. & Maniscalco, R. 1993. Monotonic and cyclic loading behavior of two sands at small strains. *Geotechnical Testing Journal* 16(4): 409-424.
Ray, R.P. 1983. Changes in shear modulus and damping in cohesionless soils due to repeated loading. Ph.D. thesis, University of Michigan.
Shibuya, S., Mitachi, T., Fukuda, F. & Degoshi, T. 1995. Strain rate effects on shear modulus and damping of normally

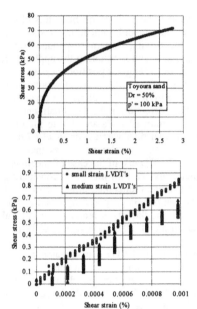

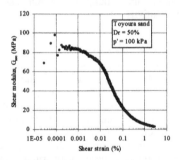

Figure 3. (a) Stress-strain curve from a monotonic torsional shear test. (b) Comparison between stress-strain curves obtained by both small and medium shear strain LVDT's

Figure 4. Shear modulus versus shear strain

consolidated clay. *Geotechnical Testing Journal* 18(3): 365-375.
Shibuya, S., Tatsuoka, F., Abe, F., Teachavorasinskun, S. & Park, C.S. 1991. Elastic properties of granular materials measured in the laboratory. *Proc. Tenth European Conf. on SMFE, Firenze* 1: 163-166.
Simpson, B., O'Riordan, N.J. & Croft, D.D. 1979. A computer model for the analysis of ground movements in London clay. *Geotechnique* 29(2): 149-175.
Stokoe, K.H.II, Hwang, S.K., Lee, J.N.-K. & Andrus, R.D. 1995. Effects of various parameters on the stiffness and damping of soils at small to medium strains. Keynote Lecture, *Proc. First Int. Conf. on Pre-failure Deformation Characteristics of Geomaterials, Sapporo, Japan 1994.* 2:785-816. Rotterdam: Balkema.
Tatsuoka, F., Lo Presti, D.C.F. & Kohata, Y. 1995. Deformation characteristics of soils and soft rocks under monotonic and cyclic loads and their relationships (State of the art). *Proc. Third Int. Conf. Recent Adv. Geotech. Earthquake Eng. and Soil Dyn, St. Louis, Missouri.* 2: 851-879.
Tatsuoka, F., Ochi, K., Fujii, S. & Okamoto, M. 1986. Cyclic undrained triaxial and torsional shear strength of sands for different sample preparation methods. *Soils and Foundations* 26(3): 23-41.

Advanced Laboratory Stress-Strain Testing of Geomaterials, Tatsuoka, Shibuya & Kuwano (eds),
© 2001 Taylor & Francis, ISBN 90 2651 843 9

Application of an elasto-viscous model to one-dimensional consolidation of clay under cyclic loading

T. Moriwaki & Y.E. Supranata
Hiroshima University, Higashi-Hiroshima, Japan

M. Saitoh
Osaka City, Osaka, Japan

M. Hashinoki
Japan Highway Public Corporation, Hikone, Japan

ABSTRACT: In order to clarify the consolidation behavior of clay ground under cyclic loading, a series of one-dimensional consolidation tests was carried out using inter-connected oedometers. A clay sample reconsolidated at high temperature was used along with a clay sample reconsolidated at room temperature to investigate cyclic effects on the consolidation properties of clay with a high-grade structure. A series of numerical analyses was also carried out using an elasto-viscous model, which can explain the various types of time-dependent behavior during the one-dimensional consolidation process under static loading. Applicability of this elasto-viscous model to the one-dimensional consolidation behavior of saturated clay under cyclic loading is discussed.

1 INTRODUCTION

The consolidation behavior of clay ground under cyclic loading is different from that under static loading. However, the mechanism of cyclic consolidation has not been clarified as well as that of static consolidation because cyclic consolidation behavior is influenced in a complex way by many factors such as the cyclic period, the magnitude and pattern of the cyclic load, the initial consolidation conditions, the sensitivity of clay, etc. In this paper, a series of one-dimensional consolidation tests under static and cyclic loading was carried out using inter-connected oedometers in order to clarify the consolidation behavior of every element at different locations inside a clay layer. A clay sample reconsolidated at high temperature was used along with a clay sample reconsolidated at room temperature to investigate cyclic effects on the consolidation properties of saturated clay with a high-grade structure.

Yoshikuni et al. (1994) have proposed a non linear elasto-viscous model to explain various kinds of time-dependent behavior during the one-dimensional consolidation process under static loading, including creep in both the normal and overconsolidated stages, relaxation of effective stress under undrained conditions, strain rate dependency of consolidation yield stress in the constant rate of strain consolidation test, and quasi-overconsolidated behavior. Applicability of this elasto-viscous model to the one-

dimensional consolidation behavior of saturated clay under cyclic loading is also discussed in this paper.

2 CONSOLIDATION TEST

2.1 *Sample preparation and test procedure*

The clay sample used in this study was an alluvial marine clay known as Maizuru clay. The physical properties of this clay were w_L=83.3%, w_P=31.1%, I_P=52.2 and G_S=2.737. In order to maintain sample homogeneity, a natural sample was completely remolded in a mixer, screened through a 0.420mm sieve and finally mixed into slurry with a water content twice its liquid limit. The slurry was placed in a vacuum and periodically agitated to remove entrapped air. It was then poured into a 250mm diameter stainless steel mold up to 200mm thick, and initially consolidated under the self-weight of the slurry for one day. Consolidation pressures of 2.0, 7.8, 29.4, 49.0 and 98.0kPa were subsequently applied to the samples at both 20℃ and 70℃ under one-dimensional conditions. The clay samples reconsolidated at 20℃ and 70℃ will hereafter be called the RT sample and the HT sample respectively. The consolidation period under the consolidation pressures of 2.0, 7.8, 29.4 and 49.0kPa was one day, while the consolidation period under the final consolidation pressure of 98.0kPa was determined by three times the t_E method.

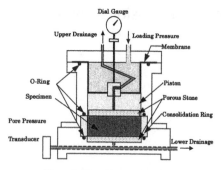

Figure 1. Closed type oedometer

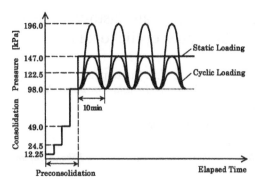

Figure 3. Loading pattern

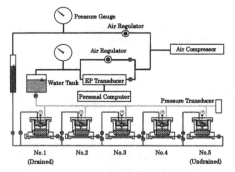

Figure 2. Inter-connected consolidation test apparatus

A series of consolidation tests under static and cyclic loading was carried out for both the RT and HT samples by using a set of inter-connected oedometers in order to clarify the internal behavior of the clay under cyclic loading. Figures 1 and 2 show a sketch of a single closed type oedometer and the inter-connected consolidation test apparatus, which consisted of five inter-connected closed type oedometers. Each specimen was 60mm in diameter and 20mm in height. Each sub-specimen prepared from the clay samples mentioned above was set in an oedometer and was preconsolidated stepwise from 12.2kPa to 98.0kPa under a back pressure of 98.0kPa, permitting single drainage for every separate sub-specimen.

After preconsolidation at 98.0kPa, the five sub-specimens were connected through a path of pore water as shown in Fig. 2, and were then subjected to a static consolidation pressure of 147.0kPa ($\Delta p =$ 49.0kPa) and one-way cyclic consolidation pressures whose maximum amplitudes, Δp_{cyc}, were 24.5, 49.0 and 98.0kPa. The maximum values of the cyclic consolidation pressure were 122.5, 147.0 and 196.0kPa, as shown in Fig. 3. The cyclic period was 10min and both the static and the cyclic consolidation tests were continued 10,000min.

2.2 Test results and discussion

The relationships between the strain and the elapsed time on the RT and HT samples under static and cyclic loading are shown in Figs. 4 and 5. In the case of static loading, although the accumulation of the strain begins from the drainage side (No.1), every curve of the strain versus elapsed time relationship finally converges in a single line as shown in Fig.4. In the case of cyclic loading, however, the strain of the sub-specimen on the drainage side is still larger than that of the other sub-specimens even when the consolidation progresses sufficiently as shown in Fig.5. Since the final strain of the sub-specimen on the drainage side under cyclic loading is almost equal to that under static loading, and the strain of the sub-specimens on the inside undrained side is smaller than that under static loading, the average strain for all sub-specimens under cyclic loading is thus smaller than that under static loading. Comparing Figs. 4 and 5, it was also found that consolidation progress is slower under cyclic loading than under static loading. This is due to the difference in loading duration, i.e., the loading duration where excess pore water pressure is generated is shorter for cyclic loading than for static loading because of the existence of the unloading duration in cyclic loading.

Comparing the difference between the RT sample and the HT sample, it was apparent that the strain of the RT sample was larger than that of the HT sample under both static and cyclic loading. This is considered to be due to the difference in the void ratios of the RT sample and the HT sample at the start of both static and cyclic loading. The void ratio of the RT sample at the start of static and cyclic loading was 1.65, and was more than 1.56 for the HT sample. These results differ from a previous study (Moriwaki et al., 1999), in which the strain of the HT sample was larger than that of the RT sample. It was pointed

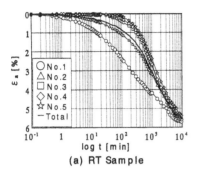

(a) RT Sample

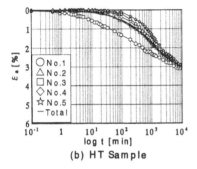

(b) HT Sample

Figure 4. Strain ~ time curves under static loading

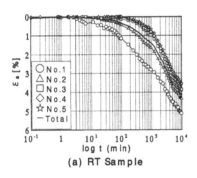

(a) RT Sample

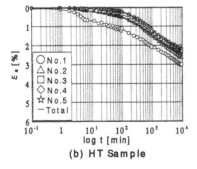

(b) HT Sample

Figure 5. Strain ~ time curves under cyclic loading

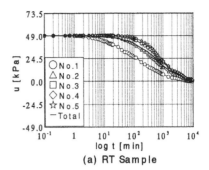

(a) RT Sample

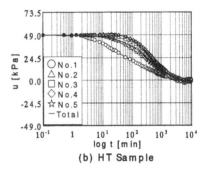

(b) HT Sample

Figure 6. Pore water pressure ~ time curves under static loading

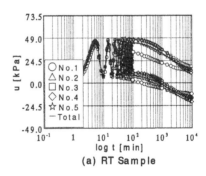

(a) RT Sample

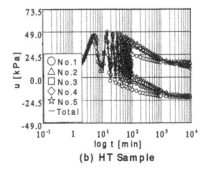

(b) HT Sample

Figure 7. Pore water pressure ~ time curves under cyclic loading

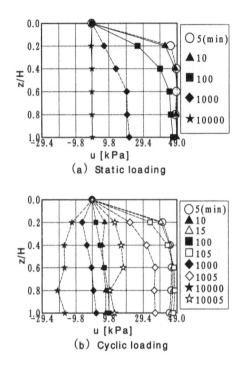

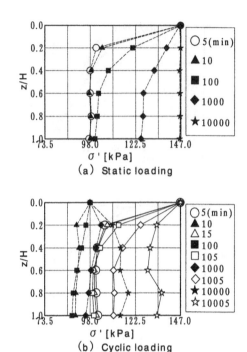

Figure 8. Isochrones of pore water pressure for RT sample

Figure 9. Isochrones of effective stress for RT sample

out that the high temperature reconsolidation has opposing effects in that the void ratio increases due to development of cementation but decreases due to acceleration in the secondary consolidation (e.g. Moriwaki et al., 1995). Moreover, the former is common in cases where the consolidation pressure is small and the later is common in cases where the consolidation pressure is large. As the reconsolidation pressure of 98.0kPa in this test is larger than the 49.0kPa of the previous study, secondary consolidation is accelerated and a strong dense structure is formed in the HT sample. Therefore, the strain of the HT sample under static and cyclic loading is smaller than that of the RT sample.

From the relationship between the excess pore water pressure and the elapsed time as shown in Figs. 6 and 7, it was found that the excess pore water pressure on the undrained side had almost dissipated in the later stages of static consolidation. This implies that the accumulation of strain in the later stages of consolidation is caused by secondary consolidation. On the other hand, it is difficult to analyze simply the excess pore water behavior, because its pressure varies cyclically with time. The absolute values of the excess pore water pressure on loading and unloading are almost equal after the elapsed time of 10,000min and therefore its behavior is considered

as having reached a state of equilibrium in the later stages of cyclic consolidation. However, the tendency for the accumulation of strain to decrease in the later stages of consolidation cannot be seen in the case of cyclic loading. Strain under cyclic loading continues to be accumulated in the later stages of consolidation due to cyclic effects. Since the drainage distance from the sub-specimen on the undrained side is 10cm, the pore water pressure cannot be dissipated in a cyclic period of 10min and therefore should have the same amplitude as the cyclic load. However, the amplitude of the excess pore water pressure is smaller than that of the cyclic load, as shown in Figs. 6 and 7. This is due to a time lag in the measurement system of the excess pore water pressure as mentioned below.

Figures 8 and 9 show the isochrones of the excess pore water pressure and the effective stress of the RT sample. As mentioned above, because the excess pore water pressure after the elapsed time of 10,000min under static loading is fully dissipated on the inside of the specimen, effective stress reaches the value of the static consolidation pressure. Therefore, although the progression of the strain of each sub-specimen is dependent on the difference in drainage distance, the strain of each sub-specimen attains the same value in the end. On the other hand,

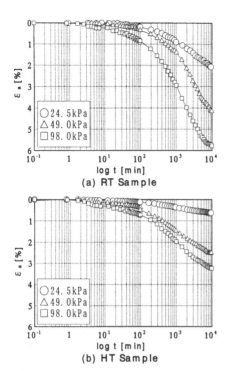

Figure 10. Average strain ~ time curves
under cyclic loading

the excess pore water pressure on the inside of the specimen after the elapsed time of 10,000min under cyclic loading is not dissipated, remains at about half the initial value, and reaches a state of equilibrium. Therefore, the strain of the sub-specimens on the inside undrained side under cyclic loading becomes less than that under static loading because their effective stresses cannot reach the consolidation pressure value. However, the strain of the sub-specimen on the drainage side under cyclic loading reaches the same value as the final strain under static loading because its effective stress does reach the consolidation pressure value.

The relationships between the average strain for all sub-specimens and the elapsed time in cases where the magnitude of the cyclic loads were different are shown in Fig. 10. In spite of the difference in magnitude of the cyclic load, the strain of the RT samples is larger than that of the HT samples. This is due to the fact that the void ratios of the RT samples at the start of cyclic loading were larger than those of the HT samples as mentioned above. In those tests, the cyclic consolidation stage is carried out normally a consolidated region because the specimens have already been consolidated to the reconsolidation pressure of 98.0kPa. However, when the

cyclic load is small, there is no clear end to primary consolidation and the consolidation curve is similar to that of an overconsolidated region under static loading, as pointed out in general (e.g. JGS, 2000).

3 NUMERICAL ANALYSIS

3.1 Elasto-viscous consolidation theory

Having summarized the previous studies on compression and consolidation of clayey soils and interpreted the available information from a rheological stand point, Yoshikuni et al. (1994) proposed a model for non-linear elasto-viscous liquids. The rheological equation of the proposed model is:

$$-\frac{de}{dt} = m_v \frac{d\sigma'}{dt} + \frac{\sigma'}{\eta} \quad (1)$$

where e is the void ratio, σ' is the effective stress, m_v is the coefficient of volume compressibility in the overconsolidated region, and η is the coefficient of viscosity. m_v and η are given by

$$m_v = 0.434 C_\gamma \frac{1}{\sigma'} \quad (2)$$

$$\log \eta = \frac{e_0 - e}{C_\alpha} + \frac{C_\beta - C_\alpha}{C_\alpha} \log\left(\frac{\sigma'_0}{\sigma'}\right) + \log \eta_0 \quad (3)$$

$$C_\alpha = \frac{-de}{d \log t}, \quad C_\beta = \frac{-de}{d \log \sigma'}, \quad C_\gamma = \frac{-de}{d \log \sigma'} \quad (4)$$

where C_α is the gradient of creep lines on the $e \sim \log t$ plane, C_β is the gradient of equi-creep rate lines on the $e \sim \log \sigma'$ plane and C_γ is the gradient of unloading-reloading lines on the $e \sim \log \sigma'$ plane. Although the values of C_α, C_β and C_γ have to be determined by using the irrecoverable change in the void ratio, i.e., the viscous component Δe^v, they are virtually equal to the coefficient of secondary compression, C_α, the compression index, C_c, and the swelling index, C_s, which are defined by the total change in the void ratio, respectively.

The one-dimensional consolidation equation based on the proposed elasto-viscous model is given by

$$\frac{\partial u}{\partial t} = \frac{(1+e_0)k}{\gamma_w m_v} \frac{\partial^2 u}{\partial z^2} + \frac{(1+e_0)}{\gamma_w m_v} \frac{\partial k}{\partial z} \frac{\partial u}{\partial z} + \frac{dp}{dt} + \frac{\gamma' z + p - u}{m_v \eta} \quad (5)$$

where u is the excess pore water pressure, k is the coefficient of permeability and p is the applied consolidation pressure. The first and second terms in Eq. (5) are the drain terms and take a negative value when consolidation progresses, the third term is the surcharge term and is positive on loading and negative on unloading, while the fourth term is the relaxation of effective stress and always takes a positive value. The change in the excess pore water pressure in the consolidation process is interpreted as the resultant phenomena of the four terms.

3.2 Analytical results and discussion

The consolidation equation of Eq. (5) can be numerically solved by the finite difference method, in which time t differentiation is approximated by forward finite differences and depth z differentiation by central finite differences (Yoshikuni et al., 1995a).

A series of numerical analyses was carried out to discuss the applicability of the proposed elasto-viscous consolidation theory and to investigate the consolidation behavior of clay under cyclic loading. A clay specimen with a thickness of 10cm was divided into 50 elements with the nodal point of No.1 on the drainage side and No.51 on the undrained side. The time difference interval was set to satisfy the following condition and gradually increased with time:

$$\frac{k}{\gamma_w m_v} \frac{\Delta t}{\Delta z^2} < \frac{1}{6} \tag{6}$$

In the calculation, the coefficient of permeability, k, was changed as follows:

$$\log k = \log k_0 - \frac{e_0 - e}{C_k} \tag{7}$$

where C_k is the gradient of the coefficient of permeability lines on the $e \sim \log k$ plane. The consolidation pressure under cyclic loading was varied with time as follows:

$$p = \left(p_0 + \frac{\Delta p_{cyc}}{2}\right) + \frac{\Delta p_{cyc}}{2} sin\left(2\pi\frac{t}{T} + \frac{3\pi}{2}\right) \tag{8}$$

where p_0 is the preconsolidation pressure, Δp_{cyc} is the cyclic consolidation pressure added to the preconsolidation pressure, and T is the cyclic period. The consolidation parameters used in this analysis are listed in Table 1. These parameters were determined from a conventional consolidation test and a long term consolidation test (Yoshikuni et al., 1995b).

Figures 11, 12, 13 and 14 show the comparisons of the analytical results with the test results for the RT sample under a cyclic consolidation pressure of 98.0kPa. Comparing the average strain for all sub-specimens versus the elapsed time relationships shown in Fig. 11, the average strain in the analysis is smaller than that in the test. Viewed in detail, although there is a tendency for the strain of the sub-specimens on the inside undrained side to be slightly smaller in the analysis than in the test, the value of the final strain in the analysis is almost equal to that in the test. The strain of the sub-specimen on the

Table 1. Consolidation parameters

	RT sample	HT sample
C_α	0.0122	0.0120
C_β	0.614	0.498
C_γ	0.117	0.128
C_k	0.324	0.324
e_0	1.65	1.56
η_0 ($kPa \cdot min/cm^2$)	1.95×10^7 (at $\sigma_0' = 98.0$kPa, $e_0 = 1.437$)	1.82×10^7 (at $\sigma_0' = 98.0$kPa, $e_0 = 1.423$)
$\dot{e}_0$ (1/min)	5.382×10^{-6} (at $e_0 = 1.65$)	5.035×10^{-6} (at $e_0 = 1.56$)
k_0 (cm/min)	5.13×10^{-8} (at $e_0 = 1.0$)	5.40×10^{-8} (at $e_0 = 1.0$)

drainage side is larger than that on the inside undrained side in the test, but is almost equal to that on the inside undrained side in the analysis. Therefore, it can be concluded that the reason why the average strain in the analysis is different from that in the test is due to the fact that the accumulation of strain of the sub-specimen on the drainage side cannot be calculated by the analysis. In the cyclic consolidation analysis, the elasto-viscous model is needed to express the cyclic effects which cause more accumulation of strain than under static loading. The rate of consolidation, including the secondary consolidation stage, in the analysis agrees with that in the test, and the applicability of the elasto-viscous model is considered to be satisfactory.

In the dissipation behavior of the excess pore water pressure shown in Fig. 12, the pressure in the analysis varies cyclically, having the same amplitude as the cyclic load, i.e., the excess pore water pressure varies from 0kPa to 98.0kPa at the start of loading and the center of the cycle decreases with time while keeping the same amplitude. In the later stages of consolidation, it varies from −49.0kPa to 49.0kPa, as 0kPa is the center of the cycle, and reaches a state of equilibrium. The dissipation of the excess pore water pressure is faster as the distance to the drainage boundary decreases, but there is no difference in the final behavior of the excess pore water pressure on the inside of the specimen. On the other hand, although the behavior of the excess pore water pressure in the test is practically similar to that in the analysis, the magnitude of the amplitude in the test is smaller than that in the analysis. The excess pore water pressure in the test does not reach the value of the cyclic load of 98.0kPa in the first cycle and gradually increases to 98.0kPa after 40~50min. The magnitude of the amplitude of the excess pore water pressure should coincide with that of the cyclic load

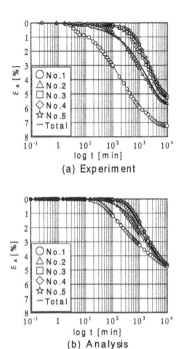

(a) Experiment

(b) Analysis

Figure 11. Strain ~ time curves for RT sample
under cyclic loading (Δp_{cyc}=98.0kPa)

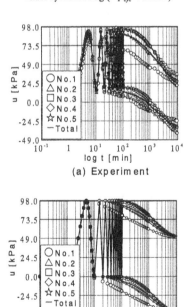

(a) Experiment

(b) Analysis

Figure 12. Pore water pressure ~ time curves for RT sample
under cyclic loading (Δp_{cyc}=98.0kPa)

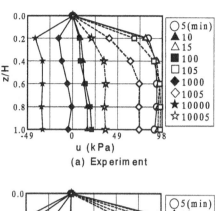

(a) Experiment

(b) Analysis

Figure 13. Isochrones of pore water pressure for RT sample
under cyclic loading (Δp_{cyc}=98.0kPa)

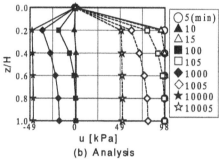

(a) Experiment

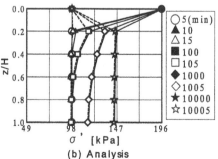

(b) Analysis

Figure 14. Isochrones of effective stress for RT sample
under cyclic loading (Δp_{cyc}=98.0kPa)

because, as mentioned above, there is no time to dissipate the excess pore water pressure on the undrained side within one cycle. It can be considered from these results that the reason why the magnitude of the amplitude of the excess pore water pressure in the test is smaller than that of the cyclic load is due to the time lag in the measurement system of the pore water pressure.

In the analysis shown in Fig. 13, the excess pore water pressure on loading is dissipated from the drainage side, the negative value of the excess pore water pressure on unloading increases, and their distributions on loading and unloading become symmetric with respect to the zero axis in the later stages of consolidation. The final values of the excess pore water pressure on loading are almost equal in the region of $z/H \geq 0.2$ and the excess pore water pressure remains at about half the cyclic load on loading in the later stages of consolidation.

It was also found from Fig. 14 that the effective stress in the analysis hardly varies during one cycle and gradually increases with time from the drainage side. In the later stages of consolidation, the values of the effective stress in the analysis are almost equal in the region of $z/H \geq 0.2$ and increase by about half of the cyclic load. On the other hand, the effective stress in the test varies at all measured points during one cycle as shown in Fig. 14(a). However, because the variation in the strain of each sub-specimen corresponding to the variation in the effective stress is not measured, this variation in the effective stress is considered to be an apparent variation caused by the time lag in the measurement system of the pore water pressure as mentioned above.

4 CONCLUSIONS

The main conclusions drawn from this study can be summarized as follows:

1) The excess pore water pressure on the inside of the specimen under cyclic loading was not fully dissipated and remained at about half the applied cyclic load even in the later stages of consolidation. Therefore, because the effective stress cannot reach the value of the applied consolidation pressure, the strain of the clay samples reconsolidated at both high and room temperatures under cyclic loading became smaller than under static loading.

2) The strain of the HT sample is smaller than that of the RT sample under both static and cyclic loading. This was considered to be due to the difference in the void ratios of the RT sample and the

HT sample after reconsolidation. In high temperature reconsolidation under large consolidation pressure, secondary consolidation was accelerated and a strong dense structure was formed in the HT sample.

3) In the analysis based on the elasto-viscous model the calculated strain on the inside of the specimen under cyclic loading agreed with that measured in the test. However, the calculated strain on the drainage side of the specimen in the analysis was smaller than that measured in the test. In order to predict precisely the cyclic consolidation behavior, the elasto-viscous model is needed to express the cyclic effects which cause more accumulation of strain than under static loading.

4) The progress of consolidation is slower under cyclic loading than under static loading. The rate of consolidation under cyclic loading, including the secondary consolidation stage, in the analysis agreed with that in the test. The applicability of the elasto-viscous model for cyclic consolidation is considered to be satisfactory.

REFERENCES

Moriwaki, T., Saitoh, M. and Im, E.S. (1999). One-dimensional consolidation properties of a saturated clay under cyclic loading, Poster Session Proceedings of the 11th ARC on SMGE, pp.25-26.

Yoshikuni, H., Kusakabe, O., Hirao, T. and Ikegami, S. (1994). Elasto-viscous modelling of time-dependent behaviour of clay, Proc. of the 13th ICSMFE, Vol.1, pp.417-420.

Yoshikuni, H., Kusakabe, O., Okada, M. and Tajima, S. (1995a). Mechanism of one-dimensional consolidation, Proc. of the International Symposium on Compression and Consolidation of Clayey Soils, Vol.1, pp.497-504.

Yoshikuni, H., Okada, M., Ikagami, S. and Hirao, T. (1995b). One-dimensional consolidation analysis based on an elasto-viscous liquid model, Proc. of the International Symposium on Compression and Consolidation of Clayey Soils, Vol.1, pp.505-512.

Moriwaki, T., Yoshikuni, H., Nagai, O. and Nago, M. (1995). Cyclic consolidation characteristics of clay reconsolidated at high temperature, model, Proc. of the International Symposium on Compression and Consolidation of Clayey Soils, Vol.1, pp.135-142.

Japanese Geotechnical Society (2000). Test method for one-dimensional consolidation properties of soils using incremental loading, Method and explanation for laboratory soil testing, pp.348-388.

Advanced Laboratory Stress-Strain Testing of Geomaterials, Tatsuoka, Shibuya & Kuwano (eds),
© *2001 Taylor & Francis, ISBN 90 2651 843 9*

Effects of stress path on the flow rule of sand in triaxial compression

H.Nawir, R.Kuwano & F.Tatsuoka
Department of Civil Engineering, University of Tokyo, Japan

ABSTRACT: To investigate the viscous effects on the flow characteristics of sand, several creep tests were performed on Toyoura sand at a selected stress state during otherwise monotonic anisotropic triaxial compression at a constant stress ratio (i.e., R= σ_v'/σ_h'= 3.8) and monotonic shearing along several stress paths with positive rates of stress ratio. The direction of irreversible strain increment vector was noticeably different between anisotropic compression and shearing. When the creep deformation started following anisotropic compression, the direction of irreversible strain increment started rotating towards the one during monotonic shear loading. Such a noticeable rotation as above did not take place as a creep test commenced following monotonic shear loading. The test results suggest that the so-call plastic potential concept, which gives a unique direction of plastic strain increment vector that is independent of stress path and strain rate, is not relevant to describe the actual flow characteristics of sand. It is argued that such behaviour as observed in the present study is a result of an complicated interaction between two types of yielding (shear and compression), which may be subjected to different viscous effects.

1 INTRODUCTION

Time effects on stress-strain behaviour of geomaterials consist of mainly two components; ageing (changes of material properties with time) and viscous effects (loading rate effects, creep and stress relaxation). Noticeable viscous effects on the stress-strain behaviour of sand were observed in plane strain and triaxial compression tests (e.g., Murayama et al. 1984; Lade et al. 1998; Matsushita et al. 1999; Tatsuoka et al. 2000a, 2001). This factor plays an important role in modelling the behavior of sand. A new promising three-component model (i.e., the TESRA model) was proposed (e.g., Tatsuoka et al. 2000b, 2001; Di Benedetto et al. 2000). However, this model in the present version can simulate only the shear deformation characteristics along a fixed stress path (i.e., a constant confining pressure) and without referring to the flow rule (i.e., the characteristics of irreversible strain increment vector).

When constructing a constitutive model for the stress-strain behaviour of sand under general stress conditions, the following characteristics should be defined: 1) the basic structure how to decompose the stresses and strains into respective components; 2) the yielding characteristics with a yield function; 3) the flow rule; and 4) the hardening rule, describing the development of stress due to irreversible straining. With respect to the issue 1), it is herein assumed that a given total strain increment is decomposed into elastic and irreversible (or inelastic) components. The irreversible strain component is due to the visco-plastic property of

sand, and an irreversible strain increment cannot be linearly decomposed into plastic and viscous components (Tatsuoka et al. 2000a,b).

Within the framework of elasto-plasticity theory, Poorooshasb et al (1966) proposed to assume the direction of plastic strain increment vector of sand at a given stress point to be unique irrespective of instantaneous stress path direction (Figure 1). When it is the case, a plastic potential, which is a unique function of stress state, can be defined. However, Tatsuoka and Ishihara (1974) reported results from triaxial compression tests with loading and unloading along different stress paths (i.e., constant shear stress, constant axial stress, constant p', constant confining pressure, constant stress ratio and so on) that showed that the direction of plastic strain increment vector is usually not unique at a given stress state, but depends on stress path. This test result also suggested that there exist two types of yielding; shear yielding and compression yielding.

Double hardening models were proposed for sand by Lade (1977), Vermeer (1978), Molenkamp (1980), and Tatsuoka and Molenkamp (1983). According to such models, two different types of yield loci and corresponding two types of plastic potentials are defined to obtain; 1) plastic shear strain increments and associated plastic volumetric increment (i.e., dilatancy characteristics) due to shear yielding; 2) plastic volumetric strain increment due to compression yielding. It should be noted, however, that none of the previous double hardening models deals with viscous properties of the stress-strain behaviour of sand.

In order to get a deeper insight into the flow rule of sand and the viscous effects on it, a set of several triaxial compression tests were performed, in particular

1. to reconfirm whether the direction of irreversible strain increment vector is not unique at a given stress for different stress paths, and whether the irreversible strain potential does not exist; and
2. to find how the flow rule changes, or does not change, during creep at a fixed state.

The relevancy of double yielding theory will be argued based on test results.

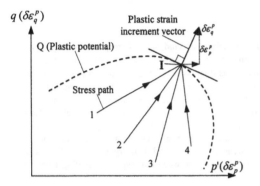

Figure 1. Illustration of plastic potential (or irreversible strain potential).

2 TEST EQUIPMENT AND PROCEDURES

2.1 Specimen preparation

Rectangular specimens, 18.2 cm in height and 11.1 cm x 11.1 cm in cross-section (Figures 2 and 3), were prepared to obtain accurate axial and lateral strains by local measurements along the lateral surface (as described later), which are free from effects of bedding errors at the top and bottom ends of specimen and those of membrane penetration at the lateral surface of specimen. This type of specimen was first used with sand by Hoque et al. (1996) and then with sedimentary softrock by Hayano et al. (1997). Medium dense specimens of Toyoura sand were prepared (Table 1), expecting more time-dependent deformation to take place than with dense specimens. Air-dried particles of Toyoura sand were pluviated through fixed in total four sieves with an opening of 1.0 mm. The specimen density was adjusted by controlling the rate of a flow from twelve small nozzles located above the multiple sieves. No other compaction efforts were employed. Void ratios between 0.691 and 0.724 were obtained. Enlarged ends having polished surfaces of stainless steel were used. The end surfaces were lubricated by using a 0.3 mm-thick latex rubber disc smeared with a 50 μm-thick layer of Dow Corning high-vacuum silicon grease (Tatsuoka et al. 1984).

Under a suction of 30 kPa, the specimen was made saturated by the dry setting method, consisting of vacuuming, flushing with de-aired water and back pressurizing (Ampadu and Tatsuoka, 1993).

2.2 Loading and pressure system

The vertical load was applied at a constant axial strain rate to the triaxial specimen by using a precise displacement-control gear system (Tatsuoka et al. 1994; Santucci de Magistris et al. 1999). By using the loading system together with an electrical pneumatic transducer for the cell pressure control, loading along a wide variety of stress path could be achieved in an automated way. With the use of a servo-motor to drive a gear system, the axial strain rate could be instantaneously varied within a range of 300 times, from 0.003%/min up to 0.9%/min. The maximum cell pressure that was controlled was 700 kPa. Creep tests could also be performed by this system.

2.3 Deformation measurement system

A pair of 12 cm-long vertical and three pairs of 8 cm-long lateral LDTs (local deformation transducers) (Goto et al. 1991; Hoque et al. 1997) were used to locally measure axial and lateral deformations respectively (Figs. 2&3). The largest strain measured with the vertical and lateral LDTs was about 6 %. The lateral LDTs were set on a pair of lateral surfaces in parallel assuming homogenous deformation of specimen in the horizontal planes in a small strain pre-peak range studied in the present study. Each pair of lateral LDTs was set at the mid-height of each one third of height (see Figs. 2&3).

2.4 Stress paths

The following four stress paths, as shown in Table 1 and Figure 4, were employed, with and without a creep stage for five hours at a common stress state (point I). The nominal axial strain rate (controlled based on external axial strains) was $\dot{\varepsilon}_v = 0.057$ %/minute $(= \dot{\varepsilon}_0)$ for all the

Figure 2. Triaxial specimen under a suction of 30 kPa, instrumented with vertical and lateral LDTs.

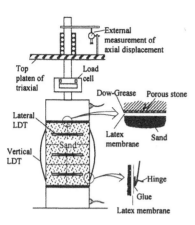

Figure 3. Triaxial compression testing system (not to scale).

stress paths. Stress path SP-1: The specimen was isotropically compressed from the common initial stress state at $\sigma_v' = \sigma_h' = 30$ kPa (point A) to 400 kPa (point B), followed by shearing at a constant effective lateral stress $\sigma_h' = 400$ kPa. Stress path SP-2: From point A, the specimen was sheared up to $\sigma_v' = 114$ kPa and $\sigma_h' = 30$ kPa (point D) and then was anisotropically compressed at a constant stress ratio R= $\sigma_v'/\sigma_h' = 3.8$ up to point I. Finally the specimen was sheared to the ultimate conditions at $\sigma_h' = 400$ kPa. Stress path SP-3: From an isotropic stress state at $\sigma_v' = \sigma_h' = 600$ kPa (point C), the specimen was sheared at a stress increment ratio of dp'/dq= 0.15 to q= 450 kPa (point E). Then, the q value was maintained constant until point F, from which the specimen was sheared at p' = 773 kPa until point I. After creep at point I, loading was restarted to the ultimate state (point J). The stress path during each test was controlled without taking into account changes in the cross-sectional area to achieve stable stress state control for a long duration. So, as approaching to the ultimate state, the actual stress path deviated to some extent from the intended one. Stress path SP-4: From point G at q = 850 kPa, the specimen was loaded at a constant q (= 850 kPa) to point H and then loaded at a constant dp'/dq (=-0.47) until point I. After this over-

consolidation stress history and creep loading at point I, the specimen was sheared at a constant σ_h' (= 400 kPa) to the ultimate state (point K). In these tests, several creep tests were performed at several stress states (each for five hours) as listed in Table 1, in addition to the one at the common point I.

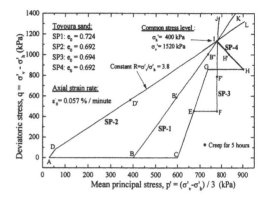

Figure 4. Stress paths (measured) that were followed to investigate the viscous deformation characteristics.

3 TEST RESULTS

3.1 Locally measured strains

Figure 5 shows the relationships between the average of the vertical strains locally measured with a pair of vertical LDTs and the lateral strain measured with each lateral LDT from a monotonic loading test along stress path SP-1a (without any creep). The average lateral strain is also shown. The following trends of behaviour may be seen:

1) The scatter in the lateral strain measured with three pairs of lateral LDTs is not significant at small strains, showing rather uniform deformation of specimen without significant end restraint effects.
2) During isotropic compression, the average lateral strain is slightly larger than the average axial strain, showing some degree of inherent anisotropy.

Table 1. List of tests

Test name (stress path)	Initial void ratio, e_0 (at point A)	Stress path before reaching the common stress state (I)	Stress path after I	Stress points for creep
SP-1	0.724	ABB'B"	K	B', B" and I
SP-1a	0.709	AB	K	-
SP-2	0.692	ADD'	K	D' and I
SP-2a	0.707	AD	L	-
SP-3	0.694	ABCEFF'	J	E, F, F' and I
SP-4	0.695	ABCEGHH'	K	H, H' and I

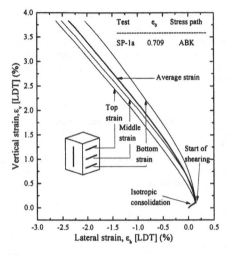

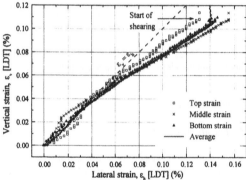

Figure 5. Drained triaxial compression test along stress path SP-1a (without creep) on saturated Toyoura sand; relationships between ε_v and ε_h; a) overall; and b) during isotropic consolidation.

3.2 Stress-strain characteristics

The relationships between the deviator stress q and the total and irreversible shear strains from all the tests with a single or multiple creep stages are shown in Figures 6a & 6b. The irreversible shear strains were obtained as follows:

$$\gamma^{ir} = \int d\gamma^{ir}; \qquad d\gamma^{ir} = d\gamma - d\gamma^e;$$

$$d\gamma^e = d\varepsilon_v^e - d\varepsilon_h^e;$$

$$d\varepsilon_v^e = \frac{1}{E_v} \cdot d\sigma_v - 2 \cdot \frac{v_{hv}}{E_h} \cdot d\sigma_h; \qquad (3.1)$$

$$d\varepsilon_h^e = \frac{1-v_{hh}}{E_h} \cdot d\sigma_h - \frac{v_{vh}}{E_v} \cdot d\sigma_v$$

where E_v, E_h, v_{hv}, v_{vh} and v_{hh} are the stress state-dependent elastic deformation parameters, each of which is a specific function of instantaneous stress state, according to the hypo-elastic model for sand (Hoque and Tatsuoka,1998). For example, $E_v = E_0 \cdot (\sigma'_v / \sigma'_o)^{m_v}$,

where E_0 is the value of E_v when $\sigma'_v = \sigma'_o$; and m_v is the constant (equal to 0.49 for Toyoura sand).

The following trends of behaviour may be seen from this figure:
1) As the stress state approached the failure state, the creep deformation became more obvious; and
2) Immediately after loading was restarted at the original axial strain rate following a creep stage, the q ~ γ relationship exhibited nearly elastic behavior for a noticeable stress range until a clear yield point was reached. Then, the q ~ γ behavior rejoined the original relation that would have been obtained by continuous loading without an intermission.

3.3 Direction of irreversible strain increment vector

Figure 7 compares the total and irreversible strain paths from continuous loading tests without any intermission for a creep test along stress paths SP-1 and SP-2. It is apparent that the directions of both total and irreversible strain increment vectors at the same stress state (i.e., point I) are utterly different for the two stress paths. This result indicates that the irreversible strain increment vector could not be unique at a given stress state, but it could largely depend on the stress path currently followed. Consequently, it can be concluded that the irreversible strain potential does not exist.

Figures 8 and 9 compare the total and irreversible strain paths from all the tests with a creep test at point I (and at other points) performed along the four stress paths. The difference in the direction of total strain increment vector between stress path SP-2 and the other stress paths is larger than that in the direction of irreversible strain increment vector. This is due to different amounts of elastic strain increment involved in the total strain increment for the different stress paths. The following trends of behaviour can be seen:
1) Confirming the results presented in Figure 7, the direction of irreversible strain increment during continuous loading at the same stress state was not unique. However, when compared with the one along stress path SP-2, the direction was not very different among stress paths SP-1, SP-3 and SP-4, where the increasing rate of dq/dp' was positive immediately before reaching the stress point I.
2) After the start of creep stage following continous loading at a constant strain rate along stress path SP-1, SP-3 and SP-4, the direction of irreversible strain increment vector rotated only slightly. The rotation was also small when loading was restarted at the original strain rate along either of stress paths SP-1 or SP-3.
3) On the otherr hand, the direction of irreversible strain increment vector rotated largely when a creep stage started following continuous loading along SP-2.
However, the rotation was very small when loading at the original constant axial strain rate was started along stress

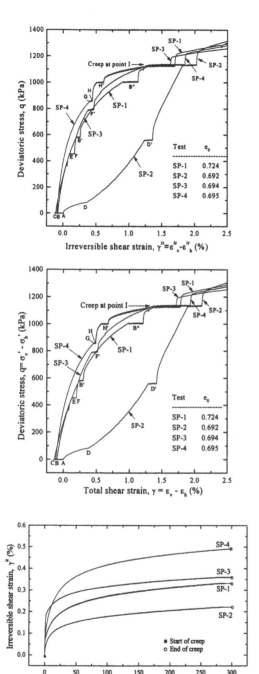

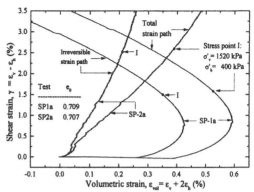

Figure 7. Total and irreversible strain paths during continuous loading at a constant axial strain rate along stress paths SP-1 & SP-2.

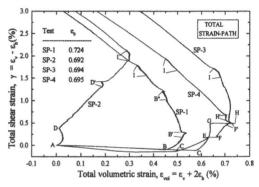

Figure 8. Total strain paths from tests with creep stages.

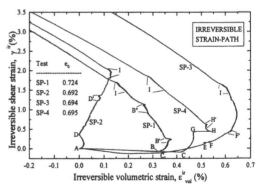

Figure 9. Irreversible strain paths from tests with creep stages.

Figure 6. Drained triaxial compression tests along different stress paths with creep on saturated Toyoura sand; relationships a) between q and γ; b) between q and γ^{ir}; and c) details of creep behaviour at stress point I ($\sigma_h' = 400$ kPa and $\sigma_v' = 1,520$ kPa).

path IK, which is the same as the one along stress path SP-1.

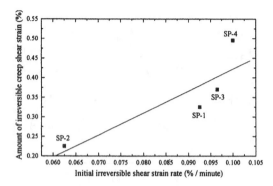

Figure 10. Relationship between the initial irreversible shear strain rate and the total amount of irreversible creep for a period of five hours.

3.2 Amount of creep shear strain

Figure 6c compares the time histories of shear strain at the creep stage at the same stress state (point I). Figure 10 shows the relationship between the irreversible creep shear strain and the initial shear strain rate at the start of creep stage. The largest creep shear strain was obtained for stress path SP-4, where p' decreased while q increased until point I and the smallest value was obtained for stress path SP-2, where both p' and q increased at the lowest rate of dq/dp'(i.e.=0). The magnitude of creep shear strain for stress paths SP-1 and SP-3 was between the two above. This phenomenon indicates that for the same creep period at the same stress state, the amount of creep shear strain apparently depends on the previous stress path before the start of creep. It is known, however, that for the same stress path before the start of creep stage, the amount of creep shear strain increases with the increase in the initial shear strain rate at the start of creep, which should be essentially the same with the shear strain rate during monotonic loading immediately before arriving at the stress state for the creep test (Tatsuoka et al. 2000b).

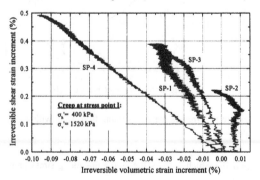

Figure 11. Summary of creep deformation at the same stress state (point I).

On the other hand, for the same axial strain rate during continuous loading, the initial irreversible shear strain rate at the start of creep stage was larger in the order of stress paths SP-4, SP-3, SP-1 and SP-2. Therefore, it is possible that the observed dependency of the magnitude of creep strain on the stress path before the creep stage was actually due to different initial irreversible shear strain rates at the start of creep. Further tests with largely different shear strain rates at the start of creep stage following continuous loading along the same stress path will be necessary to confirm whether such a relationship as above is independent of stress path for continuous loading.

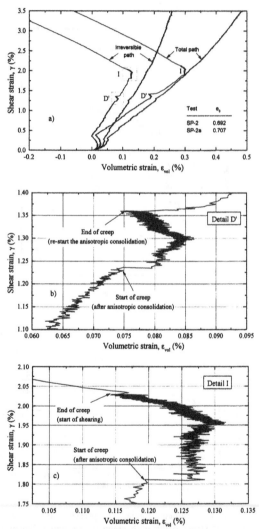

Figure 12. Details of irreversible strain paths before, during and after a creep stage during otherwise loading at a constant axial strain rate along stress path SP-2.

292

Table 2. Summary of possible double yielding concept

	K-consolidation at a constant $\dot{\varepsilon}_v$	Initial stage of creep	Later stage of creep	Shearing at a constant σ_h
$(d\varepsilon_{vol}^{e})_{comp}$	Positive (dominant)	0.0	0.0	Positive (not dominant)
$(d\varepsilon_{vol}^{ir})_{comp}$	Positive (dominant)	Existing (positive); diminishes at a relatively fast rate	May have stopped occurring (positive)	Positive (not dominant)
$(d\varepsilon_{vol}^{ir})_{shearing}$	Negative (masked)	Existing (negative); diminishes at a relatively slow rate	Could be still existing	Negative (dominant)
$(d\varepsilon_{vol})_{total}$	Positive	Positive, or a small negative value	Negative	Negative

4 TWO MECHANISMS OF YIELDING

Figure 11 summarizes the creep deformation at point I by re-plotting the data from the same origin at the start of creep. A large rotation in the direction of irreversible strain increment vector was observed only immediately after the loading mode was switched from monotonic loading at a constant axial strain rate at a constant stress ratio (stress path SP-2) to a creep test, as seen from Figure 9. From Figure 11 it can also be seen that as the creep strain increased, the effect of the previous stress path decreased and the direction of creep strain increment vector became rather unique.

Figure 12 shows details of strain path around and at two stress points D' and I along stress path SP-2. It may be seen from Figure 12b that soon after the loading was restarted at the original strain rate at a constant stress ratio after creep at stress point D', the direction of irreversible strain increment vector returned to the original one after exhibiting a large rotation again. That is, as anisotropic compression loading was resumed, the volumetric deformation property changed from a dilative one during the creep stage into a contractant one again.

The behaviour described again could be interpreted based on the double-yielding concept, as summarized in Table 2. In this interpretation, it is assumed that the viscous effects diminish faster with compression yielding than with shear yielding. It appears that compression yielding with positive irreversible volumetric strain is dominant only along stress path SP-2, in which the rate of increase in p' is particularly large. It could be understood based on this concept why the rotation of irreversible strain increment after creep started was smallest with stress path SP-4, where the component of compression yielding was unlikely to be included during monotonic loading.

5 CONCLUSIONS

With respect to the viscous properties of the stress-strain behavior of sand along general stress paths in triaxial compression, the following conclusions were obtained:

1. At least at high stress ratios, the irreversible strain potential does not exist; i.e., the direction of irreversible strain increment vector during continuous loading could strongly depend on stress path, in particular between different stress paths with and without a large increasing rate of q/p'
2. The direction of creep strain increment vector could be initially very different at the start of creep stage at a given stress state, affected by the stress path before the start of creep. However, it gradually becomes unique, becoming essentially the same as those by shear yielding.
3. It seems that the magnitude of creep shear strain and associated volumetric strain depends on the initial shear strain rate.
4. It seems that the yielding of sand consists of two types of yielding; shear yielding and compression yielding. It is necessary to define the shapes of yield locus and the hardening functions for each type of yielding. It will also be necessary to find whether an irreversible strain potential function can be defined for each type of yielding.

ACKNOWLEDGEMENTS

The authors are deeply grateful for the help provided by Mr. Ishihara, M. and Mr. Uchimura, T. in performing the tests. The experimental work presented here was performed at Geotechnical Engineering Laboratory, Civil Engineering Department, University of Tokyo, Japan.

REFERENCES

Ampadu,S.K. and Tatsuoka,F. (1993), "A hollow cylinder torsional simple shear apparatus capable of a wide range of shear strain measurement", *Geotechnical Testing Journal, ASTM*, Vol.16, No.1, pp.3-17.

Di Benedetto,H., Tatsuoka,F. and Ishihara,M. (2000): Time-dependent deformation characteristics of sand and their constitutive modelling, *Soils and Foundations* (submitted).

Hayano,K., Sato,T. and Tatsuoka,F. (1997), "Deformation characteristics of a sedimentary softrock from triaxial compression tests rectangular prism specimens", *Géotechnique*, Vol.47, No.3, Symposium In Print, pp.439-449.

Hoque,E., Tatsuoka,F. and Sato,T. (1996), "Measuring anisotropic elastic properties of sand using a large triaxial specimen", ASTM Geotechnical Testing Journal, ASTM, Vol.19, No.4, pp.411-420.

Hoque,E., Sato,T. and Tatsuoka,F. (1997), "Performance evaluation of LDTs for the use in triaxial tests", *Geotechnical Testing Journal, ASTM*, Vol.20, No.2, pp.149-167.

Hoque,E., and Tatsuoka,F. (1998), "Anisotropy in the elastic deformation of materials", Soils and Foundations, *Soils and Foundations*, Vol.38, No.1, pp.163-179.

Goto.S, Tatsuoka,F., Shibuya,S., Kim,Y.-S., and Sato,T.(1991), "A simple gauge for local small strain measurements in the laboratory", *Soils and Foundations*, Vol.31, No.1, pp.169-180.

Lade, P.V., and Liu, C.T. (1998). "Experimental study of drained creep behavior of sand", *Journal of Engineering Mechanics*, ASCE124(8): 912-920.

Matsushita,M. et al (1999). "Time effects on the pre-peak deformation properties of sands", Proc. Second Int. Conf. on Pre-Failure Deformation Characteristics of Geo-materials, IS Torino '99 (Lo Presti eds.), Balkema, Vol.1, pp.681-689.

Molenkamp, F. (1980), "Elasto-plastic double hardening model MONOT", *Delft Soil Mechanics Laboratory, Co.218595*.

Murayama,S., Michiro,K., and Sakagami,T. (1984). "Creep characteristics of sands", *Soils and Foundation*, Tokyo, Japan, 24(2), pp.1–15.

Poorooshasb, H.B., Holubec, I., and Sherbourne, A.N. (1966), "Yielding and flow of sand in triaxial compression: Part I", *Canadian Geotechnical Journal* 3(4), 179-90.

Poorooshasb, H.B., Holubec, I., and Sherbourne, A.N. (1967), "Yielding and flow of sand in triaxial compression: Part II and III", *Canadian Geotechnical Journal* 4(4), 376-97.

Santucci de Magistris,F., Koseki,J., Amaya,M., Hamaya,S., Sato,T. and Tatsuoka,F. (1999), "A triaxial testing system to evaluate stress-strain behaviour of soils for wide range of strain and strain rate", *Geotechnical Testing Journal, ASTM*, Vol.22, No.1, pp.44-60.

Tatsuoka, F., and Ishihara, K.(1974). "Yielding of sand in triaxial compression", *Soils and Foundation*, Tokyo, Japan, 14(2), 63-76.

Tatsuoka, F., and Molenkamp, F. (1983), "Discussion on yield loci for sands", *Mechanics of Granular Materials: New Models and Constitutive Relations*, Elsevier Science Publisher B.V., pp. 75-87.

Tatsuoka,F., Molenkamp,F., Torii,T., and Hino,T. (1984), "Behavior of lubrication layers of platens in element tests", *Soils and Foundations*, Vol.24, No.1, pp.113-128.

Tatsuoka,F., Sato,T., Park,C.-S., Kim,Y.-S., Mukabi,J.N. and Kohata,Y. (1994), "Measurements of elastic properties of geomaterials in laboratory compression tests", *Geotechnical Testing Journal, ASTM*, Vol.17, No.1, pp.80-94.

Tatsuoka,F., Santucci de Magistris,F., Hayano,K., Momoya,Y. and Koseki,J. (2000a): "Some new aspects of time effects on the stress-strain behaviour of stiff geomaterials", Keynote Lecture, *The Geotechnics of Hard Soils – Soft Rocks, Proc. of Second Int. Conf. on Hard Soils and Soft Rocks*, Napoli, 1998 (Evamgelista and Picarelli eds.), Balkema, Vol.3, pp.1285-1371.

Tatsuoka,F., Ishihara,M., Di Benedetto,H. and Kuwano,R., (2000b): Time-dependent deformation characteristics of geomaterials and their simulation, *Soils and Foundations* (submitted).

Tatsuoka,F., Uchimura,T., Hayano,K., Di Benedetto,H., Koseki,J. and Siddiquee,M.S.A. (2001); Time-dependent deformation characteristics of stiff geomaterials in engineering practice, the Theme Lecture, *Proc. of the Second International Conference on Pre-failure Deformation Characteristics of Geomaterials, Torino, 1999*, Balkema (Jamiolkowski et al., eds.), Vol. 2 (to appear).

Vermeer, P.A. (1978), "A double hardening model for sand", *Géotechnique*, Vol. 28, No.4, pp.413-433.

294

Advanced Laboratory Stress-Strain Testing of Geomaterials, Tatsuoka, Shibuya & Kuwano (eds),
© 2001 Taylor & Francis, ISBN 90 2651 843 9

Simulation of viscous effects on the stress-strain behaviour of a dense silty sand

F. Santucci de Magistris
Department of Geotechincal Engineering, University of Napoli "Federico II", Italy

F. Tatsuoka
Department of Civil Engineering, University of Tokyo, Japan

M. Ishihara
Public Work Research Institute, Ministry of Construction, Tokyo, Japan

ABSTRACT: The viscous aspect of the stress-strain behaviour of a compacted silty sand obtained from a special undrained triaxial compression test is analyzed. During otherwise strain-controlled monotonic loading, the strain rate was changed stepwise many times by about two orders of magnitude and several undrained creep tests were performed. The viscous response of specimen to such loading history depends on the geomaterial type. With this silty sand, at low strain levels, the stress is rather a unique function of instantaneous irreversible strain and its rate, while at larger strains, the viscous effect decays with strain, which is at a higher rate at larger strains. It is shown that among several three-component rheology models, the so-called general TESRA model is most relevant to simulate the observed stress-strain behaviour of the studied silty sand. The influence of the parameters controlling the model behaviour is analyzed.

1 INTRODUCTION

The time effect on the stress-strain behaviour of geomaterials is very important in predicting ground deformation and structural displacement in many geotechnical projects. For such predictions, it is necessary to be able to simulate the stress-strain-time behaviour of geomaterial subjected to various loading histories, including; a) sustained load (i.e., creep deformation); b) fixed displacement (i.e. stress relaxation); c) restarting of monotonic loading after ageing; and d) monotonic or cyclic loading at different strain rates. It seems that realistic and comprehensive constitutive models that can predict a wide variety of time-dependent stress-strain behaviour of geomaterials have not been developed. This may be because the general framework of the time-dependent deformation property of geomaterials has not been obtained, since in most of the previous experimental studies, monotonic loading tests at different constant strain rates, creep tests and relaxation tests were performed rather separately (Santucci de Magistris & Tatsuoka 1999; Matsushita et al. 1999).

In view of the above, the authors and their colleagues performed a comprehensive series of laboratory stress-strain-time tests on a wide variety of geomaterials (Tatsuoka et al. 1999a,b,c, 2000a,b, 2001). In these tests, various loading histories including a) - d) described above were applied systematically to each single specimen using an advanced triaxial apparatus (Tatsuoka et al. 1994; Santucci de Magistris et al. 1999). This apparatus is

reliable and robust, having a very high-level long-term stability, allowing performing a wide range of stress and strain paths.

In the present study, triaxial compression tests were performed on a particular type of Italian silty sand (Metramo silty sand) using such a triaxial apparatus (Santucci de Magistris et al. 1998, 1999; Santucci de Magistris & Tatsuoka 1999). In this paper, the test results are analyzed and simulated based on a set of non-linear three-component rheology models that were proposed recently (Di Benedetto et al. 2000; Tatsuoka et al. 2000a&b).

2 A BRIEF REVIEW OF RECENT RELATED STUDIES

Tatsuoka et al. (1999a&b, 2000a) showed that the stress-strain behaviour at strains less than about 0.001 %, for which the elastic Young's modulus E_0 or shear modulus G_0 is usually defined, could noticeably depend on the strain rate. It was found (cf. figure 4.1 of Tatsuoka et al. 2000a) that the strain rate-dependency of E_0 is a function of geomaterial type and, the dependency reduces as the stiffness increases and as the strain rate increases. This behaviour was interpreted on the base of the Elastic Limiting Line (Tatsuoka & Shibuya 1991), which is the upper bound of the stress-strain relationships at different strain rates. Therefore, when the strain rate is larger than a certain limit, the same initial stiffness could be obtained by all but different types of stress-strain tests performed at different strain rates under otherwise the same testing conditions. Di Benedetto & Tatsuoka (1997) showed that a simple linear three-component rheology model could simulate the observed strain-rate dependency of the stress-strain behaviour at these small strains.

Based on results from a comprehensive series of laboratory stress-strain-time tests, Tatsuoka et al. (2000a) pointed out the followings:

1) The time-dependent deformation behaviour of a wide variety of geomaterials observed in shear tests (e.g., triaxial compression tests) is similar in nature to that observed in oedometer tests on soft clay, sand and rock in that:
a) a high-stiffness zone develops in the stress space by ageing (i.e., drained creep) with the initial stiffness immediately after the restart of loading being close to the elastic modulus; and
b) when loading continues further at the original loading rate, the stress-strain state may overshoot the primary stress-strain curve that is obtained by continuous loading without an intermission of ageing. The overshooting may be temporary or persistent depending on the geomaterial type, perhaps due to different effects of densification, structuration, thixotropic hardening and cementation developed during the ageing stage.

2) With clays and sedimentary soft rocks, the stress-strain behaviour tends to be controlled uniquely by instantaneous strain rate, especially when far before the failure state. This property is called "the isotach property".

3) With poorly-graded sands, the stress-strain behaviour is essentially independent of instantaneous strain rate as far as the strain rate has been constant for some large strain range. Despite the above, immediately after the strain rate is increased and decreased stepwise, the stress-strain state temporarily overshoots and undershoots the stress-strain curve that is obtained by loading at a constant axial strain rate. In addition, a considerable amount of creep deformation and stress relaxation is observed. This property has been called the TESRA property (Temporary Effects of Strain Rate and Acceleration; Di Benedetto et al. 2000; Tatsuoka et al. 2000b).

Santucci de Magistris et al. (1999) and Santucci & Tatsuoka (1999) studied in detail the stress-strain-time property of Metramo silty sand. Metramo silty sand is a crushed granite from South Italy, having a maximum particle diameter $D_{max}= 2$ mm, a uniformity coefficient $Uc \approx 400$ and a clay fraction of 16 % with $w_L = 35.4$ % and $I_P = 13.7\%$. They found that the viscous property of Metramo silty sand is of hybrid nature, influenced by both those of clay and sand. That is, at relatively small strains, the stress is rather a unique function of instantaneous strain and strain rate (i.e., the isotach property) as with reconstituted soft clays. At relatively large strains, on the other hand, the TESRA property becomes dominant. Santucci de Magistris & Tatsuoka (1999), Di Benedetto et al. (2000) and Tatsuoka et al. (2000a&b) proposed a set of constitutive models to simulate a wide variety of viscous aspect of the stress-strain behaviour of geomaterials. The framework of the models is the general three-component model, which consists of an elastoplastic component EP1 that is connected in series to a set of another elastoplastic component EP2 and a non-linear viscous component V connected to each other in parallel. In the models that are referred to in this paper, the component EP1 has been simplified into a hypoelastic component and the component EP2 has been simplified into a purely inelastic part. Consequently, irreversible strains take place only in components EP2 and V.

3 CONSTITUTIVE MODELING

According to these three-component models, the total strain rate $\dot{\varepsilon}$ is decomposed into elastic component $\dot{\varepsilon}^e$ and irreversible components $\dot{\varepsilon}^{ir}$ as:

$$\dot{\varepsilon} = \dot{\varepsilon}^e + \dot{\varepsilon}^{ir} \tag{1}$$

The elastic component can be obtained by a relevant hypoelastic model (e.g. Tatsuoka & Kohata 1995; Tatsuoka et al. 1999a&b), such as:

$$\dot{\varepsilon}^e = \frac{\dot{\sigma}}{E^e(\sigma)} \tag{2}$$

where $E^e(\sigma)$ is the elastic Young's modulus, which is a function of instantaneous stress state (and others). Tatsuoka et al. (2000a) showed that a further linear decomposition of irreversible strain into plastic (time-independent and irreversible) and viscous (time-dependent and irreversible) components is not relevant for geomaterials, since they are not independent of each other. The total stress σ is decomposed into non-viscous and viscous stress components as:

$$\sigma = \sigma^f(\varepsilon^{ir}) + \sigma^v(\varepsilon^{ir}, \dot{\varepsilon}^{ir}, others) \tag{3}$$

The $\sigma^f - \varepsilon^{ir}$ relation, called the reference relationship, can be obtained by fitting the experimental data with any relevant non-linear model. The following empirical function was used by Di Benedetto et al. (2000) and Tatsuoka et al. (2000b) to simulate stress-strain relations with the tangent modulus consistently decreasing with strain throughout the test:

$$y = y_0 + A_1(1 - e^{-x/b_1}) + A_2(1 - e^{-x/b2}) \tag{4}$$

where y and x represent the stress and the total strain; and the value $A1+A2$ is the stress at an infinite strain. Irreversible strains can be obtained by Eqs. (2) and (4). However, Eq. (4) was not relevant to simulate the stress ratio $\eta = q/p'$-strain relationships in the present case, because the measured relations had a peak before the end of each test. In this case, the following more flexible one, a forth order polynomial function, was used:

$$y = y_0 + \sum_{i=1}^{8} A_i x^{i/2} \qquad (5)$$

The five parameters (y_0, A_1, t_1, A_2 & t_2) of Eq. (4) and the nine (y_0, A_1-A_8) of Eq. (5) could be obtained empirically by the least square fitting.

The viscous stress component σ^v is obtained in several different ways as shown below.

New isotach model: σ^v is a unique function of instantaneous irreversible strain and its rate, independent of the previous stress history. The following specific form (Tatsuoka et al. 2000a) was used in the present study:

$$\sigma^v(\varepsilon^{ir}, \dot{\varepsilon}^{ir}) = \sigma^f(\varepsilon^{ir}) \cdot g_v(\dot{\varepsilon}^{ir}) \qquad (6)$$

$$g_v(\dot{\varepsilon}^{ir}) = \alpha \cdot \left[1 - \exp\left\{1 - \left(\left|\dot{\varepsilon}^{ir}\right|/\dot{\varepsilon}_r^{ir} + 1\right)^m\right\}\right] \qquad (7)$$

where α, m and $\dot{\varepsilon}_r^{ir}$ are the positive material parameters, which could be determined by a trial and error procedure. The viscous function $g_v(\dot{\varepsilon}^{ir})$ is always positive, having the lower and upper bounds, zero and α. With this model, as far as $\dot{\varepsilon}^{ir}$ is positive, σ^v is always positive and $\sigma^f(\varepsilon^{ir})$ represents the lower bound for all possible stress-strain relationships represented by:

$$\sigma = \sigma^f(\varepsilon^{ir}) \cdot \left[1 + g_v(\dot{\varepsilon}^{ir})\right] \qquad (8)$$

As underlined before, this model is relevant to reconstituted soft clays, especially at states far before the failure condition.

TESRA model: The new isotach model is not able to simulate the temporary viscous effect that is typically observed with poorly-graded sand. The TESRA model was then proposed to simulate the temporary viscous property. With this model, the viscous stress component σ^v is obtained as:

$$\sigma^v = \int_{\tau=\varepsilon_1^{ir}}^{\varepsilon^{ir}} [d\sigma^v]_{(\tau,\varepsilon^{ir})} = \int_{\tau=\varepsilon_1^{ir}}^{\varepsilon^{ir}} \left[d\left\{\sigma^f \cdot g_v(\dot{\varepsilon}^{ir})\right\}\right] \cdot g_{decay}(\varepsilon^{ir} - \tau)$$

$$= \int_{\tau=\varepsilon_1^{ir}}^{\varepsilon^{ir}} \left[\left(\frac{\partial \sigma^f}{\partial \varepsilon^{ir}}\right) \cdot g_v(\dot{\varepsilon}^{ir}) + \sigma^f \cdot \left(\frac{\partial g_v(\dot{\varepsilon}^{ir})}{\partial \dot{\varepsilon}^{ir}}\right) \frac{\ddot{\varepsilon}^{ir}}{\dot{\varepsilon}^{ir}}\right]_{(\tau)} \cdot g_{decay}(\varepsilon^{ir} - \tau) \cdot d\tau$$

$$\qquad (9)$$

where $[d\sigma^v]_{(\tau,\varepsilon^{ir})}$ is the viscous stress increment generated when $\varepsilon^{ir} = \tau$; and $g_{decay}(\varepsilon^{ir} - \tau)$ is the decay function. Di Benedetto et al. (2000) proposed:

$$g_{decay}(\varepsilon^{ir} - \tau) = r_1^{(\varepsilon^{ir} - \tau)} \qquad (10)$$

where r_1 is the decay parameter that is constant ($0 \leq r_1 \leq 1$). When r_1 approaches to 1.0, the viscous stress does not decay and the model returns to the new isotach model. Similarly with the parameters of the viscosity function g_v, the decay parameter r_1 could be determined by a trial and error procedure.

General TESRA model: To properly simulate the overall stress-strain behaviour of Metramo silty sand (as well as some clays and some well-graded gravels), it is necessary to generalize the TESRA model by modifying Eq. (10) as follows (Santucci de Magistris & Tatsuoka 1999; Tatsuoka et al. 2000b):

$$g_{decay}(\varepsilon^{ir}, \varepsilon^{ir} - \tau) = r(\varepsilon^{ir})^{(\varepsilon^{ir} - \tau)} \qquad (11)$$

Unlike the TESRA model with $r = r_1$ or the new isotach model with $r = 1.0$, the decay parameter r is not constant but changes with the strain level as:

$$r(\varepsilon^{ir}) = r_1 \qquad\qquad \text{at } \varepsilon^{ir} = 0$$

$$r(\varepsilon^{ir}) = \frac{r_1 + r_f}{2} + \frac{r_1 - r_f}{2} \cdot \cos\left[\pi \cdot \left(\frac{\varepsilon^{ir}}{c}\right)^n\right]$$

$$\qquad\qquad\qquad \text{for } 0 \leq \varepsilon^{ir} \leq c \qquad (12)$$

$$r(\varepsilon^{ir}) = r_f \qquad\qquad \text{for } \varepsilon^{ir} \geq c$$

The positive parameters ($0 \leq r_f \leq 1$, n, c) could be determined by a trial and error procedure. When $r_f = 1.0$, the model returns to the new isotach model. The general TESRA model behaves as the TESRA model after ε^{ir} exceeds any chosen value c.

4 EXPERIMENTAL PROCEDURES AND APPARATUS

To examine whether either of the models presented above could be relevant to Metramo silty sand, several triaxial compression tests were performed. Cylindrical specimens (12.5 cm high and 5 cm diameter) were obtained by trimming a block of dense material prepared with the modified Proctor compaction energy and various initial water contents. Only the result from one particular test on a specimen prepared at an initial water content of $w = 7.4$ %, corresponding to an initial dry unit weight of 19.24 kN/m³, will be analyzed in the following. It has been confirmed that the conclusions obtained with this test are also relevant to those from all the others performed under similar conditions. After made saturated in the triaxial cell, the specimen was isotropically consolidated at a constant stress rate up to an effective confining pressure $\sigma_c' = 392.4$ kPa and aged for 27 hours. The specimen was then subjected to undrained TC loading. The axial strain rate

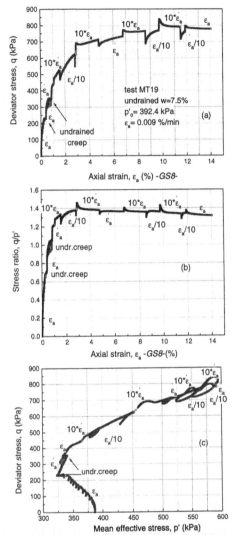

Figure 1. Undrained triaxial compression test on Metramo silty sand: (a) stress-strain relationship; (b) stress ratio-strain relationship; and (c) effective stress path.

nal axial strain throughout each test. This record was directly used to obtain the time history of axial strain rate at large strains where the readings of the sensitive transducers (LDTs and 2 mm-gap sensors) had become out of range.

Figures 1a & b show the relationship between the deviator stress q or the stress ratio q/p' and the axial strain ε_a from the axial displacement of the specimen cap measured with a proximity transducer. The effects of bedding error, which could be significant on the small strain stiffness, were ignored in the present analysis dealing with the behaviour at large strains. The effective stress path is presented in Fig. 1c. The following trends of behaviour may be seen:

1) Upon the restart of loading at the original strain rate following each creep stage, the material shows a very stiff response. Clear yielding is then exhibited before rejoining the original primary stress-strain curve that would have been obtained if the loading had been continuous.

2) When the axial strain rate is suddenly increased, the stress-strain relationship and the effective stress path exhibit a jump in the slope; i.e. the tangent Young's modulus increases suddenly to a value close to the instantaneous elastic stiffness, as observed upon the restart of loading following a period of creep. Then, the specimen exhibits a clear yielding, followed by a tendency to rejoin the primary stress-strain relationship. When the axial strain rate is decreased suddenly, the opposite behaviour is observed.

3) At small strains, the primary stress-strain relationship is apparently different for different constant strain rates, but it becomes similar for different strain rates at larger strains.

4) It appears that at the initial stage of loading, the stress-strain relationship tends to be controlled by the instantaneous strain rate. At the later stage, the relationship tends to become independent of the instantaneous strain rate as far as the strain rate has been constant for some strain range.

These results show that the general TESRA model is most relevant to this test result.

5 ANALYSIS OF THE DEVIATOR STRESS AND AXIAL STRAIN RELATION

First, the reference curve was carefully determined noting that its choice is essential for a better simulation by any of the models. Then, the elastic parameters were determined. To be consistent with the analyzed data, the elastic properties were detected on the base of externally measured axial strains. These procedures are described in Santucci de Magistris & Tatsuoka (1999). A parametric study was then performed to evaluate the influence of the controlling variables on the $q - \varepsilon_a$ relation for the respective model.

was increased or decreased stepwise by one order of magnitude with respect to the reference strain rate ($\dot{\varepsilon}_0 = 0.009$ %/min) during monotonic loading towards the failure state. Therefore, the overall range of strain rate was two orders of magnitude. Two undrained creep tests were executed at q= 231 kPa for 2540 minutes and q= 351 kPa for 1360 minutes. With the triaxial test system employed, the relationship between the input voltage into an AC servo-motor and the applied strain rate is highly linear (Santucci de Magistris et al. 1999). Therefore, the record of the input voltage, which was free from the scattering in the measured axial strains, was taken into advantage to evaluate the time history of exter-

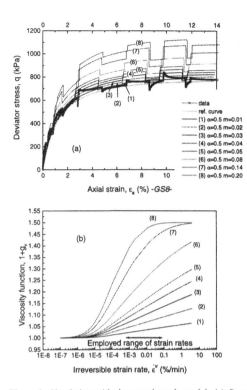

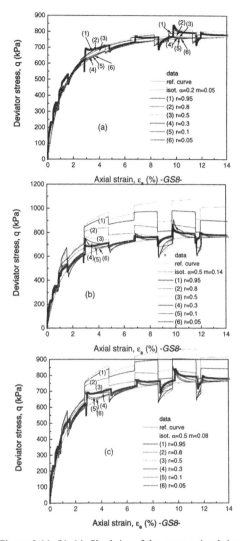

Figure 2. Simulation with the new isotach model: (a) Stress-strain relationship and, (b) viscosity function versus the irreversible strain rate.

Simulation by the new isotach model: The most sensitive parameter for the viscous function is m in the present case, because the values of α and $\dot{\varepsilon}^{ir}$ have effects only when the strain rate becomes extremely large or small, which is essentially out of the range encountered in the present case (see Figure 2b). Therefore, fixed values $\dot{\varepsilon}^{ir} = 0.0001$ %/sec and $\alpha = 0.5$ were used, while the parameter m was changed from 0.01 to 0.20. In Figure 2a, the simulation is compared with the experimental result. The following trends of behaviour may be seen:

1) As the parameter m increases, the increasing rate of σ^v with the increase in $\dot{\varepsilon}^{ir}$ becomes larger. This is particularly the case at larger strains.

2) Simulations with any value of m cannot be comparable with the measured overall stress-strain behaviour. That is, when the parameter m is selected to best fit the initial part of the experimental result, the simulated stress value largely overestimates the measured values at large strains. Conversely, if the parameter m is chosen to match the experimental result at large strains, the model largely underestimates the effects of $\dot{\varepsilon}^{ir}$ on σ^v at small strains. Therefore, it cannot properly simulate the creep behaviour, seriously underestimating the creep strain.

After all, the new isotach model is not relevant to

Figures 3 (a), (b), (c). Simulation of the stress-strain relationship using the TESRA model.

simulate the overshooting and undershooting behaviour observed in the experiment.

Simulation by the TESRA model: The parameter r (Eq. 12) controls the shape of the decay function; i.e., as r becomes smaller from 1.0, the decreasing rate of σ^v becomes larger, and vice versa. The simulation was made by changing the parameter r between 0.05 and 0.95 with relevant values of α for the viscous function (Figures 3a, b & c). The simulation using $r = 1.0$ (i.e. the new isotach model) forms the upper bound of those with smaller values of r. The following trends of behaviour may be noted:

1) The TESRA model can simulate some of the ma-

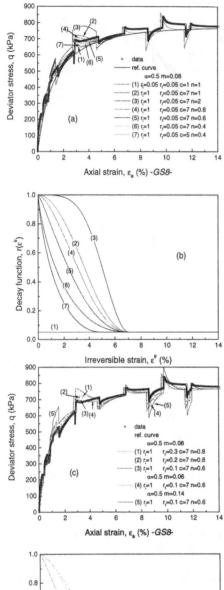

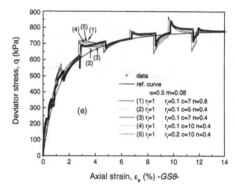

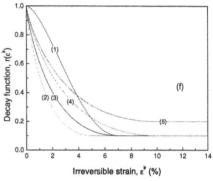

Figure 4. Simulation using the general TESRA model: (a), (c), (e) stress-strain relationship, and (b), (d), (f) variation of the decay function with the irreversible strain rate.

jor features of the experimental data when proper values of the parameters are used.

2) The TESRA model with $\alpha = 0.2$ and $m = 0.05$ (Figure 3a) underestimates the effect of the change in $\dot{\varepsilon}^{ir}$ on σ^{v} for any value of r. For this reason, creep strains are largely underestimated.

3) The TESRA model with $\alpha = 0.5$ and $m = 0.14$ (Figure 3b) simulates better the creep deformation. In this case, however, the model generally overestimates the amount of overshooting and undershooting of stress at large strains for any value of r. When r is equal or lower than 0.5, however, the shape of the simulated stress-strain curve is quite close to the experimental result at large strains.

4) The TESRA model with $\alpha = 0.5$ and $m = 0.08$ (Figure 3c) simulates much better the experimental result when using a proper value of r, while creep deformation is only slightly underestimated.

Even with the best combination of the values of the parameters, the TESRA model cannot simulate the measured overall stress-strain behaviour with the same level of accuracy from the start of loading to large axial strains, say 14 %.

Simulation by the general TESRA model: As in the previous case, the two major parameters controlling the viscosity function were fixed as $\alpha=0.5$ and $m=0.08$, except where otherwise indicated. The results of the simulation are shown in Figure 4. The following trends of behaviour can be seen:

1) Effects of the parameters c and n of the decay function were examined fixing r_i and r_f, except one case (Figure 4a). The simulation with $r_i=0.05$ means a TESRA simulation. The strain limit c, after which the decay function became constant and equal to r_f, was set to 7 %, except for one case. Apparently, $c=5$ % is too small.

2) Figure 4b shows how the shape of the decay function changes by the changes in the parameter n with $c=7$ %. As n decreases, the rate of decay in σ^v becomes faster, making the stress-strain relation rejoin faster the reference stress-strain relationship after a step change in the strain rate.

3) The effects of the two parameters r_f and n are examined in Figures 4c and d.
   a) As the parameter r_f increases, σ^v decreases at a smaller rate, particularly after ε^{ir} becomes larger than c. It seems that $r_f=0.3$ is too large.
   b) Fixing the values of the parameters of the decay function to the proper values, the effects of the parameters of the viscosity function were re-examined (Figure 4c). It was found that by using somehow different value of m (= 0.14), the creep strain could be better simulated, but the simulation of the overall strain-strain curve became worse.

4) With the fixed values, $\alpha=0.5$ and $m=0.08$, the parameters of the decay function were re-evaluated to find the best fitting (Figures 4e and f). It seems that the combination ($\alpha=0.5$, $m=0.08$, $r_i=1.0$, $r_f=0.2$, $c=10$ and $n=0.4$) is best, as shown in Figure 5.

In summary, only the general TESRA model with suitable values of the parameters can simulate properly the observed overall stress-strain behaviour of Metramo silty sand.

6 ANALYSIS OF THE STRESS RATIO Q/P' AND AXIAL STRAIN RELATIONSHIP

Based on the above, the observed q/p' vs. ε_a relationship was simulated only by the general TESRA model, as follows:

1) The reference curve was determined by a trial and error procedure.

2) The elastic properties were determined referring to the initial part of the q/p' vs. ε_a experimental curve, again based on externally measured axial strains.

3) A try and error procedure was applied to find the best values of the parameters for the viscous and the decay functions, as shown in Figure 6. With some inaccuracy allowed, a combination ($\alpha=0.4$,

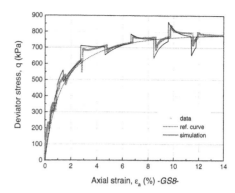

Figure 5. Best simulation with general TESRA model using $\alpha=0.5$, $m=0.08$, $r_i=1.0$, $r_f=0.2$, $c=10$, and $n=0.4$.

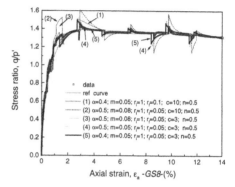

Figure 6. Simulation of the stress ratio- axial strain using the general TESRA model.

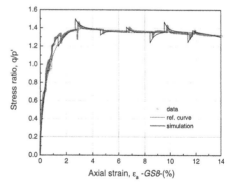

Figure 7. Best simulation with general TESRA model using $\alpha=0.4$, $m=0.05$, $r_1=1.0$, $r_2=0.05$, $c=3$, and $n=0.5$.

$m=0.05$, $r_i=1.0$, $r_f=0.05$, $c=3$ and $n=0.5$) was found most relevant (see Figure 7).

It is worth to underline that the relevant values of the parameters are utterly different between the q vs. ε_a and q/p' vs. ε_a relation. Clearly, the latter shows the TESRA behaviour from a smaller strain.

7 CONCLUSIONS

A compacted silty sand exhibited the following peculiar viscous behaviour in undrained triaxial compression tests including undrained creep stages:
1. Immediately after loading was restarted following a period of creep, the material showed a very stiff and nearly elastic response, followed by a clear yielding.
2. Immediately after the strain rate was increased stepwise, a very stiff and nearly elastic response was observed, followed by a clear yielding. The opposite behaviour was observed upon a step decrease in the strain rate.
3. At small strain levels, the stress-strain relationship tended to be controlled by the instantaneous strain rate, while at the later stages, the stress-strain relationship tended to be independent of instantaneous strain rate as far as the strain rate had been kept constant for some strain range.
4. The above-mentioned behaviour could be simulated by the general TESRA model with proper values of the parameters. According to this model, the viscous stress decays with the irreversible strain and the decay rate increases with the irreversible strain. More study will be necessary to find the relevant structure of this model.

ACKNOWLEDGMENT

Dr. Santucci de Magistris was invited at the University of Tokyo by the Japan Society for the Promotion of Science under "the JSPS Postdoctoral Fellowship Program for Foreign Researchers". The support received was highly appreciated.
Authors wish to thanks staff and graduate students for the assistance received in performing experimental tests.

REFERENCES

Di Benedetto, H. & Tatsuoka, F. 1997. Small strain behaviour of geomaterials: modelling of strain effects. *Soils and Foundations* 37(2):127-138.
Di Benedetto, H., Tatsuoka, F. & Ishihara, M. 2000. Time-dependent deformation characteristics of sand and their constitutive modelling, *Soils and Foundations* (submitted).
Matsushita, M., Tatsuoka, F., Koseki, J., Cazacliu, B., Di Benedetto, H. & Yasin, S.J.M. 1999. Time effects on the pre-peak deformation properties of sands. *Proc. Second Int. Conf. on Pre-Failure Deformation Characteristics of Geomaterials, IS Torino '99* (Jamiolkowski et al., eds.), 1: 681-689. Rotterdam: Balkema.
Santucci de Magistris, F., Sato, T., Koseki, J. & Tatsuoka, F. 1998. Effects of strain rate and ageing on small strain behaviour of a compact silty sand. *Proc. of Second Int. Conf. on Hard Soils and Soft Rocks* (Evangelista & Picarelli eds.), 1:843-851. Rotterdam: Balkema.
Santucci de Magistris, F., Koseki, J., Amaya, M., Hamaya, S., Sato, T. & Tatsuoka, F. 1999. A triaxial testing system to evaluate stress-strain behaviour of soils for wide range of strain and strain rate. *Geotechnical Testing Journal, ASTM* 22(1):44-60.
Santucci de Magistris, F. & Tatsuoka, F. 1999. Time effects on the stress-strain behaviour of Metramo silty sand. *Proc. Second Int. Conf. on Pre-Failure Deformation Characteristics of Geomaterials, IS Torino '99* (Jamiolkowski et al. eds.), 1:491-555. Rotterdam: Balkema.
Tatsuoka, F. & Shibuya S. 1992. Deformation characteristics of soils and rocks from field and laboratory tests. *Keynote Lecture for Session No.1, Proc. of the 9th Asian Regional Conf. on SMFE, Bangkok,* 1991, 2: 101-170.
Tatsuoka, F., Sato, T., Park, C.-S., Kim, Y.-S., Mukabi, J.N. & Kohata, Y. 1994. Measurements of elastic properties of geomaterials in laboratory compression tests. *Geotechnical Testing Journal ASTM* 17(1):80-94.
Tatsuoka, F. & Kohata Y. 1995. Stiffness of hard soils and soft rocks in engineering applications. *Keynote Lecture, Proc. of Int. Symp. on Pre-Failure Deformation of Geomaterials, IS-Hokkaido* (Shibuya et al., eds.) 2: 947-1063. Rotterdam: Balkema.
Tatsuoka, F., Modoni, G., Jiang, G.L., Anh Dan, L.Q., Flora, A., Matsushita, M. & Koseki, J. 1999a. Stress-strain behaviour at small strains of unbound granular materials and its laboratory tests. *Keynote Lecture, Proc. of Workshop on Modelling and Advanced testing for Unbound Granular Materials, January, 1999, Lisboa* (Correia ed.), 17-61. Rotterdam: Balkema.
Tatsuoka, F., Jardine, R.J., Lo Presti, D.C.F., Di Benedetto, H. & Kodaka, T. 1999b. Characterising the Pre-Failure Deformation Properties of Geomaterials. *Theme Lecture for the Plenary Session No.1, Proc. of XIV IC on SMFE, Hamburg, September 1997.* 4:2129-2164. Rotterdam: Balkema.
Tatsuoka, F., Santucci de Magistris, F., Momoya, M. & Maruyama, N. 1999c. Isotach behaviour of geomaterials and its modelling. *Proc. Second Int. Conf. on Pre-Failure Deformation Characteristics of Geomaterials, IS Torino '99.* (Jamiolkowski et al., eds.), 1:491-499. Rotterdam: Balkema.
Tatsuoka, F., Santucci de Magistris, F., Hayano, K., Momoya, Y. & Koseki, J. 2000a. Some new aspects of time effects on the stress-strain behaviour of stiff geomaterials. *Keynote Lecture, Proc. of Second Int. Conf. on Hard Soils and Soft Rocks, Napoli, 1998* (Evangelista and Picarelli eds.). 3:1285-1371. Rotterdam: Balkema.
Tatsuoka, F., Di Benedetto, H., Ishihara, M. & Kuwano, R. 2000b. Time-dependent deformation characteristics of geomaterials and their simulation, *Soils and Foundations* (submitted).
Tatsuoka, F., Uchimura, T., Hayano, K., Di Benedetto, H., Koseki, J. & Siddiquee, M.S.A. 2001. Time-dependent deformation characteristics of stiff geomaterials in engineering practice. *Theme Lecture, Proc. of the Second International Conference on Pre-failure Deformation Characteristics of Geomaterials, Torino, 1999* (Jamiolkowski et al. eds.), Vol. 2 (in press) Rotterdam: Balkema.

Advanced Laboratory Stress-Strain Testing of Geomaterials, Tatsuoka, Shibuya & Kuwano (eds),
© *2001 Taylor & Francis, ISBN 90 2651 843 9*

Interrelationship between the metastability index and the undrained shear strength of six clays

S.Shibuya & T.Mitachi
Hokkaido University, Sapporo, Japan

M.Temma
Geo-Research Institute, Osaka, Japan

T.Kawaguchi
Hakodate National College of Technology, Hakodate, Japan

ABSTRACT: The behaviour of natural sedimentary clay is different from the comparative behaviour of a reconstituted sample due primarily to in-situ ageing. In this paper, the metastability of Holocene clays from six nations was examined by performing a series of undrained triaxial tests on each pair of natural and reconstituted samples. The undrained shear strength c_u of the source ground was obtained by shearing each natural sample subjected to anisotropic consolidation with in-situ effective overburden pressure. The c_u value of the natural samples reflecting ageing effects was examined with respect to the metastability index, $MI(G)$, proposed by Shibuya (2000). A method for distinguishing the primary source of clay ageing; i.e., secondary compression or interparticle bonding, is also proposed.

1 INTRODUCTION

Natural sedimentary clays become stiffer and stronger as they age. In this paper, the term "ageing effects" applies to any changes in the mechanical properties of natural sedimentary clay since deposition. For example, an ageing effect is evident in the behaviour of the undrained shear strength, c_u. The ratio of the c_u to the consolidation stress σ_{vc} under one-dimensional compression is not unique for any given clay, i.e. the c_u/σ_{vc} value differs between natural and reconstituted samples, and also between two comparative laboratory samples subjected to recompression at and far beyond the in-situ effective overburden stress, $\sigma'_{v(in-situ)}$. As σ_{vc} increases, c_u/σ_{vc} of the natural sample subjected to undrained shear under specific conditions approaches a lower bound defined by the behaviour of the reconstituted sample (see, for example, Hanzawa, 1989).

Soil ageing involves 'structuration' induced by long-term creep (i.e., secondary compression) and/or formation of interparticle bonding in fabrics. However, the designation of a soil micro-structure to an explanatory mechanism is seemingly qualitative, implying that this kind of examination is unlikely to provide much in predicting the kinds of different mechanical behaviour which reflect the numerous micro-structures and fabrics of a given clay. Alternatively, Mitchell (1993) has proposed a measure termed the metastability index, MI, by which clay structuration may be assessed in a quantitative manner. The reliability of the original MI, which makes use of the 'yielding' characteristics of clay in an oe-

dometer test, suffers from the fact that the yield stress is greatly influenced by the rate of compression as well as by sample disturbance (Shibuya et al, 2001). Accordingly, Shibuya (2000) has recently proposed an alternative measure $MI(G)$, defined by considering the difference in density between non-aged reconstituted and aged natural samples with a common elastic shear modulus, G.

In this paper, the ageing effects on the c_u are intensively examined in the laboratory through international geotechnical site investigation, and the results are discussed within the framework of the metastability concept.

2 CLAY SAMPLES TESTED

Table 1 shows the basic engineering properties of the clay samples at seven sites in six nations. It should be mentioned that all the samples originate from Holocene sediments that have never been subjected to geological overconsolidation. In addition, the sampling depths from the ground surface ranged from 7m to 14m, above which the seasonal fluctuation of the effective overburden pressure is significant. Accordingly, the over-consolidation ratio, OCR, examined in oedometer tests was close to unity for all the samples. Note also that the plasticity index spans a wide range, from low-plastic Drammen clay to extremely high-plastic Bangkok clay at Nong Ngu Hao (NNH). More details of sampling sites in Ariake (JPN), Bangkok (Suttisan), Bangkok (NNH), Bothkennar (UK), Drammen (Norway),

Louiseville (Canada) and Pusan (Korea) have been described in the literature by Tanaka et al (1996), Shibuya and Tamrakar (1999), Balasubramaniam et al (1996), Hight et al (1992), Lunne at al (1976), Lefebvre et al (1994) and Shibuya et al (2001), respectively.

All the samples were retrieved from each source ground by using a fixed-piston thin wall sampler, implying high-quality in terms of both strength and compressibility (Tanaka et al, 1996). In addition, the natural samples tested were each recovered from the middle part of the sampling tube. In addition, care was taken to minimize potential sources of sample disturbance during transportation and trimming. Accordingly, the disturbance of these natural samples was probably the least in routine practice, suggesting that the in-situ structure seems well-preserved .

The natural samples were each 'reconstituted' in the laboratory by one-dimensional reconsolidation of slurry with an initial water content of twice the liquid limit to about 100 kPa of the vertical preconsolidation pressure over a period of ten days. The mechanical behaviour of the reconstituted samples, therefore, appears relatively free from ageing effects.

3 TRIAXIAL TEST

A fully-digitized triaxial apparatus (Shibuya and Mitachi, 1997) was employed for obtaining the stress-strain and strength of the natural and reconstituted samples. In this paper, the symbol G_{max} applies exclusively to the elastic shear modulus of laboratory samples obtained at very small strains of about 0.001%. It should be mentioned that the G_{max} is independent of the shearing rate as well as the type of loading. As long as a quality sample is tested in the laboratory under conditions similar to those in-situ, the G_{max} value of the laboratory sample coincides closely with the comparative G_f from the in-situ seismic survey (e.g., Tatsuoka and Shibuya, 1992, Shibuya et al, 1992).

A cylindrical specimen with dimensions of 5cm in diameter and 10cm in length was recompressed in a triaxial cell by using a constant axial stressing rate of 0.2 kPa/min. During the recompression to $\sigma'_{v(in-situ)}$, quasi-K_0 conditions were imposed on each sample by keeping the radial strain within ±0.02%. At certain stages in the course of recompression, the undrained elastic Young's modulus, E_{max}, was measured directly from the cyclic stress-strain curve obtained using a cyclic strain amplitude of about 0.001%. The G_{max} value was then estimated using the following equation based on assuming the isotropic/elastic properties with the undrained Poisson's ratio, ν_u, to be 0.5;

$$G_{max}=E_{max}/2(1+\nu_u)=E_{max}/3 \qquad (1)$$

Taking of ν_u to be 0.5 seems reasonable since the pore pressure parameter, B, when measured prior to undrained shearing, was in excess of 0.96 for all the samples (Toki et al, 1994).

Each sample was subjected to undrained shear by using an axial straining rate of 0.1%/min until the axial strain, ε_a, reached 15%.

4 METASTABILITY INDEX, $MI(G)$

The framework of critical state soil mechanics employs the principle that the effective stress of a given saturated clay, exhibiting no sign of fabric bonding, is uniquely determined by its water content. Mikasa (1964) independently modified this by incorporating an additional factor of post-depositional structuration. Recently, it has been shown that Mikasa's premise is also applicable to G_{max} (Shibuya, 2000), therefore:

$\sigma' = F_1$ (w, post-depositional structure) , and

$$G_{max} = F_2(w, post-depositional\ structure) \qquad (2)$$

Since the effective stress as well as the G_{max} of a non-aged reconstituted clay are both a monotonous and smooth function of the current water content, the water content of the non-aged reconstituted clay is expressed as:

$$w = \Phi_1(\sigma') , \text{ and } w = \Phi_2(G_{max}) \qquad (3)$$

The above-equation means that the G_{max} value of a reconstituted clay is uniquely determined by w, which is in turn uniquely related to σ'.

In comparing the mechanical behaviour of different clays worldwide, it is convenient to introduce the liquidity index, I_L, defined as follows:

$$I_L= (w_n-w_p) / (w_L - w_p) \qquad (4)$$

in which w_n, w_L and w_p denote the in-situ water content, and the water contents at the Atterberg limits, respectively. Note that the I_L for clays is by definition similar to the relative density for sands. By using the I_L, the behaviour of clays from different origins showing a wide spread of intrinsic properties is compared on a common cooking table (refer to Mitchell, 1993).

The concept of metastability, together with the metastability index, $MI(G)^{IL}$, applicable to the behaviour of normally consolidated clay can be conveniently explained the using illustrations shown in Fig.1. In this figure, the (I_L, σ'_v, G) states of the natural sample in-situ and in the laboratory are denoted by points I and L, respectively. The α^*-α^* line corresponds to the normal compression (NC) curve of a non-aged reconstituted sample subjected to one-

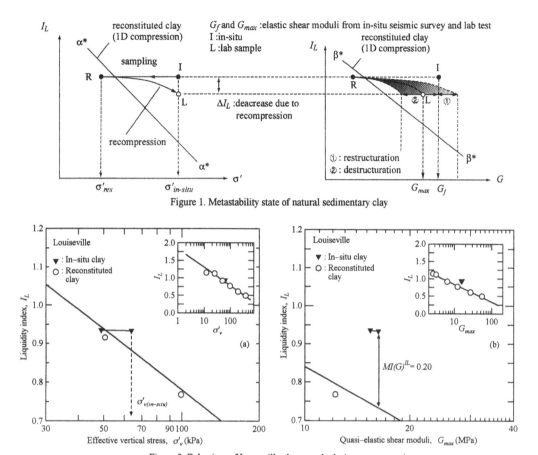

Figure 1. Metastability state of natural sedimentary clay

Figure 2. Behaviour of Louseville clay sample during recompression

dimensional compression. Similarly, the curve β^*-β^* represents the corresponding G_{max}. As shown later on, the in-situ state of the natural NC samples (i.e., point I) stays unwaveringly above the α^*-α^* and β^*-β^* lines The metastability is described by referring to this characteristic interrelationship between natural (aged) and reconstituted (non-aged) samples. The in-situ state at point I is metastable, since it migrates towards the lower bounds, i.e., the α^*-α^* and β^*-β^* lines, as σ'_v gradually increases beyond $\sigma'_{v(in\text{-}situ)}$ (Shibuya, 2000).

Alternation in water content is unlikely to occur with decent sampling in saturated clay ground, however, the sample is under residual effective stress, σ'_{res}, associated with the release of in-situ total stress (i.e., the path from I to R). With one-dimensional recompression at $\sigma'_{v(in\text{-}situ)}$, the state of ($I_L$, σ'_v, G_{max}) denoted by point L may not rejoin point I. However, it may be possible to minimize the divergence between these two points in careful testing on a high-quality sample.

When the β^*-β^* line is available, the $MI(G)^{IL}$ value of NC clay may be defined using the following

equation as the difference in the I_L between natural and reconstituted samples at the relevant G:

$$MI(G)^{IL} = (I_L - I_L^{\beta^*\text{-}\beta^*})_G \qquad (5)$$

The usefulness of $MI(G)$ for assessing in-situ structuration/destructuration has been successfully demonstrated by Shibuya (2000).

5 TEST RESULTS

Some of the engineering properties are summarized in Table 1. Two examples of variations in the G_{max} (see Eq.1) and σ'_v values with I_L during recompression to $\sigma'_{vc}=\sigma'_{v(in\text{-}situ)}$ are shown in Figs.2 and 3, for Louiseville and Bangkok clays, respectively. The undrained effective stress paths during recompression and undrained shear are shown in Fig.4, where the variations in deviator stress, q, and mean effective stress, p', are in each test normalized by $\sigma'_{vc}(=\sigma'_{v(in\text{-}situ)})$. Fig.5 shows the relationship between q, and ε_a at the undrained shearing stage. The secant

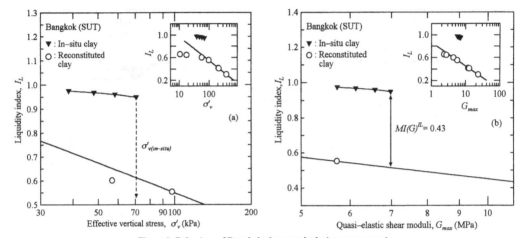

Figure 3. Behaviour of Bangkok clay sample during recompression

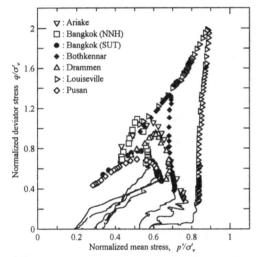

Figure 4. Effective stress paths of seven clays in triaxial test

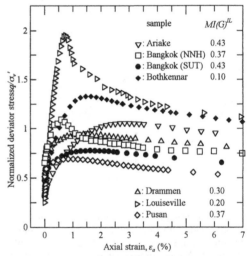

Figure 5. Stress-strain behaviour of seven clays in undrained triaxial test

Young's modulus E_{sec} when normalized by E_{max} is plotted against ε_a in Fig.6.

The following characteristics may be noted for their stress-strain and strength behaviour:

i) the alternation of the G_{max} against σ'_v during recompression to $\sigma'_{vc}=\sigma'_{v(in-situ)}$ was substantially more significant in the behaviour of Bangkok clay, whereas the G_{max} of Louiseville clay remained almost unchanged at a relatively larger value of about 15 MPa, as compared to the 7 MPa of Bangkok clay (see Figs.2 and 3),

ii) the effective stress paths of Louiseville and Bothkennar clays at an early stage of undrained shear exhibited a sharp rise to a peak (see Fig.4),

iii) the axial strain at peak conditions was remarkably small, less than 1% for Louiseville, Bothkennar and Bangkok (NNH) clays, and the stress-strain relationship showed a distinctive post-peak softening (see Fig.5),

iv) the rate of stiffness decay over the pre-peak region was exceptionally slow in the behaviour of Louiseville clay (see Fig.6).

6 DISCUSSION

A summary of the relationship between c_u σ_{vc} and $MI(G)^{IL}$ has been given in Table 1, and the results are plotted in Fig.7. The clays from six nations have

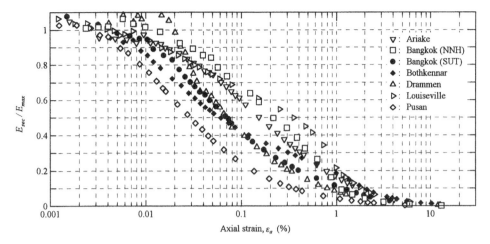

Figure 6. Secant Young's modulus against axial strain in undrained triaxial test

all been found to be in a metastable condition by showing that $MI(G)^{IL}$ is positive, bearing in mind that non-aged reconstituted clay exhibits a conditions where $MI(G)^{IL}=0$. The results appear to corroborate the ability of the $MI(G)^{IL}$ measure to quantify post-depositional ageing of natural clays (see Eq.2).

In addition, we have seen that these clays may be categorized into two groups. One group of clays (i.e., Louisville and Bothkennar clays) refers to those in which the effects of interparticle bondings are predominant in governing the stress-strain and strength behaviour. In the other group, the existence of interparticle bonding is not evident, so long-term creep induced during secondary compression may be an alternative prevailing factor in ageing in their mechanical behaviour. As also stated by Shibuya (2000), the development of bondings is strongly suggested in the behaviour of Louisville and Bothkennar clays, i.e., the independence of the G_{max} from σ'_v (see Fig.2), the sharp rise of the undrained effective stress path to a peak followed by considerable softening (see Fig.4), and the slow rate of stiffness decay over the pre-peak region (see Fig.6).

Apart from Louisville and Bothkennar clays, the $MI(G)^{IL}$ value was spread over a narrow range between 0.33 and 0.43. The c_u/σ_{vc} value also stayed over a narrow range between 0.34 and 0.53. Conversely, the $MI(G)^{IL}$ value of these cemented clays was relatively small, whereas the c_u/σ_{vc} value was distinctly large in triaxial compression. As suggested by Leroueil and Vaughan (1990), the inclusion of interparticle bonding seems to prevent volumetric strain from generating, and to exhibit brittle stress-strain behaviour. Conversely, the effects of secondary compression bring about increases in both stiffness at small strains and peak-strength at intermediate strains.

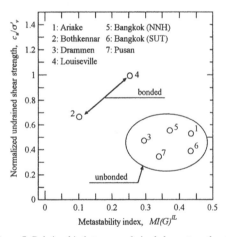

Figure 7. Relationship between undrained shear strength ratio and metastability index $MI(G)^{IL}$ of seven clays

7 CONCLUSIONS

1) A metastability index $MI(G)^{IL}$ capable of quantifying post-depositional ageing was successfully obtained by performing an undrained triaxial test on a pair of high-quality natural and reconstituted samples of soft clays worldwide.

2) Normally-consolidated Holocene clays from six nations were all considered metastable due to the metastability index $MI(G)^{IL}$ being positive in value.

3) The prevailing source of soil ageing (i.e., long-term creep or interparticle bonding) in a given natural sedimentary clay could be identified by observing the characteristics of the relationship between $MI(G)^{IL}$ and c_u/σ_{vc}. For the natural clays showing no sign of interparticle bonding, the $MI(G)^{IL}$ value was

spread over a narrow range of between 0.33 and 0.43. The c_u/σ_{vc} value also spread over a narrow range from 0.34 to 0.53. Conversely, the $MI(G)^{IL}$ value of the weakly bonded clays (i.e., Louiseville and Bothkennar clays) was relatively small, whereas the c_u/σ_{vc} value was much larger.

4) The inclusion of interparticle bonding seems to prevent volumetric stain from generating, and also to exhibit brittle stress-strain behaviour. Conversely, the effects of secondary compression bring about increases in both stiffness at small strains and peak-strength at intermediate strains.

ACKNOWLEDGEMENTS

The authors are grateful to the Geotechnical Investigation Laboratory at the Port and Harbour Research Institute headed by Dr H. Tanaka for their co-operation in soil sampling. The triaxial test on samples of Bangkok clay was carried out at the Asian Institute of Technology where the first author had warm encouragement from and stimulating discussion with Professors A. Balasubramaniam and D. T. Bergado.

REFERENCES

Balasubramaniam, A.S., Bergado, D.T., Long, P.V., Ashford, S. and Noppadol, P. (1996): Development of ground improvement techniques in the Bangkok plain, Proc. 12th SE Asian Geotech. Conf., Kuala Lumpur, Vol. 2, pp. 71-92.

Hight,D.W., Bond,A.J. and Legge,J.D. (1992): Characterization of the Bothkennar clay: an overview, Géotechnique, Vol.42, No.2, pp.303-347.

Lefebvre, G., Leboeuf, D., Rahhal, M.B., LaCroix, A., Warde, J. and Stokoe, K.H. (1994): Laboratory and field determinations of small-strain shear modulus for a structured Champlain clay, Canadian Geotechnical Journal, Vol.31, No.1, pp.61-70.

Leroueil, S. and Vaughan, P.R. (1990): The general and congruent effects of structure in natural soils and weak rocks, Geotechnique, Vol.40, No.3, pp.467-488.

Lunne, T.O., Eide, J. and de Ruiter, J. (1976): Correlations between cone resistance and vane shear strength in some Scandinavian soft to medium stiff clays, Canadian Geotechnical Journal, No.13-4, pp.430-441.

Mikasa, M. (1964): A classification chart for engineering properties of soils, Tsuchi-to-Kiso, JGS, Vol.12, No.4, pp.17-24 (in Japanese).

Mitchell, J.K. (1993): Fundamentals of soil behavior, Second Edition, John Wiley and Sons, New York.

Shibuya, S., Tatsuoka, F., Teachavorasinskun, S., Kong, X.J., Abe, F., Kim, Y.S. and Park, C.S. (1992): Elastic deformation properties of geomaterials, Soils and Foundations, Vol.32, No.3, pp.26-46.

Shibuya, S. and Mitachi, T. (1997): Development of a fully digitized triaxial apparatus for testing soils and soft rocks, Geotechnical Engineering, Vol.28, No.2, pp.183-207.

Shibuya, S. and Tamrakar, S.B.(1999): In-situ and laboratory investigations into engineering properties of Bangkok clay,

Characterization of Soft Marine Clays (Tsuchida, T. and Nakase A. edns), Balkema, pp.107-132.

Shibuya, S.(2000):Assessing structure of aged natural sedimentary clays, Soils and Foundations, Vol.40, No.3, pp.1-16.

Shibuya, S., Mitachi, T., Tanaka, H., Kawaguchi, T. and Lee, I.M. (2001): Measurement and application of quasi-elastic properties in geotechnical site chracterisation, Theme Lecture, Proc. of 11th Asian Regional Conference on SMGE, Seoul, Balkema, Vol.2 (in print).

Tanaka, H., Sharma, P., Tsuchida, T. and Tanaka, M. (1996): Comparative study on sample quality using several types of samplers, Soils and Foundations, Vol.36, No.2, pp.57-68.

Tatsuoka, F. and Shibuya, S. (1992): Deformation characteristics of soils and rocks from field and laboratory tests, Keynote Paper, Proc. of 9th ARC on SMFE, Vol.2, pp.101-170.

Toki, S., Shibuya, S. and Yamashita, S. (1994): Standardization of laboratory test methods to determine the cyclic deformation properties of geomaterials in Japan, Pre-failure Deformation of Geomaterials (Shibuya, S. et al edns), Balkema, Vol.2, pp.741-784.

Advanced Laboratory Stress-Strain Testing of Geomaterials, Tatsuoka, Shibuya & Kuwano (eds),
© 2001 Taylor & Francis, ISBN 90 2651 843 9

Case study on the practical use of elastic modulus in deformation analysis of soft clay ground

S.B. Tamrakar
Graduate School of Engineering, Hokkaido University, Japan

S. Shibuya
Graduate School of Engineering, Hokkaido University, Japan

T. Mitachi
Graduate School of Engineering, Hokkaido University, Japan

ABSTRACT: In this paper, the results of case studies in predicting deformation behavior of Bangkok (BKK) clay underneath a test embankment and in the course of deep excavation work are reported. Prior to performing elasto-plastic FE analysis by using a modified Cam clay model (MCC), the required engineering properties were all obtained from in-situ and laboratory tests, including the variation of clay stiffness with depth and with strain. Coupled consolidation analysis for embankment loading and short-term undrained analysis for the excavation work were carried out. The effects of the elastic deformation moduli, drainage boundary conditions (one-way and two-way drainage) and coefficient of permeability were each examined on the deformation behavior of Bangkok clay. Recommendations are made for the appropriate values of elastic Young's modulus and permeability in analyzing the deformation behavior of Bangkok clay.

1 INTRODUCTION

A fundamental difficulty involved in predicting the deformation behavior of soft clay ground due to construction work such as embankment and excavation works, is attributed to the complexity of the deformation characteristics of the soil in question, which vary with location and time. The elastic stiffness of the ground, for example, varies with the stress level, and also with the strain level. In a rigorous prediction, the variations in the elastic moduli with stress and strain ought to be properly accounted for in the form of a non-linear soil model. However, in a practical prediction using a packaged FE program incorporating the properties of isotropic linear elasticity, the variation in elastic stiffness with depth, and hence with stress, may be conveniently accounted for by dividing a single stratum into multiple layers, each having a fixed elastic modulus value. The strain-level dependency of the elastic modulus may also be considered by choosing a single value for the elastic modulus to match the induced ground strain level. However, such case histories are rare in the literature (Anderbrooke et al., 1997).

It is a common practice in South East Asia to determine the elastic stiffness on the basis of the profile of the undrained shear strength, S_u, with depth, using the empirical relation $E_u = \alpha S_u$ (S_u: undrained shear strength and E_u: undrained elastic Young's modulus). Earlier researchers recommended using $E_u = 70 \sim 250 S_{uFVS}$ (S_{uFVS}: uncorrected field vane shear strength) (Balasubramaniam and Brenner, 1981) and

$E' = 15 \sim 40 S_{uFVS}$ (Bergado et al., 1990) for the embankment analysis and $E_u = 200 \sim 500 S_u$ (Bowels, 1988) and $E_u = 280 \sim 350 S_u$ (soft), $E_u = 1200 \sim 1600 S_u$ (stiff) (Hock, 1997) for the excavation analysis, respectively. Note that E' stands for the drained elastic Young's modulus. Conversely, Simpson (2000) has recently suggested employing a specific value for E' equal to around half the elastic Young's modulus, E'_{max}, noting that E'_{max} stands for Young's modulus at strains of about 0.0001%. This E'_{max} value may be obtained either through an in-situ seismic survey or through an undrained triaxial test. Akino (1990), Tatsuoka et al. (1992) and Sugie et al. (1999) have also suggested using 0.3 to 0.7 times the value of E'_{max} for excavation works in Japan with retaining structures. But case histories in which the effects of elastic moduli are examined in an explicit manner have not yet been reported widely, in particular, those dealing with the deformation behavior of soft clay ground.

In this paper, two case histories concerning the deformation behavior of soft clay ground in BKK due to the loading of the test embankment and due to deep excavation with a retaining diaphragm wall are reported. Coupled consolidation analysis was carried out for the embankment, whereas undrained short-term analysis was performed on the excavation works. The study was mainly carried out to observe the effects of elastic stiffness (60 times $S_{uFVS(cor)}$ and $E'_{max}/2$) on the prediction of the deformation behavior of BKK clay.

In the case of the embankment analysis, a MCC model was employed in all cases and the effects of the drainage boundary conditions (one-way and two-way) and the coefficient of permeability ($k_{measured}$, $10 \times k_{measured}$, $50 \times k_{measured}$ and $100 \times k_{measured}$) were all examined. In the case of excavation analysis, however, the results of the elastic, MCC and CCM models were compared.

Finally, the predicted results from the embankment and excavation works were compared with the measured data in terms of on-soil/in-soil ground deformation and pore water pressure over time.

2 GEOTECHNICAL SITE INVESTIGATION

A comprehensive site investigation was carried out by performing both field and lab tests at the NNH site, which is 40 km south east of central BKK, and at the Sutthisan site, which is in central BKK, in December 1996 and in November 1997, respectively. Construction of a new international airport is planned at the NNH site. Test results obtained from site investigations at the Sutthisan site were applied to an excavation analysis of the R-station, which is located 3 km to the north of the Sutthisan station site. Both the Sutthisan site and the R-station belong to the northern section of the Metropolitan Rapid Transit Authority (MRTA) project of BKK. It was assumed that the profile as well as the properties of the subsoil in central BKK were similar over a narrow area. Hence, the properties obtained at the Sutthisan site were applied to the excavation analysis of the R-station.

Site investigations included the performance of both field (field vane shear, seismic cone and piezocone) tests and laboratory (1-D consolidation and undrained triaxial) tests. Laboratory tests were performed on high-quality samples retrieved using a fixed-piston thin wall sampler. The stiffness variation with strain over a wide strain range from the undrained triaxial test and consolidation parameters from the oedometer test were manifested in the laboratory tests, while the small-strain shear modulus from the seismic cone, the ground water profile from the piezocone and the undrained shear strength from the field vane tests were obtained from the in-situ tests.

2.1 Properties of the NNH Site and the Test Embankment

The soil profile, together with the basic properties of the subsoil at the NNH site are shown in Figure 1. The weathered clay serves as crust down to about 2 m in depth. So-called soft BKK clay extends from 2 m to 15.5 m in depth. Up to 11.5 m in depth, the clay is soft and some shells and organic matter are present. From 11.5 m to 15.5 m in depth, there is a transition in soil color with thin lenses of sand and/or silt. In this soft marine clay layer, natural water content, w_n, is almost close to the liquid limit, w_L, of about 100%.

The plastic limit, w_P, lies within a range of 30 to 40%. The sensitivity ratio of this soft clay varies from 2 to 6. A stiff clay layer extends from 15.5 m to 21 m in depth. A dense sand layer extends down from about 22 m in depth and consists of fine to medium sand.

The ground water level (GWL) measured in December 1996 for this site was 0.6 m below the ground surface. The piezometric pressure measured below about 6 m in depth was non-hydrostatic, and it was noted that it reduced equal to atmospheric pressure at the sand layer below 21 m in depth (see Shibuya et al., 1998).

The test embankment studied here was designated as Test embankment IV (AIT, 1973). The height of the test embankment at completion of construction was 2.9 m with dimensions of 100 m × 40 m, and a 10 m berm built on one side. A fine sand was used as the fill material and the average unit weight of this was taken as 17.4 kN/m^3.

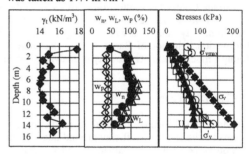

Figure 1. Basic soil properties at the NNH site.

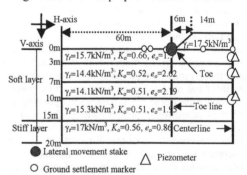

Figure 2. Instrumentation and soil layering (NNH).

A multi-stage embankment loading was carried out over a construction period of 56 days. Observation of the foundation behavior as well as that of the test embankment was made over a period of 200 days after the completion of the construction of the embankment. Instrumentations are shown in Figure 2. A total of five ground settlement markers (at 8 m, 15 m, 24 m and 27 m distance from the centerline of the embankment) and vertical settlement markers (at 2 m, 5 m and 10 m depths along the centerline of the embankment) were installed to measure the ground surface settlement and the vertical settlement along the centerline, respectively. One lateral stake was employed at the toe of the embankment to measure the lateral movement and four piezometers (at 2 m, 5 m, 10 m and 16 m depths from the ground surface) were installed along the centerline of the embankment to measure the total pore water pressure with time.

2.2 Soil Properties of the Sutthisan Station Site and Excavation at the R-station Site

The basic soil properties of the Sutthisan station site are shown in Figure 3. The subsoil stratigraphy consists of approximately 22 m thick layers of soft and stiff clay resting over the Bangkok aquifer. Weathered crust extends down to 4 m with a w_n varying from 47% to 54%. A soft clay layer extends from 4 m to 15 m. It contains shells at some depths. In this clay layer, w_P remains unchanged at around 25% and w_L varies from 60% to 100%. Stiff clay about 7 m thick lies below this soft clay layer, i.e., from 15 m to 22 m in depth. This layer has a w_P of around 25%, whereas its w_L decreases with depth down to 60%. Beyond 22 m in depth, there lies a dense sand layer. The w_n in the upper soft clay (i.e., from 5 m to 10 m in depth) was close to the w_L, whereas it decreased beyond 10 m in depth and approached the w_P. The sensitivity ratio of this soft clay ranged from 3 to 6.

The GWL, as measured on Nov. 1997, was 1 m below the ground surface. Due to excessive pumping of underground water from aquifers, a non-hydrostatic distribution of the pore water pressure was seen below 7 m in depth. At 22 m in depth, the measured pore water pressure was equal to the atmospheric pressure.

The perimeter of the R-station, i.e., a diaphragm wall (DW), was constructed by connecting the cast-in-situ reinforced concrete panels in series. A top-down bottom-up excavation method was adopted. The DW was 1 m thick and extended down to 39 m in depth. The external width of the total excavation was 25 m. The depth of the excavation reached 22.1 m in depth, i.e., 0.1 m deep into the sand layer. The length of this station box was 226.8 m. Here, the length of the excavation was far longer than its width.

The following are the sequences of the excavation simulated in the analysis:

1) Surcharge loading,
2) DW panel construction,
3) Surcharge unloading,
4) 1st stage excavation (excavate up to 1.8 m in depth),
5) Temporary strut installation (H-beam 30 cm × 30 cm),
6) 2nd stage excavation (excavate up to 4 m in depth)
7) Roof slab concreting (0.9 m in thickness)
8) Removal of temporary strut,
9) 3rd stage excavation (excavate up to 9.2 m in depth)
10) Retail slab concreting (0.7 m in thickness),
11) 4th stage excavation (excavate up to 14.4 m),
12) Concourse slab concreting (0.7 m in thickness),
13) 5th stage excavation (excavate up to 22.1 m in depth) and
14) Base slab concreting (1.75 m in thickness).

The excavation work for the installation of the DW was started in September 1998. The 2nd stage excavation was completed in December 1998. Finally, the base slab concreting was completed in October 1999.

The field instrumentations employed at the R-station are shown in Figure 4. Five surface settlement markers were installed for measuring the ground surface settlement behind the DW. Two inclinometers were installed, one at the DW panel to measure the horizontal movement of the DW and the other at a distance of 17 m from the centerline of the excavation towards the retained side to measure the ground movement in a horizontal direction. Both the inclinometers were extended down to 45 m in depth. One Vibrating Wire piezometer was installed 18.3 m from the centerline of the excavated portion behind the DW (5.3 m from the DW) to measure rapid change in total pore water pressure distribution with time.

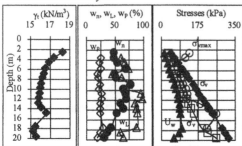

Figure 3. Basic properties of the Sutthisan site.

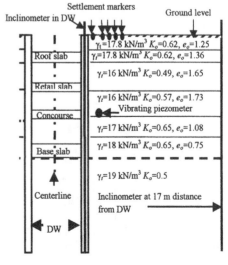

Figure. 4 Instrumentations and structures.

2.3 Soil Parameters from In-situ and Lab Tests

The properties of the sub-soil employed for the FE analysis of the NNH site and the R-station are shown in Figures 5 and 6, respectively. Common to both sites, the *OCR* value at the weathered crust is relatively high due to seasonal fluctuation of the GWL. In the soft clay layer, the *OCR* value was close to unity, suggesting a state of normal consolidation. In the stiff layer, the *OCR* value reached as high as 2.

The compression index, $\lambda(=0.434C_c)$, in the soft clay layer shows a tendency to increase gradually with depth. On the other hand, the swell index, κ $(=0.434C_s)$ remains more or less constant through depth. The λ value is smaller in the stiff clay with a value of about 0.1.

The S_u value from the FVS test with Bjerrum's correction, $S_{uFVS(cor)}$, is slightly lower than the S_u from the MTX test, S_{uMTX}. The G_{max} value from MTX test $(G_{max}=E_{max}/3)$ is lower than the comparable value from the seismic cone test, G_{SCPT}. Although the difference in the shear modulus is relatively small in the soft layer, it becomes more significant in the stiff layer. This difference may be due to sample disturbance and bedding errors, which occurred during triaxial testing. Accordingly, the G_{SCPT} value, which is free of any disturbance effects, was employed in the FE analysis as a basic shear modulus at very small strains. The drained Young's modulus is estimated using $E'=2G(1+\nu')$, where ν' stands for drained Poisson's ratio, which was taken as 0.25.

Figure 7 shows the results of the undrained MTX tests performed using the soft and stiff clay samples, each recompressed to the in-situ effective overburden pressure under K_o conditions. Note that the stress-strain non-linearity is significant over a region of small strains. The stiffness values associated with two types of stiffness parameters; *i.e.*, $E'=E'_{max}/2$ and $E'=60S_{uFVS(cor)}$, employed in the FE analysis are indicated on the stiffness decay curve of a 13.6 m sample, bearing in mind that the E'_{max} value is directly measured from the SCPT test or estimated from the results of the undrained MTX test using the relation $E'=E_u(1+\nu')/(1+\nu_u)$, where ν_u means the undrained Poisson's ratio, for which a value of 0.5 was postulated. It should be mentioned that the strain levels corresponding to $E'=E'_{max}/2$ and $E'=60S_{uFVS(cor)}$ are at approximately 0.1% and 1.0%, respectively.

3 FE ANALYSIS PERFORMED

A commercially available FE programme, SAGE-CRISP (e.g., Indraratna et al., 1992) was employed as it has often been used as a design-aid tool in geotechnical engineering for deformation analysis in many projects in South East Asia.

Plain strain conditions were assumed in both cases, considering the dimensions of the test embankment and the excavation. Coupled analysis was carried out in the analysis of the test embankment whereas a fully undrained condition was considered in the analysis of the excavation works. This was also recently postulated by Ou et al., 2000, in the case history of deep excavation works in Taipei.

As seen in Figures 5 and 6, the soil properties of the soft clay, such as its compressibility, stiffness and undrained strength, exhibit considerable variation with depth. Therefore, the soft and stiff clay in the analysis were conveniently divided into a few layers as shown in Figures 2 and 4 for the NNH site and the R-station, respectively. The initial ground conditions were assumed to have averaged values of total unit

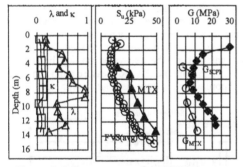

Figure 5. Soil parameters of the NNH site.

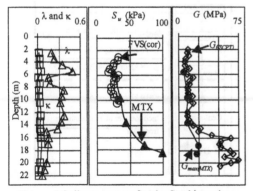

Figure. 6 Soil parameters for the Sutthisan site.

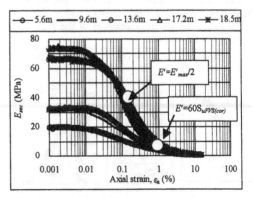

Figure. 7 Stiffness decay curve (Sutthisan).

weight, γ_t, and an initial void ratio, e_o, in each sub-layer. In a similar fashion, the coefficient of earth pressure at rest, K_o, was estimated using the following empirical expressions: $K_{NC}=1-sin\phi'_{TC}$ (Jaky, 1944) and $K_o=(OCR\ K_{NC})-[\{v'/(1-v')\}(OCR-1)]$ (Wroth, 1975), where ϕ'_{TC} is the internal angle of friction obtained from a MTX test.

The MCC model was used for the BKK site in the case of the embankment analysis, whereas elastic, CCM and MCC models were considered in the excavation analysis. On the other hand, isotropic elastic properties were considered for the sand layer, the DW and the concrete slabs. The CCM and MCC models require parameters of λ, κ, M, e_{cs}, p' and G or v'. Table 1 shows the soil parameters employed for the analysis at the NNH site. Table 2 shows the initial conditions of in-situ stresses and the pore water pressure prior to embankment loading. Similar information for the R-station site is given in Tables 3 and 4, respectively. Table 5 shows the elastic properties for the DW, the concrete slabs and the temporary strut. It should be mentioned that all the soil parameters were obtained from direct measurements at site investigation. Permeability in vertical and horizontal directions, k_v and k_h, was obtained from oedometer and dissipation tests using a piezocone.

As shown in Figure 7, the E' value equal to $E'_{max}/2$ corresponds to a strain level of about 0.1%, whereas the empirical expression of $E'=60S_{uFVS}$ corresponds to about a 1% strain. Figure 8 shows the FE meshing adopted in the analysis at the NNH site, for which the properties of the test embankment are assumed to be isotropic-elastic. The excess pore water pressure at the top drainage boundary under both one-

Table 1. Soil parameters for FE embankment analysis.

Layer (m)	e_{cs}	k_v (m/s)	k_h (m/s)	λ	κ	$S_{u(FVS)}$ (kPa)	$E'_{max}/2$ (MPa)
0~3	2.68	$1.44x^{-10}$	$1.06x^{-9}$	0.28	0.06	11.0	4.17
3~7	4.25	$1.44x^{-10}$	$1.06x^{-9}$	0.57	0.09	10.1	7.50
7~10	4.61	$2.26x^{-10}$	$2.78x^{-10}$	0.76	0.09	16.9	6.67
10~15	3.46	$4.50x^{-11}$	$1.36x^{-10}$	0.41	0.07	27.9	12.50
15~20	1.38	$4.50x^{-11}$	$1.36x^{-10}$	0.12	0.03	75.0	31.25
For embankment: E'=10 MPa and v= 0.33							
For MCC analysis, M =1.24 (for all soil layers)							

Table 2. In-situ ground condition (NNH site).

Depth (m)	σ_v (kPa)	u_w (kPa)	σ'_v (kPa)	σ'_h (kPa)	$p'_{c(MCC)}$ (kPa)
0	0	0	0	0	57
3	47	24	23	19	42
7	105	55	49	25	56
10	147	73	74	36	78
15	224	85	139	65	129
20	309	0	309	170	253

Table 3. Soil parameters for excavation analysis.

Layer (m~m)	e_{cs}	e_{cs}	λ	κ	M	K_w (MPa)	E'(MPa) (60*S_{uFVS})	G_{SCPT} (MPa)
0~1.8	2.07	2.06	0.21	0.03	1.38	72	1.78	9.9
1.8~4	2.47	2.46	0.29	0.03	1.38	184	1.78	9.9
4~9.2	3.13	3.11	0.38	0.03	1.38	373	1.50	12.7
9.2~15	3.17	3.16	0.32	0.03	1.38	710	1.73	14.6
15~18.5	1.80	1.80	0.14	0.04	1.05	683	4.50'	59.8
18.5~22	1.25	1.25	0.09	0.02	1.05	2598	4.50'	70.8
22~80		(elastic, undrained)						1125
K_w= Bulk modulus of water = $100*K'$; K'=$(1+e)p'/\kappa$								
S_u for stiff layers are taken from MTX test, *MCC, **CCM								

Table 4. In-situ condition prior to excavation.

Depth (m)	σ_v (kPa)	u_w (kPa)	σ'_v (kPa)	σ'_h (kPa)	$p'_{c(MCC)}$ (kPa)	$p'_{c(CCM)}$ (kPa)
0.0					119.6	161.1
1.8	32.0	8.0	24.0	14.9	119.6	161.1
4.0	71.1	30.0	41.1	25.5	76.4	102.9
9.2	154.3	70.1	84.2	40.9	105.2	141.6
15.0	247.1	89.4	157.8	89.6	249.0	333.9
18.5	306.6	83.5	223.1	145.0	523.5	701.8
22.0	370.0	0.0	369.6	240.3	523.5	701.8
80.0	580	1472	892	446	N.A.	N.A.

Table 5. Properties of strut, wall and slabs.

	Thickness (m)	E (MPa)	v	γ_t (kN/m³)
Diaphragm wall	1.0	$2.80x10^4$	0.2	24
Temporary strut		$9.70x10^4$	0.2	N.A.
Roof slab	0.9	$2.30x10^4$	0.2	24
Retail slab	0.7	$2.3x10^4$	0.2	24
Concourse slab	0.7	$2.63x10^4$	0.2	24
Base slab	1.8	$1.97x0^4$	0.2	24

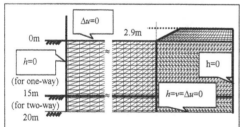

Figure 8. FE mesh and boundaries (embankment).

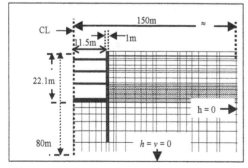

Figure. 9 FE mesh and boundaries considered.

way and two-way drainage conditions is considered equal to zero, whereas the bottom boundary in one-way drainage is considered to be at 15 m in depth and to be impermeable. In the case of the two-way drainage conditions, the bottom boundary was considered to be at 22 m in depth where the measured piezometric pore water pressure was equal to the atmospheric pressure. The meshing for the excavation is shown in

Figure 9, in which isotropic-elastic conditions are assumed for the sand layer (see also Table 3).

4 RESULTS OF FE ANLAYSIS

4.1 Test embankment

4.1.1 Effects of the Drainage Boundary

Table 6 shows the vertical and horizontal displacements predicted under one-way and two-way drainage boundary conditions using two types of stiffness parameters and the MCC model. From the table, it can be seen that a better prediction was made under two-way drainage conditions. This might be due to the fact that the non-hydrostatic distribution of pore water pressure at the site started from 6 m below the ground surface and at the top of the sand layer, i.e., below the stiff clay, the pore water pressure was equal to atmospheric pressure.

4.1.2 Effects of Elastic stiffness

Figures 10 to 14 analyze the results of the effect of elastic stiffness in the case of embankment analysis. As shown in Table 6, under both one-way and two-way drainage conditions, the use of $E'=E'_{max}/2$ grossly under-predicted the vertical as well as horizontal displacements, whereas with $E'=60S_{uFVS(cor)}$, the predicted ground and vertical settlement results were close to the measured ones (see Figures 10, 11 and 12). However, those closer predictions were still

Table 6. Effect due to stiffness and boundary conditions.

Model	E'	1-W *V (cm)	**H (cm)	2-W *V (cm)	**H (cm)
	$60S_{uFVS}$	18.90	13.33	23.88	14.79
MCC	$E'_{max}/2$	5.17	2.54	6.55	3.18

*vertical settlement at the CL on the surface (measured=50cm
**horizontal displacement at the toe point (measured=10 cm)

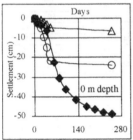

Figure 10. Settlement vs. time.

Figure 11. Settlement profile vs depth.

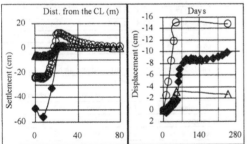

Figure 12. Ground surface settlement.

Figure 13. Horizontal displacement at toe point.

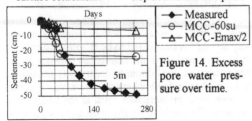

Figure 14. Excess pore water pressure over time.

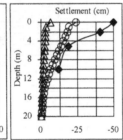

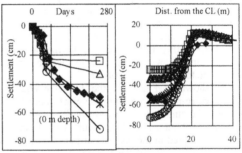

Figure 15. Settlement vs. time.

Figure 16. Ground surface settlement.

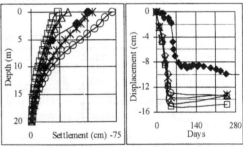

Figure 17. Settlement profile with depth.

Figure 18. Horizontal displacement at toe.

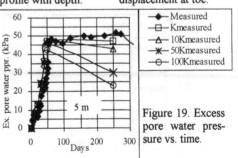

Figure 19. Excess pore water pressure vs. time.

lower than the measured ones. Figure 13 shows the horizontal displacement at the toe of the embankment. A prediction made with $E'=60S_{uFVS(cor)}$ overpredicted, while one with $E'=E'_{max}/2$ under-predicted the displacement. Excess pore pressure distribution over time is shown in Figure 14 where it can be seen that the effect of elastic stiffness was insignificant in predicting the excess pore water pressure.

4.1.3 *Effects of Permeability Coefficient Elastic stiffness*

Figures 15 to 19 show the results of the analysis made with different permeability coefficients under two-way drainage conditions and using $E'=60S_{uFVS(cor)}$. In general, with a decrease in the permeability coefficient, there was a considerable increase in the vertical settlement over time (Figures 15, 16 and 17). On the contrary, there was a decrease in the horizontal displacement with an increase in the permeability coefficients (Figure 18). Thus, it can be said that the amount of settlement over time was significantly affected by the value of the permeability coefficient selected for the analysis. From Figures 15, 16 and 17, we can say that the best simulation of the vertical settlement can be made taking the permeability coefficient as $50 \times K_{measured}$. From Figure 18, we can say that the analysis with the stiffness value, $E'=60S_{uFVS(cor)}$, over-predicted the horizontal displacement under all the permeability conditions, showing that this stiffness value was still higher.

The effect of the permeability coefficient in predicting the excess pore pressure distribution can be seen in Figure 19. Up to the construction period, the estimated maximum excess pore pressure at all the depths, 2 m, 5 m, 10 m and 16 m, was closer to the measured one. However, the dissipation rate of excess pore water pressure after the construction period of the embankment was higher in all the cases using $50 \times K_{measured}$ and $100 \times K_{measured}$ values. However, the predictions made with $K_{measured}$ and $10 \times K_{measured}$ at 5 m and 10 m depths showed that the predictions were closer to the measured ones. This suggested that the coefficient of permeability was a predominant factor in governing the settlement after completion of the construction stage. Although the dissipation rate of excess pore water pressure during the consolidation stage was slightly higher while using $50 \times K_{measured}$, it was better to use this value as a permeability coefficient since it provided closer predictions for the settlement as well as the horizontal displacements.

4.2 *Excavation*

4.2.1 *Effects of Elastic stiffness*

Figures 20 through 23 show the comparison made using different constitutive models; MCC, CCM and elastic, all using two stiffness values of $E'=E'_{max}/2$ and $E'=60S_{uFVS(cor)}$. Only the results at the 4th stage of excavation, i.e., before the base slab was cast are shown. Note that in the undrained analysis, the bulk modulus of the water, K_w, was assumed to vary with depth, i.e., 100 times that of the effective bulk modulus of the soil, K'.

Figure 20 shows a comparison between the predicted and the measured ground settlement profiles behind the DW. It should be mentioned that the measured data was corrected by taking the rate of ground subsidence as 1 cm/year, a value typically seen in central BKK. Note that there is no significant difference between the comparative results using different soil models. However, the effect of E was significant, i.e. the maximum settlement observed was around 0.5 cm when $E'=E'_{max}/2$ was employed, whereas it varied from 3.3 cm to 4.4 cm with $E'=60S_{uFVS(cor)}$. The maximum settlement of the ground was observed at around 25 m from the centerline of the excavation. The effect of the initial surcharge could be seen behind the DW, which allowed the ground to settle down even adjacent to the outer boundary. It seems that the predicted settlement with

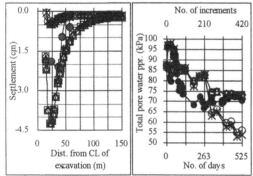

Figure 20. Surface settlement behind DW.

Figure 21. Pore water pressure comparison.

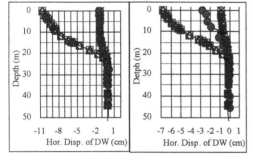

Figure 22. Horizontal displacement of DW.

Figure 23. Horizontal displacement of ground.

$E'=E'_{max}/2$ was closer to the measured profile. Figure 21 shows a similar comparison for the behavior of the total pore water pressure. The prediction of total pore water pressure was not affected by the constitutive models either. The predicted total pore water pressure at first increased to 97 kPa by using both the stiffness values (almost equal up to 260 days) but it reduced as the excavation progressed. Faster reduction in the total pore pressure was seen when using a high stiffness value down to 54 kPa than when using a low stiffness value. It should be mentioned that pore pressure response was not a critical factor in influencing the ground deformation.

Figures 22 and 23 show a comparison of the horizontal displacements of the DW and of the ground behind the DW. Again, no difference is seen among the predictions when using different constitutive models. The predicted deformation was far larger than the measurement when using $E'=60S_{uFVS(cor)}$, whereas the small amount of deformation was successfully predicted with $E'=E'_{max}/2$ (Figure 22). A similar trend can be seen in Figure 23, in which a comparison of horizontal ground deformation is shown. Thus, it could be said that the prediction with $E'=E'_{max}/2$ may be better than that with $E'=60S_{uFVS(cor)}$ in predicting the deformation of soft clay ground.

On the basis of these predictions, we can say that an elastic analysis may be sufficient to predict the deformation behavior of soft clay ground subjected to deep excavation with a rigid DW. This is because the ground strains induced refer to those below the yield point, for example, in the Cam clay model incorporated in CRISP.

5 CONCLUSIONS

The following conclusions can be made:

i) It is suggested using $E'=60 \times S_{uFVS(cor)}$ in the analysis of an embankment, whereas $E'=E'_{max}/2$ is recommended in the analysis of relatively small ground deformation associated with excavation work with a stiff retaining structure such as a DW. Also, elastic analysis is sufficient for excavation work.

ii) It is suggested considering two-way drainage boundary conditions instead of conventional one-way drainage when predicting the deformation behavior of BKK clay in coupled FE analysis.

iii) It is suggested taking the value for the coefficients of permeability as $50 \times K_{measured}$ for better prediction of ground settlement over time in the case of a coupled analysis of an embankment.

6 REFERENCES

AIT(a) 1973. Performance study of test sections for the new Bangkok airport at Nong Ngoo Hao. *Progress report submitted to Northrop airport development corporation*, Asian Institute of Technology, Bangkok, Thailand.

Akino, N. 1990. Estimation of rigidity of ground and prediction of settlement of building, *Prediction of Immediate Settlement of Building (Part 1). Journal of Structure and Construction Engineering, AIJ* 412:109-119.

Anderbrooke, T.I., D.M. Potts & A.M. Puzrin 1997. The influence of pre-failure soil stiffness on numerical analysis of tunnel. *Geotechnique* 47(3):693-712.

Balasubramaniam, A.S., & R.P. Brenner 1981. Consolidation and settlement of soft clay. *Soft clay engineering*, Brand, E.W. and R.P. Brenner, eds., Elsevier Scientific Publishing Co.

Bergado, D.T., S. Ahmeed, C.L. Sampaco & A.S. Balasubramaniam 1990. Settlement of Bangna-Bangapakong Highway on soft Bangkok clay. *Geotechnical Engineering* 116(1):136-155.

Hock, G.C. 1997. Review and analysis of ground movements of braced excavation in Bangkok subsoil using diaphragm walls. *M. Engg. Thesis*. Bangkok, Thailand.

Shibuya, S. & S.B. Tamrakar 1998. In-situ and laboratory investigations into engineering properties of Bangkok clay. In Tsuchida & Nakase (eds), *Characterization of Soft Marine Clays*: 107-132. Rotterdam:Balkema.

Simpson, B. 2000. Engineering needs. *Preprints of 2nd International Symposium on Pre-failure Deformations Characteristics of Geomaterials Keynote and Theme Lectures*: 142-157. IS-Torino 99.

Indraratna, B., A.S. Balasubramaniam & S. Balachandra 1992. Performance of test embankment constructed to failure on soft marine clay. *Journal Geotechnical Engineering, ASCE* 118 (1):12-33.

Ou, C.-Y., J.-T. Liao & W.-L. Cheng 2000. Building response and ground movements induced by a deep excavation. *Geotechnique* 30 (3):209-220.

Jaky, J. 1944. The coefficient of earth pressure at rest. *Magyar Mrenok Epitesz Kozloney*.

Sugie, S., U. Takayuki, A. Noriyuki & S. Junji 1999. Three-dimensional soil/water coupled FEM simulation of ground behavior adjacent of braced cuts. *Reports of Obayashi Corporation Technical Research Institute* 59:69-74. Obayashi Corporation, Japan.

Tatsuoka, F. & S. Shibuya 1992. Deformation characteristics of soils and rocks from field and laboratory tests. *Keynote Paper, Proc. 9th Asian Regional Conference on SMFE* 2:101-170. Bangkok.

Wroth, C.P. 1975. In-situ measurement of initial stresses and deformation characteristics. *Proc. of the specialty Conference in In-situ Measurement of Properties, ASCE*: 181-230. Rayleigh, North Carolina.

Advanced Laboratory Stress-Strain Testing of Geomaterials, Tatsuoka, Shibuya & Kuwano (eds),
© *2001 Taylor & Francis, ISBN 90 2651 843 9*

Effect of fabric anisotropy of a sand specimen on small strain stiffness measured by the bender element method

S.Yamashita & T.Suzuki
Kitami Institute of Technology, Kitami, Japan

ABSTRACT: In this study, to examine the effect of the fabric (inherent) anisotropy of a sand specimen on the initial stiffness, the shear wave velocity was measured by the bender element method on Toyoura sand and a volcanic ash sand. Two kinds of sample preparation methods were adopted: the dry-vibration (DV) method, and the MSP-F method. In the MSP-F method, the air-dry sand was pluviated into a container from a MSP-apparatus. The sand deposited in the container was frozen in a freezer. Specimens with an angle between the axial direction of the triaxial specimen and the pluviation direction of 0° and 90° were cut from the frozen sand blocks. The sand specimens were isotropically consolidated under several kinds of confining pressures ranging from 30 to 400 kPa. The shear wave velocities were measured in three different directions (VH, HH, HV-wave).

1 INTRODUCTION

To evaluate the maximum shear moduli from laboratory tests, one technique widely adopted in the last decade has been the propagation of seismic waves by means of piezoelectric transducers, called simply "bender elements", housed in a triaxial apparatus (e.g. Dyvik & Madshus 1985). In this method, the shear wave velocity propagated vertically V_{VH} is commonly measured. On the other hand, in in-situ seismic surveys, the down-hole method measures V_{VH}, whereas V_{HH} or V_{HV} is measured by the cross-hole method. Note that the first and second subscripts for V denote the directions of shear wave propagation and polarization, respectively.

The above techniques are based on the following relationships between the velocities of the seismic body waves and the shear moduli of an isotropic homogeneous elastic medium:

$$G_{max} = \rho V_S^2 \tag{1}$$

where G_{max} = the maximum shear modulus; ρ = the total mass density of the medium; and V_S = the shear wave velocity.

In the literature, the G_{HH} is fined as slightly larger than the G_{VH} even for normally consolidated soils, and the ratio of G_{HH} to G_{VH} increases from about 1.2 as the K_0-value increases (Lo Presti & O'Neil 1991, Jamiolkowski et al. 1995). In addition, the ratio of V_{HH} to V_{VH} (or V_{HV}) is 1.04 to 1.11 for two kinds of reconstituted sand specimens under isotropic con-

solidation from the bender element method (Fioravante 2000).

It would seem that this is because of the difference in the shear wave propagating direction relative to the bedding plane. In this paper, to clarify the effect of the propagating direction relative to the bedding plane on the shear wave velocity, the shear wave velocities in three different directions were measured for the specimens, with angles between the axial direction of the triaxial specimen and the pluviation direction of 0° and 90°.

2 TEST PROGRAM

2.1 Test materials

The sands used were Toyoura and Kussharo sands. Toyoura sand has been widely used for laboratory stress-strain tests in Japan. Kussharo sand is a volcanic ash sand taken from the suburbs of Tanno, Hokkaido, Japan. Kussharo sand used was graded with cut off particles more than 2 mm and less than 0.075 mm in size. The physical properties and the grain size distribution curves of these sands are shown in Table 1 and Figure 1.

Table 1. Physical properties of sands.

Sand name	ρ_s (g/cm³)	D_{50} (mm)	U_c	e_{max}	e_{min}
Toyoura	2.645	0.20	1.22	0.966	0.608
Kussharo	2.562	0.48	4.46	1.973	1.253

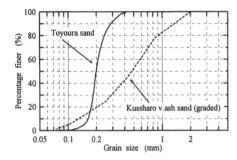

Figure 1. Grain size distribution curves of tested sands.

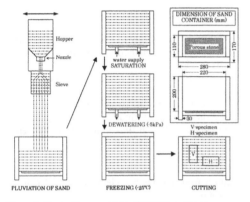

Figure 2. Preparation procedure of MSP-F specimen.

In this study, two kinds of sample preparation methods were employed. The first one was the dry-vibration (DV) method. In this method, after the air-dry sand was pluviated into a mold from the nozzle of a tube, the specimen was compacted by vibrating the mold using an electric vibrator until a desired density was attained. The second method was the MSP-F method. In this method, the air-dry sand was pluviated into a container from a multiple sieving pluviation (MSP) apparatus, as shown in Figure 2. The sand deposited in the container was permeated by water and thereafter unsaturated at suction induced by no volume expansion due to freezing. For facilitation of specimen forming, the sand deposited in the container was frozen in a freezer. Specimens with an angle between the axial direction of the triaxial specimen and the pluviation direction of 0° (V-specimen) and 90° (H-specimen) were cut from the frozen sand blocks.

Specimens of 70 mm in diameter and 100 mm in height were reconstituted by the DV method, whereas those of 70 mm in diameter and 70 mm in height were reconstituted by the MSP-F method. Toyoura sand specimens were reconstituted to a relative density D_r ranging from 40 % to 80 %. For Kussharo sand specimens, the D_r ranged from 60 % to 100 %.

2.2 Test procedure

One pair of bender elements (BE_{VH}) was attached to the top cap and the pedestal, as shown in Figure 3. After the mold was removed, two pairs of bender elements (BE_{HH} and BE_{HV}) were penetrated into the lateral surface of the specimen under a partial vacuum of 30 kPa. In the case of the frozen specimen, after it was allowed to melt under a partial vacuum of 30 kPa, two pairs of bender elements were penetrated into its lateral surface. Thereafter the cell pressure was raised to 30 kPa, and carbon dioxide and deaired water were percolated through it (saturated specimen). The specimens were subsequently isotropically consolidated under effective confining pressures ranging from 30 to 400 kPa.

In the BE_{VH}, the direction of the propagation of the shear wave and the direction of particle vibration were vertical and horizontal, respectively. The propagation and vibration directions of BE_{HH} were horizontal, and those of BE_{HV} were horizontal and vertical, respectively, as shown in Figure 4.

The shear wave velocities of the three directions were measured in each isotropic consolidation state. The transmitting element was driven by ±10V amplitude waves from a generator with a single sinusoidal

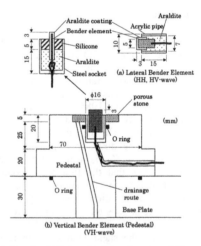

Figure 3. Details of (a) lateral bender element and (b) vertical bender element (pedestal).

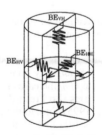

Figure 4. Arrangement of bender elements.

wave of different frequency (2 to 20 kHz) and pulse. The effective propagating distance and the arrival time of the shear wave were defined by the distance from tip-to-tip of the bender elements and the starting points of the input and received waves, respectively (Yamashita & Suzuki 2001). The shear wave velocity used was the average value in sinusoidal waves of 10, 15, 20 kHz.

3 TEST RESULTS

3.1 Effect of the propagating direction of the shear wave

Figure 5 shows the shear wave velocity V_S versus the effective confining pressure σ_c', plotted on a logarithmic scale, for three different propagation directions of shear waves on Toyoura and Kussharo dry specimens reconstituted by the DV-method. In order

to eliminate the effect of the difference in density on test results in this figure, the shear wave velocity was normalized by dividing it by the square root of the following void ratio functions: Toyoura sand $F(e) = (2.17-e)^2/(1+e)$ (Iwasaki et al. 1978), and Kussharo sand $F(e) = e^{-2}$ (Hoshi et al. 2000). The test results obtained from the saturated specimens were also plotted in this figure.

It can be seen that in Toyoura sand, the effect of the propagating direction of the shear wave is relatively small, and the V_{HH} is slightly higher than the V_{VH}. On the other hand, in Kussharo sand the V_{HH} is higher than the V_{VH} and the V_{HV}. This trend agrees with results that measured the shear wave velocity in three directions in the calibration chamber (Stokoe et al. 1995) and on the triaxial specimen (Fioravante 2000). It is considered that this is due to the effect of the fabric anisotropy of the specimen. In addition, the effect of the propagating direction on the shear wave velocity for the Kussharo specimen is larger than that for the Toyoura specimen. It would seem that this is because Kussharo sand particles are flatter than those of Toyoura sand.

On the other hand, Figure 6 shows the shear wave velocity versus the effective confining pressure on Toyoura and Kussharo saturated specimens reconstituted by the MSP-F method. As with the DV method, the V_{HH} is higher than the V_{VH} and the V_{HV}. It can also be seen that the V_{VH} is lower than the V_{HH} and the V_{HV} of the Toyoura specimen. In the case of

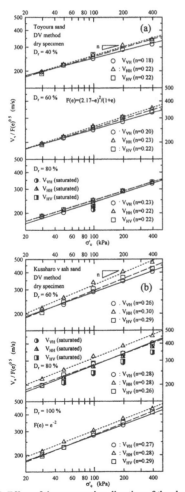

Figure 5. Effect of the propagating direction of the shear wave (DV method); (a) Toyoura sand, (b) Kussharo sand.

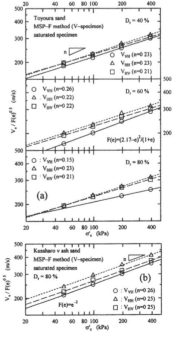

Figure 6. Effect of the propagating direction of the shear wave (MSP-F method); (a) Toyoura sand, (b) Kussharo sand.

the frozen specimen, before it was placed on a pedestal with the bender element, its top and bottom ends were cut with a groove. It would seem that the area around the vertical bender elements was disturbed by grooving. Therefore, the shear wave velocity in the vertical direction became slower than that in the horizontal direction.

It is to be noted that the shear wave velocities of the saturated specimens are lower than those of the dry specimens as shown in Figure 5. Figure 7 shows the relationships of the shear wave velocity to the effective confining pressure for the dry and saturated DV specimens and the saturated MSP-F specimen on Toyoura and Kussharo sands. It can be seen that the shear wave velocities of the dry specimens are higher than those of the saturated specimens on both sands, irrespective of the propagating direction of the shear wave. On the other hand, Figure 8 shows the relationship of the shear modulus calculated by Equation 1 (ρ = total soil density) to the effective

confining pressure. From this figure, it is found that, in the case of Toyoura sand, the maximum shear moduli are almost the same irrespective of the condition of the specimen (i.e. dry or saturated) and the sample preparation method. On the other hand, in the case of Kussharo sand the maximum shear moduli of the saturated specimens are larger than those of the dry specimens. The void ratio of the Kussharo volcanic ash sand is much larger than that of the Toyoura sand, and the particles of volcanic sand are porous. Therefore, it would seem that the shear moduli of the Kussharo saturated specimens became larger than those of the dry specimens due to the effect of coupling between soil and fluid on the shear wave velocity (e.g. Biot 1957).

3.2 Effect of the fabric anisotropy of the specimen

As mentioned above, it has been indicated that the shear wave velocity obtained from a HH-wave is

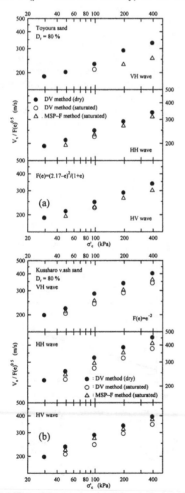

Figure 7. Effect of the sample preparation method on the shear wave velocity; (a) Toyoura sand, (b) Kussharo sand.

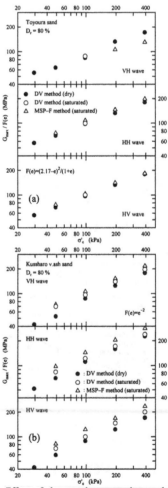

Figure 8. Effect of the sample preparation method on the maximum shear modulus; (a) Toyoura sand, (b) Kussharo sand.

higher than that from VH or HV-waves. It is considered that this is due to the effect of the fabric (inherent) anisotropy of the specimen. Thus, to further clarify the effect of the fabric anisotropy on the shear wave velocity, shear wave velocities in three different directions were measured for the specimens with an angle between the axial direction of the triaxial specimen and the pluviation direction of 0° (V-specimen) and 90° (H-specimen) cut from the frozen sand blocks (see Fig. 2).

Figure 9 illustrates the relationship of the propagating direction of the shear wave to the bedding plane. There are two kinds of H-specimens due to the difference in penetration direction of the bender elements relative to the bedding plane. The following three kinds of shear wave velocities, which were defined by the relations of the propagating direction versus the bedding plane, were measured on these specimens (see Table 2)

$-V_{VH}^{*}$ = the propagating direction of the shear wave is normal and the vibrating direction of the particles is parallel relative to the bedding plane;

$-V_{HH}^{*}$ = the propagating direction of the shear wave and the vibrating direction of the particles are parallel relative to the bedding plane; and

$-V_{HV}^{*}$ = the propagating direction of the shear wave is parallel and the vibrating direction of the particles is normal relative to the bedding plane.

Table 2. Relations between the propagating direction and the shear wave velocity.

Wave	V-specimen	H1-specimen	H2-specimen
VH-wave	V_{VH}^{*}	V_{HV}^{*}	V_{HH}^{*}
HH-wave	V_{HH}^{*}	V_{HV}^{*}	V_{VH}^{*}
HV-wave	V_{HV}^{*}	V_{VH}^{*}	V_{HH}^{*}

Figure 10 shows the shear wave velocity versus the effective confining pressure measured under the same propagating direction of the shear wave relative to the axial direction of the specimen with a different direction of bedding plane on Toyoura sand. When the propagating direction of the shear wave relative to the direction of bedding plane is the same, although the propagating direction of the shear wave relative to the axial direction of the specimen is different, the share wave velocities for propagating and vibrating parallel to the bedding plane V_{HH}^{*} are slightly higher than the V_{VH}^{*} and the V_{HV}^{*}. The V_{VH}^{*} is almost the same as the V_{HV}^{*}.

Figure 11 shows the shear wave velocity versus the effective confining pressure on Kussharo sand as in Figure 10. In the case of Kussharo sand, the V_{HH}^{*} is clearly higher than the V_{VH}^{*} and the V_{HV}^{*}, and the V_{VH}^{*} is the same as the V_{HV}^{*} as well as the Toyoura sand.

From the above, it can be concluded that shear waves propagate faster in the plane parallel to the bedding plane than in the normal one. Figure 12 shows the V_{HH}^{*} and the V_{HV}^{*} versus the V_{VH}^{*} on the MSP-F Toyoura and Kussharo specimens. It can be seen that the V_{HH}^{*} is higher than the V_{VH}^{*} with an average value of $V_{HH}^{*}/(V_{VH}^{*} = V_{HV}^{*}) = 1.05$ on

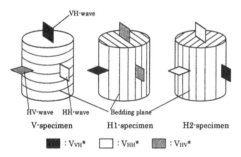

Figure 9. Relations between the propagating direction and the bedding plane.

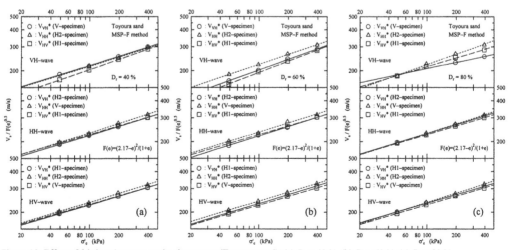

Figure 10. Effect of fabric anisotropy on the shear wave (Toyoura sand); (a) D_r = 40 %, (b) D_r = 60 %, (c) D_r = 80 %.

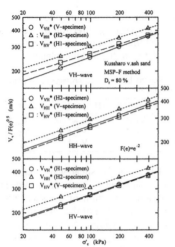

Figure 11. Effect of fabric anisotropy on the shear wave (Kussharo sand).

Toyoura sand and 1.13 on Kussharo sand. The difference between Toyoura and Kussharo sand is the difference in inherent anisotropy due to the difference in particle form.

4 CONCLUSIONS

In the frozen specimen, the area around the vertical bender elements was disturbed by grooving, therefore the shear wave velocity in the vertical direction became lower than that in horizontal direction.

The shear wave velocities of dry specimens were higher than those of saturated specimens on both sands irrespective of the propagating direction of the shear wave. On the other hand, the maximum shear moduli of the dry specimens were almost the same as the saturated specimens in Toyoura sand, whereas the maximum shear moduli of the saturated specimens were larger than those of the dry specimens in Kussharo sand due to the effect of coupling between soil and fluid on the shear wave velocity.

Shear waves propagate faster in the plane parallel to the bedding plane than in the normal one. The V_{HH}^* is higher than the V_{VH}^* with an average value of $V_{HH}^*/(V_{VH}^* = V_{HV}^*) = 1.05$ on Toyoura sand and 1.13 on Kussharo sand. The V_{VH}^* is equal to the V_{HV}^* irrespective of the inherent anisotropy and the kind of sand.

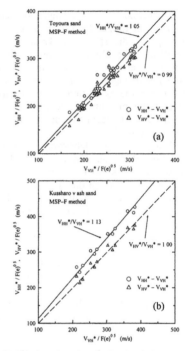

Figure 12. Relation of V_{HH}^* or V_{HV}^* to V_{VH}^*; (a) Toyoura sand, (b) Kussharo sand.

REFERENCES

Biot, M.A. 1957. Theory of propagation of elastic waves in a fluid saturated porous solid. *Journal of the Acoustical Society of America*, AIP 28(2).

Dyvik, R. & Madshus, C. 1985. Laboratory measurement of G_{max} using bender elements, *Advances in the Art of Testing Soils under Cyclic Conditions; Proc. a session sponsored by the Geotechnical Engineering Division in conjunction with the ASCE Convention, Detroit, 24 October 1985*, ASCE: 186-196.

Fioravante, V. 2000. Anisotropy of small strain stiffness of Ticino and Kenya sands from seismic wave propagation measured in triaxial testing. *Soils and Foundations*, JGS 40(4): 129-142.

Hoshi, K., Yamashita, S., Ohta, Y. & Suzuki, T. 2000. Measurement of shear wave velocity of sand and volcanic ash soil using bender elements. *Proc. of 35th Japan National Conference on Geotechnical Engineering, Gifu, 13-15 June 2000*, JGS: 271-272. (in Japanese)

Iwasaki, T., Tatsuoka, F. & Takagi, Y. 1978. Shear moduli of sands under cyclic torsional shear loading, *Soils and Foundations*, JGS 18(1): 39-56.

Jamiolkowski, M., Lancellotta, R. & Lo Presti, D.C.F. 1995. Remarks on stiffness at small strains of six Italian clays, Shibuya, Mitachi & Miura (eds), *Pre-failure Deformation of Geomaterials; Proc. Intern. Symp., Sapporo, 12-14 September 1994*. Balkema: 817-836.

Lo Presti, D.C.F. & O'Neil, D.A. 1991. Laboratory investigation of small strain modulus anisotropy in sands, Huang (ed), *Proc. ISOCCT1, Potsdam*, Elsevier: 213-224.

Stokoe, K.H.II, Hwang, S.K., Lee, N.K.J. & Andrus, R.D. 1995. Effect of various parameters on the stiffness and damping of soils at small to medium strains, Shibuya, Mitachi & Miura (eds), *Pre-failure Deformation of Geomaterials; Proc. Intern. Symp., Sapporo, 12-14 September 1994*. Balkema: 785-816.

Yamashita, S. & Suzuki, T. 2001. Small strain stiffness on anisotropic consolidated state of sands by bender elements and cyclic loading tests, *Proc. of 15th ICSMGE, Istanbul, 27-31 August 2001*. Balkema. (in press)

Advanced Laboratory Stress-Strain Testing of Geomaterials, Tatsuoka, Shibuya & Kuwano (eds),
© 2001 Taylor & Francis, ISBN 90 2651 843 9

Strength and deformation characteristics of artificially cemented clay with dispersion due to the mixing quality

N. Yasufuku, H. Ochiai, K. Omine & N. Iwamoto
Department of Civil Engineering, Kyushu University, Fukuoka, Japan

ABSTRACT: The strength and deformation modulus of artificial cemented clay can be expressed as a function of an equivalent overconsolidation ratio related to the cementation effect. A model for predicting the undrained strength of cement-treated soil with statistical scatter due to mixing quality is presented, combining cementation effects into the critical state concept. The crux of the discussion is how the statistical scatter of the undrained compressive strength and the deformation modulus due to mixing quality should be considered. Further, the applicability of the model was verified by comparing the predicted results with the results from a series of undrained triaxial tests on cement-treated clay with differing cementation and initial water content.

1 INTRODUCTION

Cement stabilization has been widely used for improving soft ground. The practical strength evaluation of cement-treated soil is generally based on unconfined compression strength, which may not include the effect of a strength increase due to overburden pressure. In recent years, it has been necessary to introduce the effect of overburden pressure into the strength of cemented treated soil for a rational prediction of the design strength. It has also been pointed out that the strength of in-situ treated soils is not necessarily coincident with that of stabilized soil in the laboratory, one reason being the differences in the mixing quality.

In this study, a model for predicting the undrained strength and deformation modulus of cement-treated soil is presented for practical application, combining cementation effects and degree of mixing quality into the critical state soil plasticity. The effect of the mixing quality on the strength and deformation modulus will be presented by utilizing the two-phase mixture model (Omine Yoshida and Ochiai, 1997). Next, the effects of the mixing qualities are discussed through parametric studies using the model. In addition, the applicability of the proposed model is verified, comparing the predicted results with the results from a series of undrained triaxial compression tests of cement-treated clay with differing cementation and initial water content. Figure 1 shows the framework of the model to be presented.

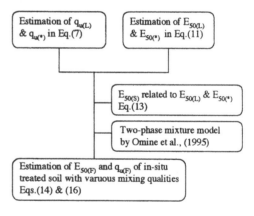

Figure 1 Framework of the model presented

2 UNDRAINED SHEAR STRENGTH AND DEFORMATION MODULUS RELATED TO CRITICAL STATE PLASTICITY

2.1 *Failure state conditions with cementation*

In order to evaluate the cementation effect on strength and deformation mobilized by adding an amount of cement, a stress parameter p_r is introduced into the p'-q space and the e-logp' space. The schematic diagram in both spaces where the cementation effect is introduced, namely p_r, is shown in Figure 2, which is considered to be an extension of the critical state concept (Yasufuku et al.,1997, Kasama et al.,2000). Noted that p' and q are deifined as

Cam-clay stress parameters, and λ^* and κ^* are defined as the slope of the equivalent normaly consolidated and swelling line in the e-ln($p'+p_r$) space, where p_r is given by the intersection of the failure line with the p'-axis. According to experimental considerations in cases where the treated original clay is the same (Morishima et al., 1999, Kasama et al, 2000, Yasufuku et al., 2000), the following experimental evidences can be made :

1) the strength parameter M in the failure state of normally consolidated region is almost constant, irrespective of the cement content, which is defined as the cement content per unitary dried soil sample weight, and the initial water content of the soil,

2) the parameter p_r , which defines the degree of cementation, is a function of the cement content independent from the initial water content of the soil,

3) when the cement content is the same, a single equivalent normally consolidated line exists, irrespective of the initial water content of the soil,

4) the failure state line in the e-ln($p'+p_r$) space becomes steeper with an increase in cementation. Based on these experimental assumptions, the failure state in the p'-q space and the e-ln($p'+p_r$) space can be shown as:

$$q = M(p'+p'_r) \quad (1); \quad e = \Gamma^* - \lambda^* \ln(p'+p'_r) \quad (2)$$

where, M is represented by a function of ϕ' such as $M = 6\sin\phi' / (3 - \sin\phi')$. Figure 3 shows the typical e-ln($p'+p_r$) relationships obtained from the isotropic

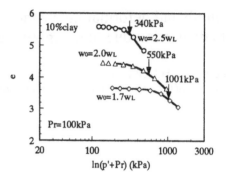

Figure 3 Typical e-ln(p'+Pr) relationships

consolidation tests of cemented clay with three different kinds of initial water content, in which all the samples have a 10 % cement content. The clear yield stresses due to cementation, which depend on the initial water content, can be seen from this figure, although one of the specimens have any stress history. Thus, as the definition of such consolidated yield stresses due to cementation seems to be somewhat different from the original, it will be called an apparent consolidated yield stress in this study. The stress regions in this space up to and exceeding the apparent consolidated yield stress are analogically named as the apparent over-consolidated and normally consolidated regions, respectively. Further, the over-consolidation ratio R^* reflecting the cementation effect is also defined as follows:

$$R^* = \frac{p'_y + p'_r}{p'_c + p'_r} \quad (3)$$

where, p_y' and p_c' are the apparent consolidated yield stress and the current isotropic consolidated stress, respectively. R^* in Eq.(3) is related to the OCR originally defined by p_y / p_c such that:

$$R^* = \frac{OCR + p'_r / p'_c}{1 + p'_r / p'_c} \quad (4)$$

Figure 4 shows the relationship between R^* and OCR as a function of p_r'/p_c' with previous experimental data (Yasufuku et al., 2000), where the effect of p_r'/p_c' on the R^*-OCR relationship can be clearly understood.

2.2 Undrained strength associated with cementation

In order to evaluate the cementation effect on the strength mobilized by adding an amount of cement, the following undrained shear strength s_u will be used in this study:

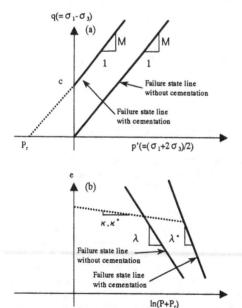

Figure 2 Schematic diagram of failure state line introducing cementation effect

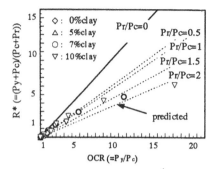

Figure 4 Relationship between R^* and OCR

$$s_u = (\sigma_1 - \sigma_3)_{peak}/2 \qquad (5)$$

where, σ_1 and σ_3 are defined as the maximum and minimum principal stresses at the peak stress ratio state. Based on critical state modeling of clay with cementation, a simple equation for the undrained effective stress path has been given as follows (Kasama, Ochiai & Yasufuku, 2000):

$$\frac{p' + p_r}{p'_0 + p_r} = \left\{ \frac{M^2}{\eta^{*2} + M^2} \right\}^{\frac{\lambda^* - \kappa^*}{\lambda^*}} \qquad (6)$$

where, p'_0 is an isotropic consolidated yield stress and η^* is defined as $\eta^* = q/(p' + P_r)$. When considering Eq.(6) as related to critical state plasticity, the model for evaluating the undrained strength s_u of the cement-treated clay in Eq.(5) in over-consolidated and normally consolidated stress regions is:

$$s_u = \frac{1}{2} M (p'_c + p'_r) \left(\frac{R^*}{2} \right)^{\frac{\lambda^* - \kappa^*}{\lambda^*}} \qquad (7)$$

Eq.(7) can be rewritten as a form of the normalized strength increment ratio such that:

$$\frac{s_u}{p'_c} = \frac{M}{2} \left(1 + \frac{P_r}{p'_c} \right) \left(\frac{R^*}{2} \right)^{\frac{\lambda^* - \kappa^*}{\lambda^*}} \qquad (8)$$

Eq.(8) indicates that the strength increment ratio of cement-treated soils should be dependent on the degree of cementation defined by P_r. In addition, the newly presented strength increment ratio including P_r is also given by :

$$\frac{s_u}{p'_c + P_r} = \frac{1}{2} M \left(\frac{R^*}{2} \right)^{\frac{\lambda^* - \kappa^*}{\lambda^*}} \qquad (9)$$

Table 1 The parameters used

Cement (%)	W_0 (%)	λ^*	κ^*	P_y (kPa)	M	P_r (kPa)	α'
5	1.5WL	1.089	0.048	85	1.5	48	100
	2WL			32	1.5		
7	1.5WL	1.355	0.039	435	1.5	68	100
	2WL			120	1.5		
	2.5WL			42	1.5		
10	1.5WL	1.545	0.035	901	1.5	100	100
	2WL			450	1.5		
	2.5WL			240	1.5		

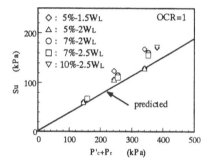

Figure 5 s_u-($p'_c + P_r$) relationships in normally consolidated regions

It is important to note that when $P_r = 0$, Eq.(9) reduces to that of the Modified Cam-clay model (Wroth, 1984). The parameters needed for prediction are λ^*, κ^*, M, P_r, and p_y which is used to determine R^* in Eq.(3).

Figure 5 shows the s_u-($p_c + P_r$) relationship of the cement-treated clay with cement content in the normally consolidated regions, where p_c is an isotropic consolidation pressure before undrained shearing. The parameters used are summarized in Table 1. Note that in preparing the samples, the slurry containing Portland Cement was mixed with Ariake Clay ($w_L = 86.5\%$, $I_p = 51.3$, $\rho_s = 2.609 g/cm^3$) with a 1.5 to 2.5 w_L water content. The details of the preparation have already been reported by Kasama et al.,(2000). The predicted results agree well with the experimental data. A comparison between the predicted and experimental s_u-($p_c + P_r$) relationships of the cement-treated clay with 10% cement content in the over-consolidated regions is shown in Figure 6. In addition, the $s_u/(p_c + P_r)$-R^* relationship is shown in Figure 7. The model in Eqs.(7) to (9) can reasonably predict the experimental results, reflecting the characteristics of the over-consolidated stress regions.

2.3 *Deformation modulus* E_{50}
The relationship between unconfined compression

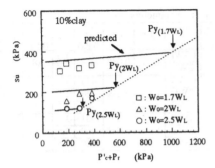

Figure 6 Predicted and experimental s_u-(p'_c+P_r) relationships

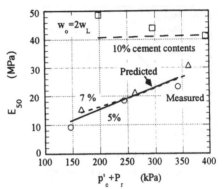

Figure 8 E_{50}-(p'_c+P_r) relationships

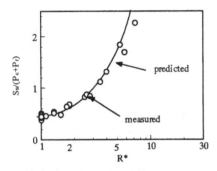

Figure 7 $s_u/(p'_c+P_r)$-R^* relationships of cemented treated-clay

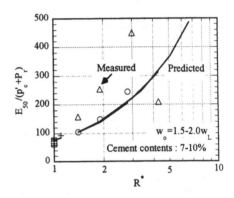

Figure 9 Depencecy of $E_{50}/(p'_c+P_r)$

strength q_u and E_{50} is expressed empirically as :

$$E_{50} = \alpha' q_u \tag{10}$$

where, q_u is defined as $q_u = 2s_u$, and α' is an experimental parameter which is practically known in practice to be in the range of 80 to 250. It has already been reported by Omine et al., (1999) that the parameter α' depends on the type of original claybeing in practice independent from the cement content, water content and curing time. When assuming that the relationship in Eq.(10) is extended to that under undrained triaxial stress conditions, E_{50} in the triaxial stress state is analogically derived from Eqs.(7), (8) and (10) such that:

$$E_{50} = \alpha' M (p'_c + p'_r) \left(\frac{R^*}{2} \right)^{\frac{\lambda-\kappa^*}{\lambda^*}} \tag{11}$$

$$\frac{E_{50}}{p'_c} = \alpha' M \left(1 + \frac{P_r}{p'_c} \right) \left(\frac{R^*}{2} \right)^{\frac{\lambda^*-\kappa^*}{\lambda^*}} \tag{12}$$

It can be seen from both equations that E_{50} is also given in terms of the current consolidation pressure, the apparent over-consolidation ratio and the degree of cementation. Figure 8 shows the characteristics of the typical predicted and measured E_{50} against (p'_c+P_r). In this prediction, α' was taken as 100. Further, Figure 9 shows the dependency of $E_{50}/(p'_c+P_r)$ on R^*. It can be seen that E_{50} in cement-treated soils is strongly dependent on the consolidated pressure and the apparent over-consolidation ratio due to cementation. The predicted results are shown by the solid lines, which roughly represent the characteristics of the measured values.

3 EVALUATION OF UNDRAINED SHEAR STRENGTH OF IN-SITU TREATED SOIL

The strength and deformation modulus of in-situ cement treated soils are considered to be influenced by the characteristics of the original soil, the scatter of the water and cement content and the degree of stirring mix rate, etc. Among such influencing factors, in this study, the effect of the stirring mix rate of the in-situ treated soil on the undrained strength and the E_{50} is discussed on the basis of the two-phase mixture model proposed by Omine et al.(1992, 1995).

3.1 Fundamental considerations

An evaluating method for the deformation modulus, E_{50}, and the unconfined compressive strength of the in-situ cement treated soils was proposed by applying the two-phase mixture model (Omine and Ochiai, 1992) to the treated soil. The basic concepts of the model can be summarized as follows:
1) The in-situ cement-treated soil is regarded as a mixture consisting of timproved and unimproved parts as the cement can not be mixed with the in-situ original soil completely.
2) The improved and unimproved parts are assumed to be an inclusion and a matrix, respectively.
3) Volume content of the improved part of the treated soil is defined as a stirring mix rate obtained by measuring the areas of the improved and unimproved parts in the cross section of the sampling specimen.
4) The stirring mix rate of the in-situ treated soil decreases with decreasing mixing level, and tthe stirring mix rate of the laboratory mixed soil is then assumed to be 100%.
5) The deformation and strength properties of the laboratory mixed soil and the in-situ original soil, and the stirring mix rate of the in-situ cement treated soil become the basic parameters of the two-phase mixture model used in this study, derived by assuming that the strain energy per unit volume in the mixture is constant.

3.2 Deformation modulus and undrained strength of in-situ cement treated soils

Based on the basic concepts mentioned above, the deformation modulus of the improved part of in-situ cement treated soil, $E_{50(s)}$, related to the E_{50} of the laboratory mixed soil, $E_{50(L)}$ and the E_{50} of the in-situ original soil, $E_{50(*)}$, is considered now to be a simple function of the stirring mix rate such that:

$$E_{50(s)} = E_{50(*)} + \left(\frac{E_{50(L)} - E_{50(*)}}{f_s} \right) \qquad (13)$$

Eq.(13) indicates that when $f_s=1$, $E_{50(s)} = E_{50(L)}$, and that $E_{50(s)}$ gradually increases with decreasing f_s value.Thus, when applying Eq.(13) to the two-phase mixture model proposed by Omine et al. (1995), the deformation modulus of the in-situ cement treated soil is represented as follows :

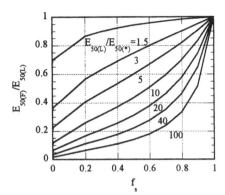

Figure 10 $E_{50(F)}/E_{50(L)}$ -f_s relationships with various $E_{50(L)}/E_{50(*)}$

$$E_{50(F)} = \frac{\{(b-1)f_s+1\}E_{50(s)}}{bf_s + \left(1-f_s\right)\left(\dfrac{E_{50(s)}}{E_{50(*)}}\right)} \;, \quad b = \left(\frac{E_{50(s)}}{E_{50(*)}}\right)^{1/2} \qquad (14)$$

where, $E_{50(F)}$: E_{50} of the in-situ treated soil, $E_{50(L)}$: E_{50} of the laboratory mixed soil, $E_{50(*)}$: E_{50} of the in-situ original soil and f_s: stirring mix rate of the in-situ treated soil. Note that $E_{50(s)}$ is defined by $E_{50(L)}$ and $E_{50(*)}$, as shown in Eq.(13). When Eq.(14) is assumed to be applicable in the triaxial stress region, both $E_{50(L)}$ and $E_{50(*)}$ can be analytically determined by Eq.(11). Thus, after calculation, $\left(E_{50(L)}/E_{50(*)}\right)$ is given by the following relationship:

$$\frac{E_{50(L)}}{E_{50(*)}} = \left(1+\frac{P_r}{p'_c}\right)\left(\frac{R^*}{2}\right)^{\frac{\lambda^*-\kappa^*}{\lambda^*}}\left(\frac{2}{OCR}\right)^{\frac{\lambda-\kappa}{\lambda}} = \frac{q_{u(L)}}{q_{u(*)}} \qquad (15)$$

In an analogical approach related to Eqs.(14) and (15), the undrained compressive strength of in-situ cement treated soils, $q_{u(F)}$, is derived as follows:

$$q_{u(F)} = \frac{\{(b-1)f_s+1\}q_{u(s)}}{bf_s + \left(1-f_s\right)\left(\dfrac{q_{u(s)}}{q_{u(*)}}\right)} \;, \quad b = \left(\frac{q_{u(s)}}{q_{u(*)}}\right)^{1/2} \qquad (16)$$

where, $q_{u(F)}$: q_u of the in-situ treated soil, $q_{u(L)}$: q_u of the laboratory mixed soil, and $q_{u(*)}$: q_u of in-situ original soil. Both $q_{u(L)}$ and $q_{u(*)}$ in Eq.(15), are calculated as twice of s_u in Eq.(7).

Figure 10 shows the relationship between the stirring mix rate and $E_{50(F)}/E_{50(L)}$. Although the E_{50} of the in-situ treated soils, $E_{50(F)}$,increases with the increase in the stirring mix rate, this tendency varies in the deformation modulus ratio in the laboratory, $E_{50(L)}/E_{50(*)}$ in Eq.(15). The same tendency can be found in the f_s- $q_{u(F)}/q_{u(L)}$ relationship as a function

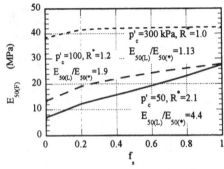

Figure 11 $E_{50(F)}$-f_s relationships

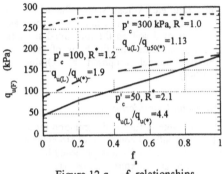

Figure 12 $q_{u(F)}$-f_s relationships

of $q_{u(L)}/q_{u(*)}$. It is clear from these figures that the deformation modulus and undrained strength of the in-situ treated soils are affected by not only the stirring mix rate but also the properties of in-situ original soil.

3.3 Example

Let us find the $E_{50(F)}$ and $q_{u(F)}$ of an in-situ cement treated soil with various stirring mix rate, f_s, over a wide stress region based on the $E_{50(L)}$, $E_{50(*)}$, $q_{u(L)}$ and $q_{u(*)}$ which are computed by the experimental parameters. The experimental parameters in this example are assumed as M=1.5, λ^*=1.355, κ^*=0.039, Pr=68.3kPa, p_{yo}=176kPa, λ=0.231, κ=0.034 and α'=150. When using these parameters, $E_{50(L)}$, $E_{50(*)}$, $q_{u(L)}$ and $q_{u(*)}$ are easily calculated from Eqs(11) and (7). Next, introducing the calculated values into Eqs.(14) and (16) with Eq.(13), the $E_{50(F)}$ and $q_{u(F)}$ with various stirring mix rates are estimated as a function of the p'_c values. Figures 11 and 12 show the $E_{50(F)}$- f_s and the $q_{u(F)}$-f_s relationships, respectively. It is found that $E_{50(F)}$ and $q_{u(F)}$ have a clear p'_c - and f_s -dependency and also the degree of f_s - dependency clearly changes with the values of p'_c.

4 CONCLUSIONS

A model for predicting the undrained strength and deformation modulus, E_{50}, of in-situ cement-treated soils with various mixing qualities was presented combining the critical state concept related to cementation effects with the two-phase mixture model presented by Omine et al., (1992, 1995). The use of this model offers potential for a better prediction of the undrained strength and the E_{50} of in-situ treated soils by taking into account the magnitude of the apparent overconsolidation ratio, cementation effect, overburden pressure and mixing qualities defined by the stirring mix rate.

ACKNOWLEDGEMENT

The authors wish to thank Mr. M. Nakasima, Mr. K. Kasama and Mr. H. Okamoto of Kyushu University for their advice and experimental support.

REFERENCES

Kasama, K., Ochiai, H. & Yasufuku, N. 2000. On the stress-strain behaviour of lightly cemented clay based on an extended critical state concept. Soils and Foundations, Vol. 40, No.5, 37-47.

Morishima, T., Ochiai, K., Yasufuku, N. & Kasama, K. 2000. Undrained strength characteristics of cemented clay paying attention to the initial water contents. Technology Reports of Kyushu University, Vol.73, No.2, 133-140 (In Japanese).

Omine, K. & Ochiai, H. 1992. Stress-strain relationship of mixtures with two different materials and its application to one-dimensional property of sand-clay mixed soils, Journal of Geotechnical Engineering, JSCE, No.448, III-19, 121-130 (In Japanese).

Omine, K., Ochiai, H. & Yoshida, N. 1995. Evaluation of Deformation-Strength Properties of In-Situ Cement-Treated Soils. Journal of Material Science, JSMS, Vol.44, No.503, 994-997 (In Japanese).

Omine, K., Ochiai, H., Yasufuku, N & Sakka, H. 1999. Prediction of strength-deformation properties of cement-stabilized soils by nondestructive testing. Proc. of 2nd Int. Symp on Pre-failure Deformation Characteristics of Geomaterials (IS Torino 99), Vol.1, 323-330.

Yasufuku, N., Ochiai, H. & Kasama, K. 1997. The dissipated energy equation of lightly cemented clay in relation to the critical state model, Proc. of 9th Int. Conf. Computer Methods and Advances in Geomechanics, Vol. 2, 917-922.

Yasufuku, N., Ochiai, H. & Kasama, K. 2000. Strength and deformation characteristics of artificial cemented clay as an equivalent overconsolidated soil. Proc. of the Symp. on Mechanics of Overconsolidated soils and ground, The Japanese Geotechnical Society, Vol.1, 19-24 (In Japanese).

Wroth, C. P. 1984. The interpretation of in situ soil tests, Geotechnique 34, No.4, 449-489.

Advanced Laboratory Stress-Strain Testing of Geomaterials, Tatsuoka, Shibuya & Kuwano (eds),
© 2001 Taylor & Francis, ISBN 90 2651 843 9

Author index

AnhDan, L.Q. 187, 209

Balakrishnaiyer, K. 195
Berilgen, M.M. 203

Cavallaro A. 15
Correia, A. 209
Cuellar, V. 275

Di Benedetto, H. 217
d'Onofrio, A. 15

Fioravante, V. 15
Furuta, I. 111

Geoffroy, H. 217
Georgiannou, V.N. 45, 227

Hashinoki, M. 279
Hatanaka, M. 111
Howie, J.A. 235

Inan, O. 203
Ishihara, M. 295
Iwamoto, N. 323

Katagiri, M. 47, 53, 245
Kawaguchi, T. 65, 303
Kita, K. 53
Kohata, Y. 65
Kongsukprasert, L. 251
Koseki, J. 111, 187, 195, 209, 259
Kurachi, Y. 259
Kuwano, J. 47, 53
Kuwano, R. 1, 53, 251, 287

Lai, C.G. 265
Lanzo, G. 15
Lo Presti, D.C.F. 15, 265

Manzanas, J. 275
Mitachi, T. 303, 309
Moriwaki, T. 279

Nakano, M. 53
Nawir, H. 287

Ochiai, H. 323
Ogata, T. 259
Omine, K. 323
Ozaidin, I.K. 203

Pallara, O. 265
Rampello, S. 15

Saitoh, K. 245
Santucci de Magistris, F. 15, 295
Sauzeat, C. 217
Shibuya, S. 1, 65, 303, 309
Shozen, T. 235
Silvestri, F. 15
Sitoh, M. 279
Supranata, Y.E. 279
Suzuki, T. 317

Tamrakar, S.B. 309
Tatsuoka, F. 1, 111, 187, 209, 251,
    287, 295
Temma, M. 303
Turco, E. 265

Uchida, K. 111

Vaid, Y.P. 235

Yamashita, S. 65, 317
Yasufuku, N. 111, 323

Printed and bound by CPI Group (UK) Ltd, Croydon, CR0 4YY

23/10/2024

01777679-0016